Artificial Intelligence to Solve Pervasive Internet of Things Issues

Artificial Intelligence to Solve Pervasive Internet of Things Issues

Edited by

Gurjit Kaur

Department of Electronics and Communication Engineering, Delhi Technological University, Delhi, India

Pradeep Tomar

Department of Computer Science and Engineering, University School of Information and Communication Technology, Gautam Buddha University, Greater Noida, India

Marcus Tanque

Department of Computer Networks and Cybersecurity Programs, Washington, DC, United States

ELSEVIER

ACADEMIC PRESS

An imprint of Elsevier

Academic Press is an imprint of Elsevier
125 London Wall, London EC2Y 5AS, United Kingdom
525 B Street, Suite 1650, San Diego, CA 92101, United States
50 Hampshire Street, 5th Floor, Cambridge, MA 02139, United States
The Boulevard, Langford Lane, Kidlington, Oxford OX5 1GB, United Kingdom

British Library Cataloguing-in-Publication Data
A catalogue record for this book is available from the British Library

Library of Congress Cataloging-in-Publication Data
A catalog record for this book is available from the Library of Congress

ISBN: 978-0-12-818576-6

For Information on all Academic Press publications
visit our website at https://www.elsevier.com/books-and-journals

Publisher: Mara Conner
Acquisitions Editor: Chris Katsaropoulos
Editorial Project Manager: Amy Moone
Production Project Manager: Swapna Srinivasan
Cover Designer: Christian J. Bilbow

Typeset by MPS Limited, Chennai, India

Contents

List of contributors

Lakshita Aggarwal
Jagan Institute of Management Studies, Rohini, New Delhi, India

Navin Ahlawat
SRM University, NCR Campus, Modinagar, India

Harshit Bhardwaj
Department of Computer Science and Engineering, University School of Information and Communication Technology, Gautam Buddha University, Greater Noida, India

Supreet Bhatnagar
Electronics and Communication Engineering Department, Chandigarh University, Mohali, India

Phillip G. Bradford
University of Connecticut Stamford, Stamford CT, United States

Dushyant Singh Chauhan
Department of Electronics and Communication Engineering, Delhi Technological University, Delhi, India

Tony Clark
School of Engineering and Applied Science, University of Aston, Birmingham, United Kingdom

Pranjit Deka
Department of Electronics and Communication Engineering, Delhi Technological University, Delhi, India

Pedro Faria
Polytechnic of Porto, Porto, Portugal

Luís Gomes
Polytechnic of Porto, Porto, Portugal

Mani Shekhar Gupta
Department of Electronics and Communication Engineering, National Institute of Technology, Hamirpur, India; Department of Electronics and Communication Engineering, Delhi Technological University, Delhi, India

Ramgopal Kashyap
Amity School of Engineering and Technology, Amity University Chhattisgarh, Raipur, India

Gurjit Kaur
Department of Electronics and Communication Engineering, Delhi Technological University, Delhi, India

Harpreet Kaur
Computer Science and Engineering Department, Chandigarh University, Mohali, India

Latika Kharb
Jagan Institute of Management Studies, Rohini, New Delhi, India

Vinay Kulkarni
Software Systems & Services Research, TCS Research, Pune, India

Michele Myauo
Microsoft, Reston, VA, United States

George S. Oreku
Tanzania Industrial Research and Development Organizational, Open University of Tanzania, Tanzania

Carlos Ramos
Polytechnic of Porto, Porto, Portugal

Mamata Rath
School of Management (IT), Birla Global University, Bhubaneswar, India

Himadri N. Saha
Department of Computer Science, Surendranath Evening College, Calcutta University, Kolkata, India

Aditi Sakalle
Department of Computer Science and Engineering, University School of Information and Communication Technology, Gautam Buddha University, Greater Noida, India

Jyotirmaya Satpathy
Academics Department National Defence Academy, Pune, India

Rajeev Kr. Sharma
SRM University, NCR Campus, Modinagar, India

Rupak Sharma
SRM University, NCR Campus, Modinagar, India

Uttam Sharma
Department of Computer Science and Engineering, University School of Information and Communication Technology, Gautam Buddha University, Greater Noida, India

Aditya Pratap Singh
Ajay Kumar Garg Engineering College, Ghaziabad, India

Anamika Singh
Gautam Buddha University, Greater Noida, India

Garima Singh
Department of Electronics and Communication Engineering, Delhi Technological University, Delhi, India

Prateek Singh
Jagan Institute of Management Studies, Rohini, New Delhi, India

Priya Singh
Indira Gandhi Delhi Technical University for Women, Delhi, India

Rashbir Singh
RMIT- Royal Melbourne Institute of Technology, Melbourne, Australia

Simar Preet Singh
Department of Computer Science and Engineering, Chandigarh Engineering College (CEC), Landran, Mohali, India

Tarana Singh
School of Information and Communication Technology, Gautam Buddha University, Greater Noida, India

Arun Solanki
School of Information and Communication Technology, Gautam Buddha University, Greater Noida, India

Akanksha Srivastava
Department of Electronics and Communication Engineering, Delhi Technological University, Delhi, India

Marcus Tanque
Independent Researcher, United States

Akash Tayal
Indira Gandhi Delhi Technological University for Women, Delhi, India

Pradeep Tomar
Department of Computer Science and Engineering, University School of Information and Communication Technology, Gautam Buddha University, Greater Noida, India

Zita Vale
Polytechnic of Porto, Porto, Portugal

About the editors

Dr. Gurjit Kaur
Associate Professor, Department of Electronics and Communication Engineering, Delhi Technological University, Shahbad Daulatpur, Delhi, India
E-mail: gurjeet_kaur@rediffmail.com

Dr. Gurjit Kaur is an Associate Professor in the Department of Electronics and Communication Engineering at the Delhi Technological University (DTU), Delhi, India. She has been a topper throughout her academic education. As a testimonial to the same, she has been awarded by the Chief Minister S. Prakash Singh Badal for being the topper in Punjab. After that, she was awarded a "Gold Medal" by the former President of India Dr. A.P.J. Abdul Kalam for being the overall topper of the Punjab Technical University, Jalandhar in B.Tech program. She also received an honor by Guru Harkrishan Education Society for being a topper among all the colleges and all the disciplines of PTU, Jalandhar. She then proceeded to PEC University of Technology, Chandigarh to complete her M.Tech in 2003 and also earned her PhD degree from Panjab University, Chandigarh in 2010 with distinction.

She has spent over 17 years of her academic career toward research and teaching in the field of Electronics and Communication in well-reputed institutes such as PEC University of Technology, Punjab University, Jaypee Institute of Information and Technology, Gautam Buddha University, and Delhi Technological University, Delhi. During this academic tenure, she has guided 4 PhD students and more than 50 M.Tech students. Her research interests mainly include optical CDMA, wireless communication system, high-speed interconnect, and IoT. She has also authored four books at the international and national level. Her three books, that is, "Green and Smart Technologies for Smart Cities," "Handbook of Research on Big Data and the IoT," and "Examining Cloud Computing Technologies Through the Internet of Things," were published by CRC & IGI-Global, International Publisher of Progressive Information Science and Technology Research in 2019, 2018, and 2017, respectively. She also authored a national level book entitled "Optical Communication," which was published by Galgotia Publications. Apart from that She contributed various book chapters in well-renowned publishers books, that is, IGI, Springer, CRC, etc. She has presented her research work as short courses/tutorials in many national and international conferences.

In recognition of her outstanding contribution as an active researcher, she received "Research Excellence Award" by Delhi Technological University in March 2020. She also received Bharat Vikas Award by Institute of Self Reliance in National Seminar on Diversity of Cultural and Social Environment at Bhubaneswar, Odisha, India and Prof. Indira Parikh 50 Women in Education Leader Award in World Education Congress, Mumbai, India in 2017. Her career accomplishments over the past 16 years include over 90 peer-reviewed publications, including 60 journal publications, 35 conference publications, and over 10 invited talks/seminars. Due to her pioneer research, her name is also listed in Who's Who in Science and Engineering Directory, United States and her biography was also published by International Biographic Center, Great Britain as 2000 Outstanding Scientists.

She also worked as a convener for two international conferences, that is, International ICIAICT 2012 that was organized by CSI, Noida Chapter and International conference EPPICTM 2012 which was held in collaboration with MTMI, United States, University of Maryland Eastern Shore, United States, and Frostburg State University, United States at School of Information and Communication Technology, Gautam Buddha University, Greater Noida, India. Till date, she has delivered many expert talks on the emerging trends in optical for all domain applications at various universities and has served as a reviewer of multiple journals such as IEEE transactions on communication, etc.

Dr. Pradeep Tomar
Assistant Professor, Department of Computer Science and Engineering, University School of Information and Communication Technology, Gautam Buddha University, Greater Noida, India
E-mail: parry.tomar@gmail.com

Dr. Pradeep Tomar is working as an Assistant Professor in the School of Information and Communication Technology, Gautam Buddha University, Greater Noida, UP, India since 2009. He earned his PhD from MDU, Rohtak, Haryana, India. Before joining Gautam Buddha University, he worked as a Software Engineer in a multinational company in Noida and as a lecturer in M.D. University, Rohtak, Haryana and Kurukshetra University, Kurukshetra, Haryana. He has excellent teaching, research, and software development experience as well as vast administrative experience at university level on various posts such as HoD Computer Science and Engineering, Chairperson Admission, Vice Chairperson, Admissions, Research Coordinator, Examination Coordinator, University Coordinator, Admission, Time Table Coordinator, Proctor, and Hostel Warden. He is also a member of Computer Society of India (CSI), Indian Society for Technical Education (ISTE), Indian Science Congress Association (ISCA), International Association of Computer Science and Information Technology (IACSIT), and International Association of Engineers (IAENG). He qualified the National Eligibility Test (NET) for Lectureship in Computer Applications in 2003, Microsoft Certified Professional (MCP) in 2008, SUN Certified Java Programmer (SCJP) for the Java platform, standard edition 5.0 in 2008, and qualified the IBM Certified Database Associate – DB2 9 Fundamentals in 2010.

Dr. Tomar was awarded Bharat Jyoti Award by India International Friendship Society in the field of Technology in 2012 and Bharat Vikas Award by Institute of Self Reliance in National Seminar on Diversity of Cultural and Social Environment at Bhubaneswar, Odisha, in 2017. He has been awarded for the Best Computer Faculty award by Government of Pondicherry and ASDF society. His biography is published in Who's Who Reference Asia, Volume II. He has been awarded distinguished Research Award from Institute for Global Business Research for his work in "A Web-Based Stock Selection Decision Support System for Investment Portfolio Management in 2018." He delivered expert talks at FDP, workshops as well as national and international conferences. He has organized three conferences: one national conference with COMMUNE group and two international conferences, in which one international ICIAICT 2012 was organized by CSI,

Noida Chapter and second international conference 2012 EPPICTM was organized in collaboration with MTMI, United States, University of Maryland Eastern Shore, United States, and Frostburg State University, United States at School of Information and Communication Technology, Gautam Buddha University, Greater Noida, India.

Apart from teaching, he is running a programming club for ICT students and he is guiding various research scholars in the areas of software engineering, reusability of code, soft computing technique, Big Data, and IoT. His major current research interest is in Component-Based Software Engineering. He is working as coinvestigator in a sponsored research project in high throughput design, synthesis, and validation of TALENs for targeted Genome Engineering, funded by Department of Biotechnology, Ministry of Science and Technology Government of India. Two books "Teaching of Mathematics" and "Communication and Information Technology" at national levels as well as three books "Examining Cloud Computing Technologies Through the Internet of Things (IoT)," "Handbook of Research on Big Data and the IoT," and "Green and Smart Technologies for Smart Cities" at international level are authored and edited by Dr. Tomar. He has also contributed more than 100 papers/articles in national/international journals and conferences. He served as a member of the editorial board and reviewer for various Journals and national/international conferences.

Dr. Marcus Tanque is a highly regarded principal technology advisor, senior cyber strategist, and an independent author. Marcus' expertise blends interdisciplinary areas—advanced, traditional, and emerging technologies. He is also a published technologist, researcher, and scholar with commended skills in various domains—IT governance, security analytics, policy, and strategic business practices. His professional skills encompass multidisciplinary research and development areas spanning digital technology, such as augmented analytics, machine learning, deep learning, quantum computing, and biometrics. His research interest involves policy management, blockchain, identity and access management, and decentralized Internet of Things devices and systems. He has consulted to many government, public, and private customers. His consultative and technical contributions bridge, multifaceted traditional, mission-centric programs, and emerging technologies. Such technological solutions involve artificial intelligence, robotics, advanced analytics, cybersecurity, and the Internet of Things.

Dr. Tanque has supported complex enterprise projects amid science and technology management. His experience in business and scientific domains encompass IT engineering, program and project management, information security and assurance, and natural language processing. He has conducted several studies in machine-to-machine and advanced database systems, computing, information systems, and cognitive science/informatics. His research work merges cyber resilience, business intelligence, cybersecurity operations, risk management, cryptographic solutions, information management, and security policy. He is a recipient of several academic awards, credentials, and professional letters of commendation.

Dr. Tanque is an active member of the Institute for Defense and Government Advancement, the Association for Computing Machinery, the Collegiate National Center for Academic Education, Committee on National Security Systems-Practice & Research, and the Institute of Electrical and Electronics Engineers. He has authored and edited scientific/technical books as well as articles on AI, ML, DL, data science, cloud, data analytics, full-spectrum cyber operations, Telecom, and related interdisciplinary areas. He received a PhD in Information Technology with a dual specialization in Information Assurance and Security, and an MS in Information Systems Engineering. In his spare time, he serves as an independent researcher/author, editorial review board member, and referee for several books, journal articles, and published chapters. He enjoys coaching and mentoring others. He is an enthusiastic reader and researcher who inspires many in academic, professional, and social communities.

Affiliations and Expertise

Independent Researcher and Associate Professor, Department of Computer Networks and Cybersecurity Programs, Washington, DC, United States

Preface

Dr. Gurjit Kaur, Dr. Pradeep Tomar and Dr. Marcus Tanque

Artificial intelligence (AI) leverages computer systems designed to support tasks that call for human interaction, intelligent design, and computational agents. AI and Internet of Things (IoT) attract the interest of researchers and practitioners. These professionals have worked on machine learning (ML), knowledge representation and reasoning (KRR), and deep learning (DL). Such solutions aim to solve issues affecting IoT devices and systems. AI involves the integration of data from IoT to the ability for devices and systems to perform automated tasks beyond human intelligence. The method integrates AI's deep insights into data provisioning, managing, security, visualization, and monitoring through augmented analytics processes. These AI capabilities support the interaction of IoT devices and systems.

This book applies AI, ML, KRR, DL, and IoT technologies. It focuses on each of these technologies' benefits, challenges, drawbacks, and trends. The book covers other emerging technologies needed for integrating AI-based solutions to solve pervasive IoT issues. It points toward the adoption and integration processes needed for sustaining AI and IoT infrastructure operations and management functions. The content comprises innovative AI and IoT concepts, theories, procedures, and methods. AI and IoT solutions have contributed to the acquisition, planning, execution, implementation, deployment, operation, and monitoring of enterprise technology assets or technological resources. The target audience of this book involves professionals, such as researchers and practitioners working in the fields of AI, ML, and IoT. These experts focus on building knowledge-based agnostic solutions for applications, devices, and systems.

This book contains 20 contributed chapters authored by experts in the field of AI, IoT, ML, DL, and KRR. The book introduces basic AI and IoT components and applications, that is, standards, legal issues, privacy, security, and ethical considerations. Detailed use cases are described, covering a variety of technological advancements, implementations, and challenges. The authors explore other technical domains, which have contributed to AI, ML, and IoT technologies. These domains have emerged over time due to disruptive technical and scientific innovations in the industry.

The book is organized as follows:

Chapter 1: Impact of Artificial Intelligence on Future Green Communication

The chapter explores AI impact on future green communication. It discusses the increase in mobile subscriptions for base stations. Base stations involve systems that require more power to operate. Minimizing the number of base stations while enhancing their energy efficiency poses significant opportunities for green energy. AI plays a crucial role in green areas, which is energy forecasting, energy-efficient, and energy accessibility. This chapter provides a brief foundation and history of AI technology and green communication roadmap. It highlights critical AI-based applications, practices, and future research directions.

Chapter 2: Knowledge Representation and Reasoning in AI-Based Solutions and IoT Applications

The chapter focuses on KRR in AI-based and IoT applications—it explains AI, KRR, and IoT disruptive evolution. Researchers and practitioners develop and integrate analytical solutions to solve pervasive issues affecting computational applications. These technological developments comprise relevant computational areas: devices, sensors, autonomous vehicles, robotics, virtual reality, and augmented intelligence. The author discusses similar technical solutions that researchers need to solve issues affecting AI, KRR, and IoT applications.

Chapter 3: Artificial Intelligence, Internet of Things, and Communication Networks

The chapter examines AI, IoT, and communication networks. AI handles connectivity, self-optimization, and self-configuration. The method assesses and predicts the current state as well as historical data needed to automate the network. Communication networks are becoming more complex to manage, due to the disruption of data, which affects device and systems connectivity. This process comprises low cost, power-efficiency, and high-performance network technologies. Incorporating AI into these networks requires automated solutions to introduce smart and intelligent decision-making processes needed for managing networked control systems.

Chapter 4: AI and IoT Capabilities: Standards, Procedures, Applications, and Protocols

The chapter analyzes AI and IoT capabilities—standards, procedures, applications, and protocols. The authors present AI-based applications, methods, standards, and protocols for interacting with IoT-based objects. Advanced AI technologies interact with IoT devices and systems—technical crossing point, mimics human intelligence interaction, collects, and processes data in real-time.

Chapter 5: Internet of Intelligent Things: Injection of Intelligence into IoT Devices

The chapter discusses the IoT seven-layer model—physical or sensor, processing and control action, hardware interface, radio frequency, section/message, user experience, application. This protocol stack discussion, hence, illustrates each layer's function and overarching operations posture. Similarly, the authors underscore the relationship between AI and IoT and other automated AI/IoT solutions.

Chapter 6: Artificial Intelligence and Machine Learning Applications in Cloud Computing and Internet of Things

The chapter examines AI, ML, IoT, and cloud computing solutions that dominate the global business and technology landscape. These technical innovations contribute to the decentralization of IoT devices and systems. The authors illustrate a detailed AI analytical review and challenges and ML solutions applied to IoT devices and systems. These solutions are essential to the technical and scientific advancement of IoT devices and systems and AI/ML solutions.

Chapter 7: Knowledge Representation for Causal Calculi on Internet of Things

The chapter introduces Pearl's model, Shafer's model, and the Halpern–Pearl's model. This chapter introduces knowledge representation for causal calculus in IoT. IoT devices and systems harness causal inference. Causal calculi are the mathematical foundations for expressing and computing causation. In contrast, causation illustrates how one event may cause another—causal systems founded in logic and probability theories. Pearl's method uses Bayesian networks on acyclic directed graphs. Shafer's method works on the dynamics of probability trees. The Halpern–Pearl model builds on Pearl's model producing two causal views. One of these views attributes causes from past events. The other ones focus on causal reasoning giving predictions.

Chapter 8: Examining the Internet of Things–Based Elegant Cultivation Technique in Digital Bharat

The chapter discusses IoT, elegant cultivation techniques in digital Bharat. These innovative technologies do farm development and management. These technological farming solutions focus on enhancing effectiveness, competence, and leverages international markets—for instance, solutions have diminished human intercession. The authors' research approach focuses on the device, system function, and applications.

Chapter 9: Machine Learning and Internet of Things for Smart Processing

The chapter discusses the relationship between ML and IoT. It defines processes and impacts these technologies present to academic, business, technical, and scientific communities. ML and IoT solutions comprise three distinct areas. These are recurrence groups, spatial channels, classifiers arrangement, and execution required to determine the best settings.

Chapter 10: Intelligent Smart Home Energy Efficiency Model Using Artificial Intelligence and Internet of Things

The chapter analyzes the design and implementation of a smart home system model to safeguard all the electrical equipment and monitoring the performance of each system installed in smart homes. These systems use AI and IoT solutions that optimize energy usage for an intelligent smart home energy efficiency model applying AI and IoT.

Chapter 11: Adaptive Complex Systems: Digital Twins

The chapter reviews the features of complex systems. It proposes solutions to support digital twins and adaptable systems needed for interacting with ML solutions. This method is presented in the chapter through simple tutorial agents' examples using ML technology. The process focuses on authors who use technology to build digital twins for supply chain networks.

Chapter 12: Artificial Intelligence Powered Healthcare Internet of Things Devices and Their Role

The chapter examines AI-based technologies, for instance, the vulnerability, threats, risks, and challenges on the Internet of Medical of Things (IoMT) devices and systems. IoMT is a healthcare domain that complements AI-solutions, IoT devices, and systems.

Chapter 13: IoIT: Integrating Artificial Intelligence With IoT to Solve Pervasive IoT Issues

The chapter examines the Internet of Intelligence of Things, AI-based solutions' integration, IoT devices, and systems. It discusses the ML-Random Forest Regression model that evaluates applicability and preventability with variance score. This process focuses on AI and IoT areas.

Chapter 14: Intelligent Energy-Oriented Home

The chapter discusses two areas—intelligent energy systems and smart homes. The foundations of these fields are presented concisely. This process entails projects on intelligent energy systems for homes, buildings, and associated business ventures.

Chapter 15: Corporate Cybersecurity Strategy to Enable Artificial Intelligence and Internet of Things

The chapter addresses cyber-adversarial system, internal and external threats, the anatomy of a cyber-attack, and financially motivated cyber-attackers. It further analyzes ideologically and politically motivated cyber-attackers. It covers many aspects of cybersecurity, that is, cyber-related laws, cybersecurity Inertia, cybersecurity IT portfolio management, and smarter cybersecurity leveraging artificial intelligence.

Chapter 16: Role of Artificial Intelligence and the Internet of Things in Agriculture

The chapter discusses how AI and IoT solutions help the software and systems engineers develop innovative capabilities for the agriculture industry. These technologies may be built with low cost and few resources.

Chapter 17: Integrating Artificial Intelligence/Internet of Things Technologies To Support Medical Devices and Systems

The chapter argues IoT/AI integration, concepts, procedures, and requirements relating to medical devices and systems. It underlines various security characteristics amid the AI/IoT integration. This process includes several changes in the ecosystem and analyses on the next generation of medical devices and systems.

Chapter 18: Machine Learning for Optical Communication to Solve Pervasive Issues of Internet of Things

The chapter analyzes ML-based solutions for optical communication and technical issues affecting physical and network layers. This process involves selected ML applications along with optical communication systems associated with physical and network layers.

Chapter 19: Impact of Artificial Intelligence to Solve Pervasive issues of Sensor Networks of Internet of Things

The chapter examines AI's impact on solving pervasive IoT-based intelligent sensors and systems issues. It presents a brief introduction of IoT devices and systems, history, characteristics, and network formation topologies. IoT sensor networks and AI-based solutions and features are further discussed in the chapter.

Chapter 20: Principles and Foundations of Artificial Intelligence and the Internet of Things Technology

The chapter explores AI and IoT technological foundations and principles. It illuminates how AI helps computers to learn from various experiences by adapting to new environments. This method includes devices and systems that perform tasks beyond human capacity. Comparably, the chapter analyzes how IoT-based technologies help objects observe, identify, simulate, and understand a situation and/or environment with limited human assistance.

The collected authors in this book explore the concepts, techniques, procedures, and implementations of these combined and/or integrated technologies.

CHAPTER

IMPACT OF ARTIFICIAL INTELLIGENCE ON FUTURE GREEN COMMUNICATION

1

Akanksha Srivastava[1], Mani Shekhar Gupta[1,2] and Gurjit Kaur[1]

[1]*Department of Electronics and Communication Engineering, Delhi Technological University, Delhi, India*

[2]*Department of Electronics and Communication Engineering, National Institute of Technology, Hamirpur, India*

1.1 INTRODUCTION

Future wireless communication networks (5G) will be highly complex and composite networks due to the integration of the different wireless and wired networks. This integration is known as heterogeneous networks (HetNets) where each network is having its different protocols and properties [1]. This combined HetNet is having various critical challenges for network scheduling, operation, troubleshooting, and managing. In the ongoing scenario the technology paradigm shifts from user-centric to device-oriented communication, which is responsible for converting the simple wireless networks into a complex form. Nowadays to justify and resolve the operational complexity of future wireless communication networks, several novel approaches like cognitive radio, fog computing, Internet of Things (IoT), and so on have become very important. The artificial intelligence (AI) is one of the most promising approaches to make the adoption of the new principles, which include learning, cognitive, and decision-making processes, for designing a strongly integrated network. Integration of AI with data analytics, machine learning, and natural language processing approach is used to improve the efficiency of the future wireless network generations. There are remarkable growth and progress in AI technology, which facilitates to overcome the problem of human resource deficiencies in many fields. Among the countries, the competition of becoming a global leader in the field of AI has officially started. Most of the countries like India France, China, Japan, Denmark, Canada, Finland, Italy, Mexico, the United Kingdom, Singapore, South Korea, North Korea, Taiwan, and the UAE, have already represented their strategies to endorse the development and usage of AI policies [2]. These countries are promoting the various tactics of the AI techniques like technical research, AI-based products, talent, and skills development, adoption of AI in private and public sector, standards and regulations, and digital infrastructure. Table 1.1 is representing the top 10 countries rankings in AI index in the year of 2018–19.

1.2 THE HISTORY OF ARTIFICIAL INTELLIGENCE

AI is one of the latest topics for research in advance wireless communication system. A very interesting fact related to this technology is that this is much older technology than you would imagine.

Artificial Intelligence to Solve Pervasive Internet of Things Issues. DOI: https://doi.org/10.1016/B978-0-12-818576-6.00001-0

Table 1.1 Top 10 Countries Rankings in Artificial Intelligence Index With score in the Year of 2018–19 [3].

Country	Ranking	Score
Singapore	1	9.186
United Kingdom	2	9.069
Germany	3	8.810
United States of America	4	8.804
Finland	5	8.772
Sweden	6	8.674
Canada	6	8.674
France	8	8.608
Denmark	9	8.601
Japan	10	8.582

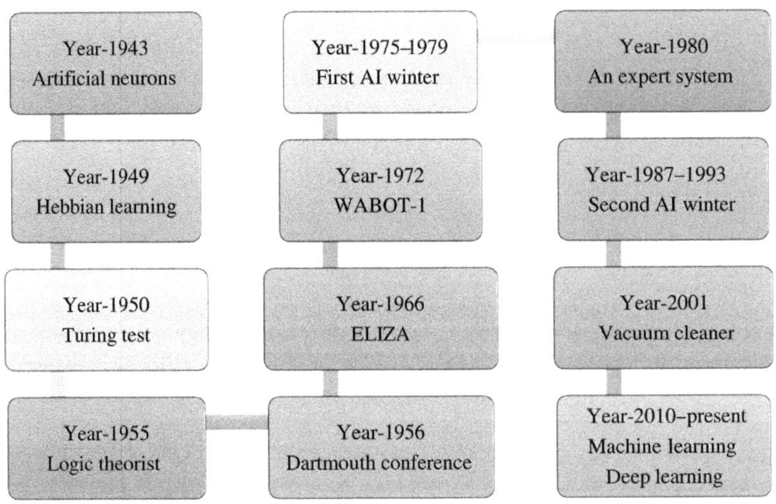

FIGURE 1.1

History of AI. *AI*, Artificial intelligence.

The concept of intelligent robots was presented by Greek myths of Hephaestus "mechanical men" and Talos "bronze man" [4]. Some important milestones of the journey of AI from an initial state to till date are represented pictorially as in Fig. 1.1.

1.2.1 THE FOUNDATION OF ARTIFICIAL INTELLIGENCE

- Artificial neurons: The artificial neurons were the first model of AI, which was proposed by Walter pits and Warren McCulloch in 1943.

- Hebbian learning: A modified rule of construction of neurons is presented by Donald Hebb in 1949. This rule is known as Hebbian learning.
- Turing test: This test can evaluate the intelligent behavior of a machine and also compare it with human intelligence. An English mathematician Alan Turing author of "Computing Machinery and Intelligence" has proposed this test in 1950.
- Logic theorist: "The first AI-based program" that was organizes by the Herbert A. Simon and Allen Newell in 1955.
- Dartmouth conference: The AI technology was the first time adopted by the American scientist John McCarthy in the academic field at this Conference in 1956.

1.2.2 PROGRESSION OF ARTIFICIAL INTELLIGENCE

After the year 1956, the researchers have invented high-level computer languages like COBOL, PASCAL, LISP, and FORTRAN. These language inventions increased the scope of AI in society [5].

- ELIZA: The first AI-based algorithm developed by Joseph Weizenbaum is known as ELIZA in 1966. This algorithm is used to solve the problems of mathematics.
- WABOT-1: Japan has constructed the first humanoid intelligent robot known as WABOT-1 in 1972.
- First AI Winter: This is the time duration (from 1975 to 1979) when the interest of AI was reduced due to the scarcity of funding, for the research of AI.

1.2.3 EXPANSION OF ARTIFICIAL INTELLIGENCE

- An expert system: After the first AI winter period, AI came back again into the light as an "expert system" in 1980. This system has ability to take decision like human expert. In this year the first national conference on AI was organized at Stanford University.
- Second AI winter: The time duration from year 1987 to 1993 was the time duration of second AI winter.
- AI in home and business: At the year 2001 first time, AI-based application, a vacuum cleaner used in the home. After that AI entered into business world companies such as Gmail, Facebook, Instagram, Twitter, and so on.

1.2.4 MODERN ARTIFICIAL INTELLIGENCE

Now AI is the most significant technology, which is used in almost all areas. The concept of machine learning, deep learning, cloud computing, and big data are just like a boom for the present scenario. Many well-known leader corporate companies like IBM, Google, Flipkart, and Amazon are focusing on AI for making their remarkable devices to provide their users with a better quality of experience (QoE). The future AI technology will be based on a high level of intelligence and amazing capacity and speed [6].

- Machine learning: Machine learning concept is one of the types of data mining techniques. Machine learning is an approach of analyzing data, absorb from that data, and then make a decision. Now, most of the big companies use machine learning for their working operations like YouTube uses machine learning to offer better suggestions to their subscribers of the movie, shows, and videos that they would like to watch.
- Deep learning: Deep learning is a subclass of machine learning. It is functioning like machine learning but it has some distinct capabilities. The key difference between machine learning and deep learning is, machine learning model requires some guidance to take accurate decision while the deep learning model does it by itself. The good example of deep learning is automatic car driving system.

1.3 A ROAD MAP OF USING ARTIFICIAL INTELLIGENCE FOR GREEN COMMUNICATION

This will be a great step to introduce AI technologies in the field of wireless communication systems. Incorporation of AI technologies in the field of signal processing and pattern recognitions has represented the amazing results [7]. Presently, the key concern of the AI technologies in wireless communication systems is to find out the accurate wireless node position, proper resources allocation and optimization, and secure data transmission without delay. However, new research is to think about how to incorporate AI schemes into wireless communication. Compared to the conventional wireless communication systems, the new AI-based wireless communication systems should have four eminent aptitudes. These aptitudes are analyzing aptitude, thinking aptitude, learning aptitude, and proactive aptitude. The new framework of AI wireless communication systems with these aptitudes is illustrated in Fig. 1.2.

1.3.1 ARCHITECTURE OF ARTIFICIAL INTELLIGENCE-BASED GREEN COMMUNICATION

The future wireless communication networks should have inherent capabilities like low-latency, ultrareliable communication and intelligently manage the resources, energy efficient, and combination of IoT devices in a real-time dynamic environment [8]. Such communication necessities and core mobile edge requirements can only be accomplished by integrating the fundamentals and principles of AI and machine across the wireless infrastructure. Fig. 1.3 represents the wireless network architecture with AI principles for a different environment. The diagram shows the integration of various latest communication technologies used for greening communication in different scenarios (urban, suburban, and rural areas).

1.3.2 OPTIMIZATION OF NETWORK USING ARTIFICIAL INTELLIGENCE

Effective data gathering and information acquisition are the most essential requirements for optimizing the future wireless communication system. To extract the relevant information from the collected data in an effective manner is under the processing of data. In the third step, researches

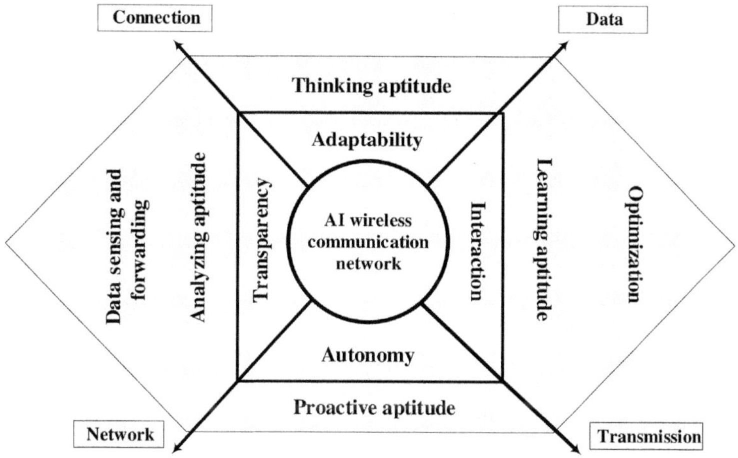

FIGURE 1.2

Framework of AI wireless communication systems with aptitudes. *AI*, Artificial intelligence.

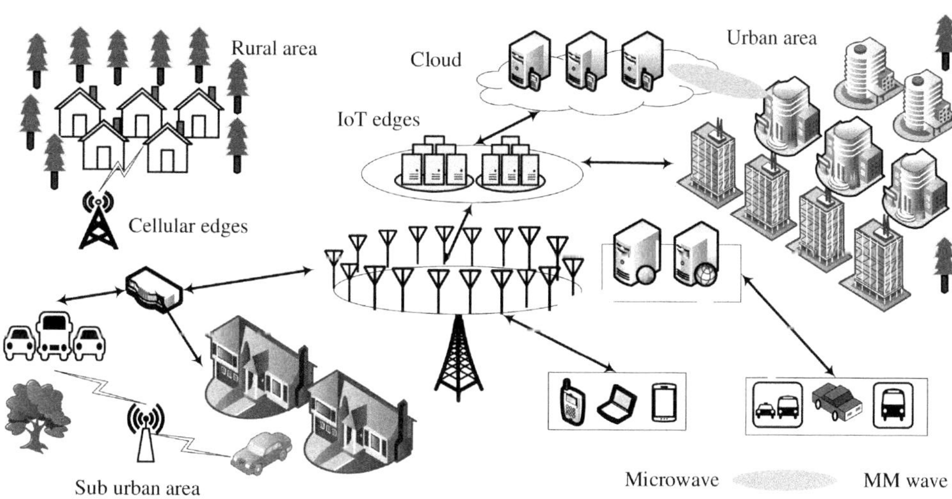

FIGURE 1.3

Energy-efficient wireless network with AI principles analyzing, cognitive, and decision making. *AI*, Artificial intelligence; *mm*, millimeter.

analyze this received information and apply various aptitudes on it. Finally, at the last step, an optimized decision is presented which converts the wireless network into an optimized network. Fig. 1.4 represents the networks optimization process to identify best network for better QoE.

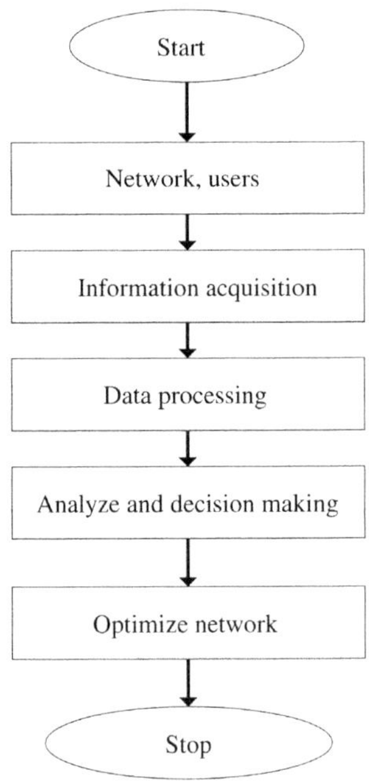

FIGURE 1.4

Network optimization by artificial intelligence technique.

1.4 KEY TECHNOLOGIES TO MAKE 5G IN REALITY USING ARTIFICIAL INTELLIGENCE

The necessity to deal with this rapid progression of wireless services has required a large research activity that explores what are the optimal options for designing of user-oriented context-aware next-generation (5G) wireless communication network. The key components for 5G are multiple input multiple output (MIMO), massive MIMO, ultradense deployment of small cells, millimeter (mm) wave communications, and device-to-device (D2D) communications have been recognized. The integration of these technologies in the wireless system with the cooperation of AI principles in the most effective manner is a challenging task for operators and researchers.

1.4.1 MULTIPLE INPUT MULTIPLE OUTPUT

This is the most promising approach to consider the development of the next-generation wireless network system. In this technique, multiple antennas are situated at both the end transmitter (source) and receiver (destination) [9]. For enhancing efficiency and reducing the errors of the network, these antennas are associated effectively [10]. This technique facilitates to multiply the capacity of the antenna more than 10 times, without increasing the power and bandwidth of the system [11]. This QoE focused approach is made it an essential element of the wireless communication network [12]. The comparison of MIMO with single input single output, multiple input single output, and single input multiple output is given in Table 1.2.

1.4.2 MASSIVE MIMO

This technique is not only energy efficient but also spectrum efficient. Massive MIMO (M-MIMO) is one of the advanced versions of technologies of MIMO having several antennas at the base station of the communication system. This technique requires shorter wavelengths (higher frequencies) because the system needs to physically pack more antennas into a small area than the other mobile networks [13,14]. The main advantage is that a base station can serve multiple subscribers simultaneously within the same spectrum. Fig. 1.5 represents the architecture of the Massive MIMO technique where ten to hundreds of antennas are serving for the communication process simultaneously.

1.4.3 ULTRADENSE NETWORK

In the new age, ultradense network (UDN) has emerged as a prominent solution to fulfilling the requirement of enormously high capacity and data rate of the 5G wireless network. Qualitatively, this network has a much higher density of radio resources than that of other existing networks in the telecommunication market [15,16]. In literature, there are various definitions of UDN suggested by various authors. In Ref. [17], the author has defined the UDN as a network where the access point and base station density exceeds the user density in a particular area. In Refs. [18,19], a UDN is considered as a network where the distance between the access points and base stations is only a few meters. The architecture of a UDN is showing in Fig. 1.6. A UDN plays a vital role in converting the communication into green communication. In this technique, the access points and base stations are presented very close distance to the mobile subscribers. The relation between the power and distance shows that the distance is directly proportional to the power. So, if the distance

Table 1.2 Comparison of Multiple Input Multiple Output (MIMO) With Single Input Single Output (SISO), Multiple Input Single Output (MISO), and Single Input Multiple Output (SIMO).

S. No.	SISO	SIMO	MISO	MIMO
1.	Simple circuitry	Known as receive diversity	Known as transmitter diversity	Improve channel capacity
2.	Diversity not required	Easy to implement	Reduce the problem of interference	Improve channel throughput
3.	Low throughput	High cost than SISO	High cost than SISO	Highest cost
4.	Channel bandwidth is limited	Problem of battery drain	Channel capacity is not improved	Complex circuitry

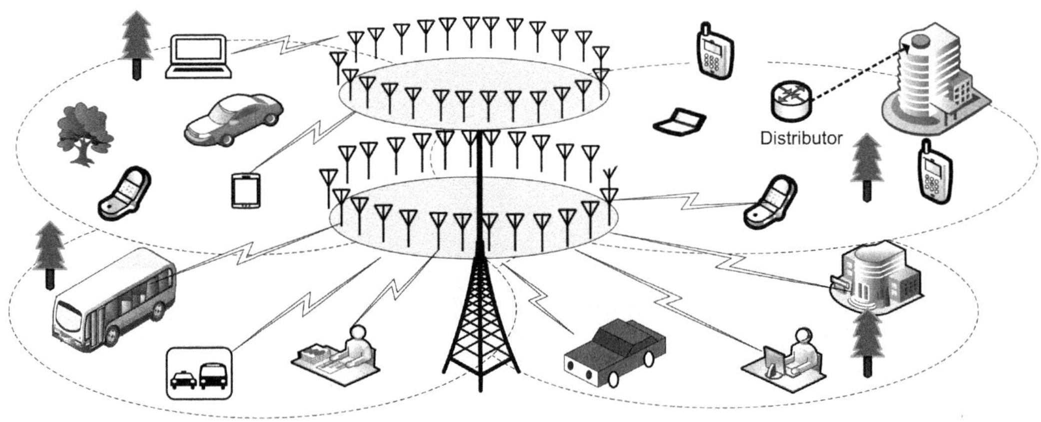

FIGURE 1.5

Massive multiple input multiple output technique in 5G network.

between the mobile subscriber and access point will reduce the power of the communication system will automatically reduce. In this way, by minimizing the power consumption a UDN promotes energy-efficient communication.

1.4.4 MILLIMETER WAVE

The mm waves are one of the most important approaches for the next generation of wireless networks. For delivering fast multimedia services, high-quality audio, video, and real-time services, a large amount of bandwidth is required. To solve this problem of spectrum scarcity, mm wavelength will be used in 5G network communication system. The signals are operating between the range of 30 and 300 GHz and being shifted to a higher spectrum. A large amount of bandwidth is offered at mm-wave frequencies as compared to the bandwidth used by 4G and earlier wireless generation networks.

1.4.5 DEVICE-TO-DEVICE COMMUNICATION

D2D communication is one of the effective technical approaches to reduce the consumption of power and improve the data transmission rate [20]. In this technique, two physically separated nearby located cellular nodes can directly communicate with each other with low transmit power and high spectrum utilization efficiency without considering the base station in the communication process showing in Fig. 1.7. The D2D communication approach is recognized as a public safety network for future wireless communication by Federal Communications Commission because of the low cost and high data rates offered by this technique.

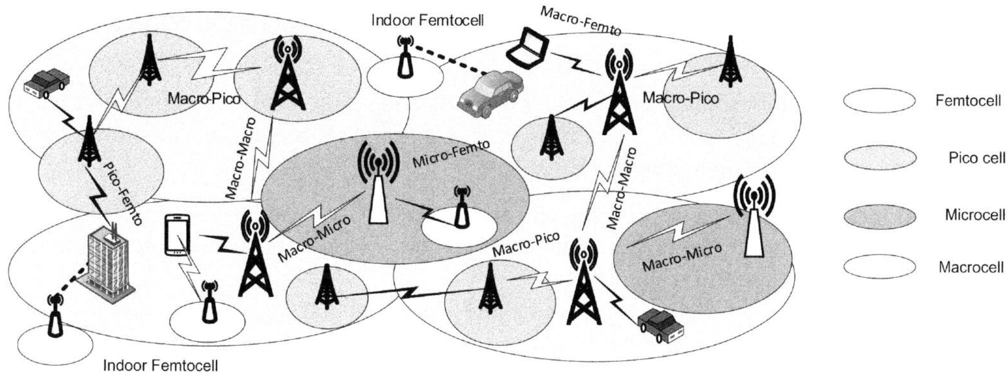

FIGURE 1.6

Architecture of ultradense network.

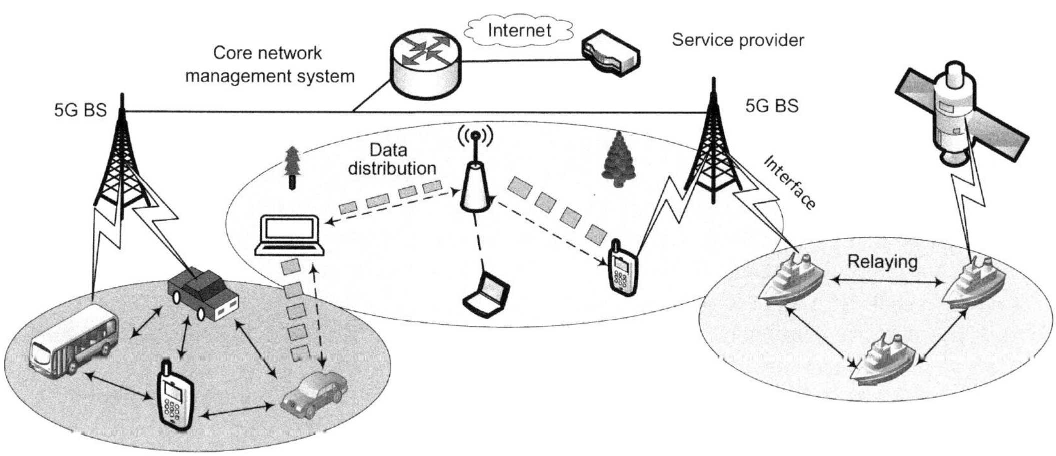

FIGURE 1.7

Architecture of device-to-device communication.

1.5 FEATURES OF ARTIFICIAL INTELLIGENCE-BASED GREEN COMMUNICATION

The present era is based on perceptional, cognitive, and computational intelligence. So, telecom researchers and operators are on the path of creation of AI-based green communication system. For the adoption of AI technology, government and other agencies are encouraging the development of AI algorithms and investing funds and resources for AI-based research activities. By these full supports, the operators have achieved success in the sequence of effective practices in several fields and accomplished productive results.

1.5.1 APPLICATION AND PRACTICES OF ARTIFICIAL INTELLIGENCE-BASED GREEN COMMUNICATION

Nowadays there are various applications of AI from collecting data to give an optimized output. The aim is to apply AI in the mobile industry is to gain a seamless network operation to improve the energy efficiency of the wireless network.

- Appling AI in the planning process: In the planning process AI is used to predict the traffic demand. In AI-driven traffic prediction there are two types of traffic tendencies short-term traffic tendency and long-term traffic tendency.
- Appling AI in network monitoring: Network monitoring and maintenance is the most complicated process. It is very difficult to analyze the requirement of customers because it dynamically changed so maintain the network according to their request is a tough process.
- Appling AI in service monitoring: To monitor the quality of service and QoE for any network is the most important task. Using AI for this purpose will give an accurate result.

1.5.2 FUTURE RESEARCH DIRECTIONS

The major research challenges are outlined in this chapter. A widespread effort is required from academia and industry in this area listed to contribute to green communication.

- Energy saving in telecommunication equipment using AI: Telecommunication systems and operators are having a large number of equipment and data centers. These data centers are made by many hardware like processing unit, input output devices that consume a large amount of power for operation. Therefore, the communication system is facing a shortage of power and energy. Various power-saving techniques based on AI like deep learning and machine learning is using to fight with this serious situation.
- Ability to improve data interaction: An AI-based system organizes the available data in a very effective manner and converts this data into relevant information.
- More effective and efficient collaboration: The benefits of AI in a collaborative manner comprise sending information more effectively at a global level. For example, real-time language translation, fast feedback, and accurate scheduling.
- Secure and seamless services: AI-based applications motivate the intelligent security management system. Based on AI services such as big data, cloud computing, and IoTs are the technologies that provide secure data transmission.

1.6 CONCLUSION

For the Information and communication technology (ICT) industry, the next-generation AI-based 5G communication network is considered as the key enabler and offers a diversity of features and services with various requirements. This chapter represents the mutual concern of the AI and next-generation wireless communication systems and technologies. The AI-based 5G wireless communication network will adopt a greater number of candidate recent technologies in the future. Therefore to manage and monitor the next-generation wireless communication, inclusion of AI with communication is very important. In this chapter, advanced wireless networks like MIMO, M-MIMO, UDN, mm wave, and D2D communication for 5G network designing and the relationship between AI and green communication are discussed. Applications and future research directions of AI-based wireless communication are also highlighted in this work. The prediction is that AI-empowered 5G communication networks will make the acclaimed ICT enabler a reality.

ACKNOWLEDGMENT

The authors would like to thank the Women Scientists Scheme-A under the Department of Science and Technology Government of India for its financial support of this work under File No: SR/WOS-A/ET-154/2017.

REFERENCES

[1] Y. Lv, H. Zhang, S. Xueming, Analysis of base stations deployment on power saving for heterogeneous network, in: 2017 IEEE 17th International Conference on Communication Technology (ICCT), IEEE, 2017, pp. 1439–1444.

[2] T. Dutton, An overview of national AI strategies. Retrieved from: <https://medium.com/politics-ai/an-overview-of-national-ai-strategies-2a70ec6edfd>, 2018, Canada.

[3] H. Miller, R. Stirling, Government artificial intelligence readiness index. Retrieved from: <https://ai4d.ai/wp-content/uploads/2019/05/ai-gov-readiness-report_v08.pdf>, 2019, Canada.

[4] AAAI, A brief history of AI. Retrieved from: <https://aitopics.org/misc/brief-history>, 2019, USA.

[5] P. Shapshak, C. Somboonwit, J.T. Sinnott, Artificial intelligence and virology-quo vadis, Bioinformation 13 (12) (2017) 410.

[6] G. Kaur, N. Gupta, Performance evaluation of one- and two-dimensional prime codes for optical code division multiple access systems, World Acad. Sci. Eng. Technol. 10 (7) (2016) 1348–1354.

[7] M.E.M. Cayamcela, W. Lim, Artificial intelligence in 5G technology: a survey, in: 2018 International Conference on Information and Communication Technology Convergence (ICTC), IEEE, 2018, pp. 860–865.

[8] A.A. Osuwa, E.B. Ekhoragbon, L.T. Fat, Application of artificial intelligence in Internet of Things, in: 2017 9th International Conference on Computational Intelligence and Communication Networks (CICN), IEEE, 2017, pp. 169–173.

[9] W. Liu, X. Li, M. Chen, Energy efficiency of MIMO transmissions in wireless sensor networks with diversity and multiplexing gains, in: Proceedings of the (ICASSP'05), IEEE International Conference on Acoustics, Speech, and Signal Processing, 2005, IEEE, 2005, vol. 4, pp. iv–897.

[10] G. Kaur, P. Goyal, High responsivity germanium on silicon photodetectors using FDTD for high speed optical interconnects, Arab. J. Sci. Eng. (2017) 1–7.

[11] Y. Gai, L. Zhang, X. Shan, Energy efficiency of cooperative MIMO with data aggregation in wireless sensor networks, in: 2007 IEEE Wireless Communications and Networking Conference, IEEE, 2007, pp. 791–796.

[12] R. Yadav, G. Kaur, Modified three dimensional multicarrier optical prime codes, Int. J. Opt. 2016 (2016) 1–7.

[13] L. Chen, F.R. Yu, H. Ji, B. Rong, X. Li, V.C. Leung, Green full duplex self-backhaul and energy harvesting small cell networks with massive MIMO, IEEE J. Sel. Areas Commun. 34 (12) (2016) 3709–3724.

[14] Y. Xin, D. Wang, J. Li, H. Zhu, J. Wang, X. You, Area spectral efficiency and area energy efficiency of massive MIMO cellular systems, IEEE Trans. Vehicular Technol. 65 (5) (2016) 3243–3254.
[15] A. Srivastava, M. Gupta, G. Kaur, Green smart cities, Green. Smart Technol. Smart Cities (2019) 1–18. Available from: https://doi.org/10.1201/9780429454837-1.
[16] X. Xu, C. Yuan, W. Chen, X. Tao, Y. Sun, Adaptive cell zooming and sleeping for green heterogeneous ultra-dense networks, IEEE Trans. Vehicular Technol. 67 (2) (2018) 1612–1621.
[17] L. Su, C. Yang, I. Chih-Lin, Energy and spectral efficient frequency reuse of ultra-dense networks, IEEE Trans. Wirel. Commun. 15 (8) (2016) 5384–5398.
[18] D. Kedia, G. Kaur, Examining different applications of cloud-based IoT, in: Examining Cloud Computing Technologies through the Internet of Things, IGI-Global, International Publisher of Progressive Information Science and Technology Research Books and Journals, 2017, pp. 1125–1146, ISBN13: 978152253445.
[19] M. Thurfjell, M. Ericsson, P. de Bruin, Network densification impact on system capacity, in: 2015 IEEE 81st Vehicular Technology Conference (VTC Spring), IEEE, 2018, pp. 1–5.
[20] M.S. Gupta, K. Kumar, Progression on spectrum sensing for cognitive radio networks: a survey, classification, challenges and future research issues, J. Netw. Comput. Appl. 143 (2019) 47–76.

CHAPTER

KNOWLEDGE REPRESENTATION AND REASONING IN AI-BASED SOLUTIONS AND IoT APPLICATIONS

2

Marcus Tanque

Independent Researcher, United States

2.1 INTRODUCTION

The chapter explores and addresses the intersection between knowledge representation and reasoning (KRR) applications, artificial intelligence (AI) solutions, and the Internet of Things (IoT) objects, notably intelligent devices and sensors. This chapter focuses on smart homes and cities. It discusses the fundamental transformations that researchers have made in the fields of KRR, AI, machine learning (ML), deep learning (DL), and IoT [1,2]. Hybrid AI-KRR capabilities interact with human-to-machine and device-to-system [3]. AI is an integrated, technology-based capability, which incorporates human intelligence (HI) and smart objects, such as devices and sensors [1]. The triad forms a single or multiple computer systems capable of performing massive workload computer processing [1]. AI-based solutions are developed to support and solve various events [3]. The studied process involves human insights, ability to ensure the performance of smart objects, and determine when these sound devices and sensors can interact. AI has reshaped the global technology landscape through its incorporated analytical capabilities [1].

2.2 BACKGROUND

In the last two decades, KRR, IoT, and AI-based capabilities have been used to support other technology domains. This technology evolution involves HI, smart devices/sensors, and reasoning systems that interact when deployed in a decentralized IoT networks [2,4]. This advance is due to continuous technological innovations involving AI, KRR, ML, DL, artificial neural networks (ANNs), IoT devices and sensors. KRR is a subclass of AI that focuses on data displaying and processing [1]. The research discusses other interdisciplinary areas, such as psychology and mathematics [1,2].

In 1955 John McCarthy coined the term AI, as a "science and engineering of making intelligent machines" [5]. In the summer of 1956 McCarthy organized a technology workshop at Dartmouth College. At that conference, other AI research leaders, namely, Allen Newell, Arthur

Artificial Intelligence to Solve Pervasive Internet of Things Issues. DOI: https://doi.org/10.1016/B978-0-12-818576-6.00002-2

Samuel, Herbert Simon, and Marvin Minsky, joined McCarthy [6]. As part of the conference proceedings, these experts presented a research project on AI topics. The researchers also received positive feedback from those in attendance. Those who attended the workshop at Dartmouth in 1956 were deemed to be one of the AI research leaders [7,8]. These researchers along with a group of students developed a computer program called "astonishing." In 1954 AI researchers conducted an array of tasks to satisfy past research work. Those projects led to successful technology breakthroughs, some of which today are called AI, KR, ANN, HI, ML, and DL [5].

In 1959 computers were able to play the "checkers strategy game" at a much-accelerated pace than an average human. Checkers' strategies consisted of information processing system-based games that researchers designed to solve issues involving the algebraic "proving logical theorems" [5].

2.3 KNOWLEDGE REPRESENTATION AND REASONING

Knowledge representation (KR) is a method that involves formalism [4]. KR is a subclass of integrated AI functions [4]. It lies in the accepted concept and design theories for information visualization, logical thinking, and processing [3]. This concept includes "semantic networks (SNNs), systems architecture, frames, rulers, and ontologies." Automated reasoning differs from "inference engines, theorem provers, and classifiers" [4].

AI is a domain that integrates technological areas namely HI, computers, and intelligent machines [3,9]. These domains range from AI intelligent systems to fused KRR capabilities required to process tasks through perceptive methods [2,4]. AI dates to the era of the Greek philosopher, Aristotle's earliest events.

Aristotle's findings examine the relationship between philosophy and logic concepts and theories [3]. Aristotle describes "reasoning" a *syllogism* [3,9]. He defines syllogism, a speculation that involves objects, ideas, factors, events, phenomena, or entities. A syllogism describes initial results in Aristotle's investigative methods [3]. Whereas the term "reasoning" has originated from the disciplines of computer science, sociology, and psychology. In contrast, logic is a process for gathering and blending methods, it operates at the heart of an assumption that man makes to ensure collaboration of two or more concepts and thoughts [9]. This concept illustrates different forms, such as providing or displaying data through analytical conclusions [9–11], as summarized by McCarthy [10]:

"A program delivers a common sense if it automatically deduces for itself a sufficiently wide class of immediate effects of anything, it is ordered and what it already knows... For a program to be capable of taking something, it must first be capable of being stated it." "Programs with Common Sense" [10].

2.3.1 KNOWLEDGE ENGINEERING

Knowledge engineering (KE) plays an essential role in applied AI research sphere. Researchers argue that an idealistic way to address, KRR, AI, KE, ML, and DL issues is by collecting and parsing data. This method ensures that researchers have the capacity they need to examine machine reasoning and how each of these systems can do specialized jobs. Thus KE is a cognitive operation that represents scientific and societal features. Such method plays a critical role in designing and supporting knowledge-based systems.

2.3.2 EXPERT SYSTEM

Expert system (ES) is an application that clinicians use to carry on medical diagnoses [12]. Edward Feigenbaum is one of the first scientists who coined the term "expert system" [12]. The first ES did not embrace a standardized functional process. Hence, the system lacked the capability researchers needed to extend their investigative scope, which subsequently contributed to the development of inference engines and software applications [12]. The original ES consisted of the following software processes: conventional and development methods, and the need for designing unique programs for tuning or streamlining the requirements for building ES. Whereas knowledge acquisition (KA) gives companies such as Andersen Consulting, the ability to secure research opportunities. KA ensures that the redesigning processes use different versions of prototypes [12].

2.4 KNOWLEDGE INFERENCE, FORWARD AND BACKWARD CHAINING, AND KL-ONE LANGUAGES

Knowledge inference, forward chaining (FC) and backward chaining (BC) or backward reasoning (BR), and KL-One languages are applications or engines that are being employed today. These applications/processes are deployed along with ES prototypes [12]. ES is a hardware that stores large volumes of datasets. The stored data contain large volumes of data or datasets about the world. On the contrary focuses on dual methods of reasoning prototypes [12]. Researchers argue that this process occurs when there is a logical description of the interface engine. The ES implementation process is extremely complicated. The logical process moves the commercial enterprise and production-based rule system prototypes [12]. BC/BR is a process that relies on artificial intelligence applications (AIAs), automated theorem evidence ES, and proof assistants. BC and FC are concepts integrated in AI, game theories, and other ES prototypes [12]. In KE, domain ontology (DO) and domain-specific ontology (DSO) describe many things, such as objects that exist around the world. Humans view ontology as a process, which involves poker domain prototypes [12]. This concept describes a method for modeling, computer hardware, notably "punched and video cards" imports, and others. In the DO, ontology meanings are interpreted through domain perceptions prototypes [12].

In 1977 Ronald J. Brachman discussed the term "KL-One" in his doctoral research—KL-One is an integrated NR system [13]. Preceding Brachman's research work, other scholars conducted more studies focusing on "representation." The research was independent of other studies that incorporated similar domains. KL-One permeates the representation of the indistinctness in SNNs. Its process describes technical data, logically, called a structured inheritance network [14,15]. This method aims to new technologies, namely epistemological level (EL). EL focuses on dealing with complex concepts, such as attributes, instantiations, descriptions, uses, and inheritances [14–16].

Brachman [17] states that, in early investigative studies, each of these concepts was illustrated as "representation systems" [18]. These systems range from traditional and technological SNNs and form networks (FNNs). The progress and fine-tuning methods have improved over the years. These schemes are used in scientific research areas, to implement new knowledge-based concepts and basic research methods for the AI society [15].

KL-One kernel points out that AI researchers rely on to formulate structured complex descriptions [15]. This knowledge-based method is applied to identify and defines the relationship between network theories [13]. SNNs or FNNs are networks that describe systems, which can be deployed to a KR environment. These systems focus on direct and undirected graphs equipped by the vertices [13]. In this context, vertices symbolize "concepts and edges." Vertices describe relationships between concepts [13]. SNNs are identified as semantic triples. The networks are used to support an array of applications, such as NLP [13].

2.4.1 TECHNOLOGICAL SINGULARITY AND RECURSIVE SELF-IMPROVEMENT

In 1959 Allen Newell and Herbert A. Simon developed KR. KR is designed to resolve complex issues affecting technology singularity and recursive self-improvement [4]. Newell and Simon developed a method to integrate plan and analyze data structures [3]. These strategies stem from outlining conceptual plans to supporting predicted outcomes. AI solutions are deployed to support general search algorithms [19]. These procedures include (***A****) frequently known as (A star). The "A star" computer algorithm was produced to support AI solutions and issues involving GPS systems. The failure of these scientific research efforts resulted in a cognitive change [3,19]. Fig. 2.1 explains an adapted-logical relationship between the triangle of cognitive science fields [3,19]. The picture in Fig. 2.1 depicts a theoretical explanation for cognitive skill fields and variables.

2.5 ARTIFICIAL INTELLIGENCE

AI provides intelligent machines with the ability to translate external data [6]. AI collects and examines datasets, to help identify, predict, and probe the origins, applicability, or qualities of data needed for processing. In AI, the process to visualize data gives analysts and scientists the ability to discover and display dataset patterns [1,6].

Decision makers rely on the collected, analyzed, and processed data to draw informed decisions about specific goals and tasks [6]. The complexity and lack of scalability affecting neural networks are different from that of the traditional systems [1,3]. In AI, regression reasoning can be insignificant, irrespective of the relationship that these variables display in a limited heuristic reason [3,10,11].

Kaplan and Haenlein [6] cite AI researchers argue that traditional computer systems are not immune to comparable functional behavior. Despite these integrated intelligent functions, AI shows pseudointelligent reasoning. It incorporates "strong and weak" behaviors within the environment—these computers often display a substantial degree of parallelism [3]. Weak AI

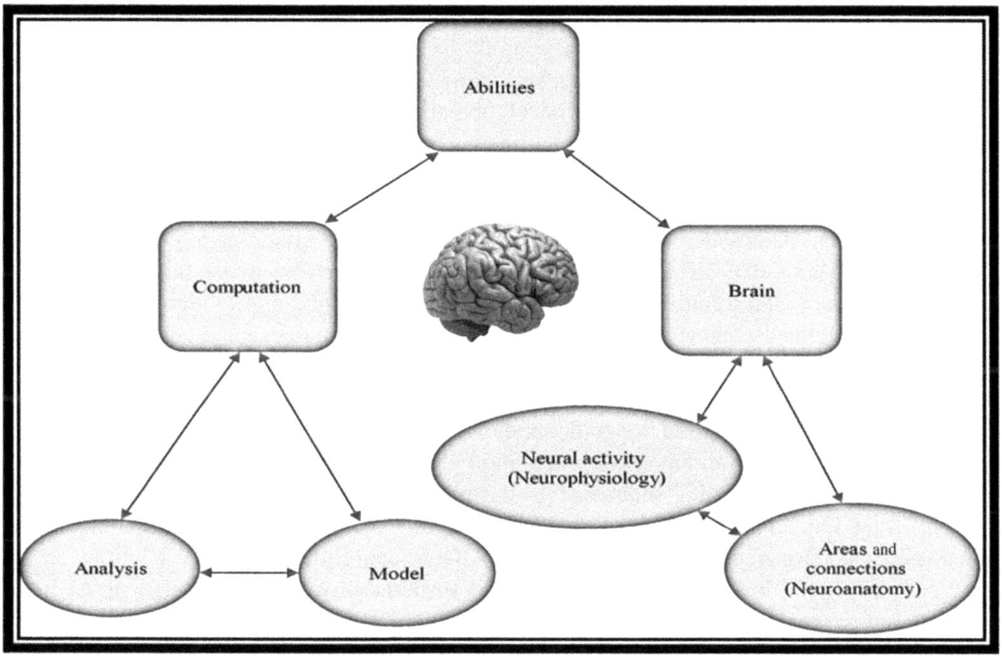

FIGURE 2.1

Cognitive science studies.

consists of information processing system-based applications that have limited interaction with other intelligent objects [5,6]. This method stems from applied knowledge and exceptional characteristics, that is, autonomous systems and heuristic ability search algorithms [10,11]. Despite the progress that scientists and researchers made, AI-based organizations view performance limitations as a challenge to AI technology. These technical constraints include "strong and weak" AI capabilities. Such restrictions are attributable to systems with autonomous interaction [3].

2.6 ARTIFICIAL INTELLIGENCE FOR INFORMATION TECHNOLOGY OPERATIONS

Artificial intelligence for IT operations (AIOPS/AIOps) was previously known as "algorithmic IT operations analytics." AIOps is a solution that provides agile and cost reduction for public and private sector. AIOps is an "umbrella term" that was coined for advanced technological domains: ML, big data analytics, and AI. Researchers developed AIOps to provide technical support for the identification, automation, and resolution of various issues affecting the IT field. In the recent decade, business, scientific, and research communities have used AIOps to monitor IT solutions while acquiring clear visibility into dependencies beyond IT systems.

Despite this adaptive terminology, researchers define AIOps a platform, which syndicates big data and AI functionalities. AIOps focuses on IT operational structure—for example, AIOps processes tasks, executes, deploys and provides continuous backup to functional areas in the enterprise. Commercial enterprise and technological functions can be deployed as a toolkit. AIOps capabilities support performance monitoring, event requirements, and business/data analytics solutions. IT service management and automation offer customers hybrid technical solutions to ensure that middle management and senior executives have the capability they need to make informed decisions. AIOps capabilities are vital to an organization's mission success and agile IT resource distribution. AIOps delivers integrated technical functions, such as propelling an organization's business capability, while providing infrastructure, continuous monitoring to support end-to-end IT solutions. Engineering science measures, automates, and monitors business practices to boost productivity and increase efficiency or streamline operational delivery timeframe. AIOps platforms are deployed as technical solutions to support enterprise integration of IT solutions. Such solutions involve monitoring, automating, and providing maintainable business/functional posture needed to sustain the holistic enterprise functioning. Managers and executives rely on these cognitive operations to make informed decisions. Organizations should invest in AIOps capabilities. These IT solutions are vital to organizations' digital business transformation and sustainability. The integration of AI into IT operations ensures that businesses have the seamless capability needed to perform at a faster rate and supply that the layer of productivity.

2.7 AI, KRR AND IoT FUNCTIONS

In 1956 John McCarthy coined the term "AI" at a conference organized by Dartmouth as part of his scientific exhibition. CS involves the following technical areas [19]: psychology, linguistics, anthropology, AI, neuroscience, and philosophy. In the same year, McCarthy devoted more time to research on parallel topics—NLP, image identification, sorting, and ML. McCarthy's research yielded significant AI results from, most of which many industries have used for many years [3,19]. Today, research and development and academic institutions are taking advantage of these AI-based results to grow operations management functions [3]: technology assets and technological resources [10]. Implementing AI and KRR capabilities gives clients the tools they need to operate effectively and efficiently in the digital era [2,4]. Despite this progress, other industries have employed similar AI/KRR processes to support their daily operations—these industries describe AI and KRR as different things [3]. Many researchers state that AI and KRR can be embedded with ML and DL to provide a hybrid functionality [3].

AI is a subclass of computer science that focuses on the study of human behavior. Researchers argue that AI does not focus on human behavior only; instead, it encompasses the interaction between humans and computer systems. The process includes complex activities that comprise awareness, perceptual experience, reasoning, mimicking, and the environment [10]. These computer systems consist of swarm intelligence (SI) networks, standard algorithms, intelligent robotic systems, and neural systems [1,2]. There is a sheer functional characteristic between traditional computer systems and AIAs. The variances between the core and established systems or the AI-based applications include, but are not limited to [2–4]:

- Sophisticated and functional features involving AI-based applications;
- A complicated step-by-step algorithmic procedure that computer systems should play along to execute/perform functions effectively; and
- The time AI computer systems need to model and process data in nontypical functional behavior.

In the 1950s "cognitive revolution" was an experimental intellectual term. This term was named as cognitive science (CS). CS involves transdisciplinary or systematic areas that emphasize the underlying research on HI and operations. These methods include the functions executed in the CS and cognitive functions of human awareness.

In the last decade, cognitive scientists developed the use of logical thinking to analyze the behavior and intelligence of the human mind [19]. Reasoning defines the computer logic needed for automating and sorting distorted scientific issues. It further describes how humans utilize the "application of instructions" amid a group of objects and subclasses of relationships [1,3]. Researchers are studying neural systems, theatrical performance, and how these operations can be enhanced to process information in real-time. This process includes language, belief, attention, abstract thought, memory, and emotion.

2.8 ARTIFICIAL INTELLIGENCE APPLICATIONS AND TOOLS

AIAs solutions are designed to solve complex problems. These products provide the capacity required to hold up the decision-making process within an organization [3]. AIAs relies on integrated applications, that is, chatters and others. The results are projected to mimic HI by providing seamless, timelessness, consistency, and cost-effectiveness unified technical and business approach [3]. AIAs are deployed to solve complex issues that may affect an organization's daily operation. Some of these problems range from system agility and holistic IT solutions that decision makers need to make informed decisions [3]. These processes are provide decision makers with the tools they need to make an informed decision, using "qualitative and quantitative" data processing methods.

Many researchers indicate that AI is a technology that its analytical methods have not fully matured [3]. AIAs continue to support different businesses while collecting, assessing, processing, monitoring, and furnishing decision makers with automated toolkit needed to solve complex issues [3]. AI analytical processes consists of algorithms, mathematical optimization, and advanced computational reasoning. These presented solutions are developed to solve AI complex issues, such as broad-based scientific solutions integration.

The following processes are built in to the artificial intelligence tool (AIT) operationalization [2,4]:

- Logical thinking
- Premises to conclusions
- Inference rule
- Planning

- Means-ends analysis
- Robotics
- Local searches in configuration space
- Learning and optimization

AITs include various research and development areas. AIT was developed to solve complex issues involving AI [2,4]. In AI, "logic" is a cognitive operation that supports KRR and problem-solving processes [3]. The advocated methods can be used interactively to support IT operations, some of which involves algorithms, planning, and logic programming. Each of these results are provided as integrated capabilities to sustain the scholarship process [3].

2.9 MACHINE LEARNING AND DEEP LEARNING

ML is part of the AI; there is a symbiotic relationship between AI, ML, and DL [20]. AI provides learning process categories, for instance, unsupervised, supervised, and reinforcement [21]. Unsupervised and supervised learning processes focus on the classification and mathematical regression theories. ML is implemented when machine classifications of things are allocated into various groupings to achieve a unique business objective.

This process applies to regression, which is a practical solution, a single machine can generate as a function. This solution produces inputs and outputs ([21,22]). How these inputs and outputs generate roles is key to delivering ML operational functions. In this context, the agent is rewarded for giving first-rate answers. The process runs agents analyzed for incorrect responses [22,23]. In the ML, agents depend on rewards and punishments to find practical strategies. Some of these concepts are designed to ensure that intelligent machines perform within assigned AI boundaries or settings [23]. For that reason, unsupervised, supervised, and reinforcement learning strategies that can be broken down within the decision theory and through the operation of judicial decisions, that is, utility and others.

2.10 ROBOTIC PROCESS AUTOMATION

Robotic process automation (RPA) is software that involves AI and ML solutions. RPA processes volumes of metadata and datasets. The term "robotic process automation" dates back to the early or mid-2000s despite its existence that goes as far as many years ago. The term spans a trio of technologies, notably screen scraping, workflow, the automated arrangement, as well as AI-based solutions [23]. Screen scraping involves machine processing and data display screen collection. Its capabilities range from traditional applications needed for data displaying through the user interface methods. The automated workflow software streamlines the process that takes for data to be sent to multiple system interfaces manually. This progress gives mechanical systems a continuous capability needed to increase the production timeframe, like efficacy, and accurateness. In IoT, this process is agile and interactive, allowing for computer systems to deliver requested or assigned services without experiencing any single point of failure [22]. The concept involves tasks that, in the past, humans used to achieve them.

In the database, AI and ML projects involve queries, computer or computerized processing, electronic recordkeeping, and other types of business transactions [23]. RPA is not an enterprise IT infrastructure enablement application. RPA can be offered as a Software as a Service (SaaS) enterprise solution to facilitate smooth deployment of technical capabilities. The technology cycle will take place without impacting IT infrastructure downtime [22].

In ML, the mathematical analysis is a branch of theoretical computer science [22,23]. The field is a computational learning theory, offering robots the capabilities they need to acquire new skills and to adapt to other environments. This unified capability supports autonomous self-exploration and social interaction involving human educationalists. Steering mechanisms, like active learning, development, and other synergic aspects can be arbitrary and repeated throughout the general learning process ([24–26]).

DL is a computer science discipline that includes other technology areas, such as deep network learning, ANNs, and cognitive computing. These three key areas are built upon computer science concepts. Vendors and researchers have used the term "cognitive computing" to exemplify a significant association between these technological areas: DL, deep network learning, and ANNs [3]. This term is generally applied and converged to support the enterprise—in part, researchers examined DL and related technologies as rapidly involving phenomena. AI objects can process data with limited human involvement ([22,23]).

2.10.1 PLANNING, SCHEDULING, AND LEARNING

Planning, programming, and learning are processes that provide intelligent agents (IAs) with the capability to determine and achieve realistic objectives.

In classical planning problems—the agents can accede that within an AI only one organization should be deployed and running at the time; such concept, yet, allows brokers to be precise about the consequences surrounding similar activities [23].

The below list explains an existing relationship between the hierarchical AI-controlled organization. Such an association includes the relationship between actuators/sensors, controlled systems, operations, and environs.

- Top level node
  - Specific goals and projects
  - Sense data and results
  - *Node (1)*
    - Actuators—actions can be embedded into the controlled system, process, or surroundings
    - Sensors—sensations from the controlled system, method, or environment into sensors
  - Node (2)
    - Actuators/sensors
      - Actions from sensors/actuators into the controlled system, method, or surroundings
      - Sensations from the controlled system, method, or environment into sensors/actuators

If a single agent or multiple brokers are not acting alone within AI, the desired process assumes that the assigned agent can reason through a layer of ambiguity. This process ensures for agents that are incapable of inept or understanding environments, to make proper estimations, or become accustomed to a specific or assigned/designated environment [23].

2.11 INTERNET OF THINGS

In 1999 Kevin Ashton coined the term "Internet of Things" after conducting several types of research at the Massachusetts Institute of Technology (MIT) Auto-ID Center. Ashton preferred to address it, "Internet for Things" [27,28]. Ashton embarked on a new research project involving the radiofrequency identification (RFID). Through such complementary research findings, Ashton was able to incorporate RFID technology into IoT-based applications. This advance allowed objects to interact and manage data or other things through a distributed or an interconnected physical network [27,29].

IoT is a dynamic network of distributed and decentralized or distributed physical objects. IoT consists of smart devices and intelligent systems [30]. These smart objects can be embedded as sensory systems—RFIDs, actuators, driverless vehicles, smart buildings, and closed-circuit television [27]. The objects range from smartphones, tablets, smartwatches, home appliances, and consumer applications, explicitly creative industries, home automation, and wearable devices [31,32]. These technological solutions can be equipped with network connectivity beyond standards. This method allows devices and organizations to collect, examine, exchange, and store data in real-time [31]. More consumers have smart devices connected to interconnected/distributed networks [32]. These objects can be interlinked via the Internet with the ability to send or receive data. These objects' embedded functions are designed for remote monitoring and supervisory [27]. According to Wigmore [32], the number of IoT objects in late years has developed significantly. This disruptive technology surge is due to a rapid increase in AI, ML, DL, commodity sensors, and others [32]. The demand for innovative methods to power the IoT devices and organizations is paramount [33]. In 1982 a group of researchers discussed IoT technology at Carnegie Mellon University. The discussion was about a technology concept on a "coke vending machine." It involved proof of concept entitled "Internet-connected appliance."

In the same year, McCarthy dedicated time to research on parallel topics, notably NLP, image identification, categorization, and ML. His research yielded significant AI results, which comprised AI analytical processes, that is, search algorithms, mathematical optimization, and advanced computational reasoning. These domains are deployed to solve AI complex issues through a wide-ranging scientific solutions integration. The proposed methods can interactively be applied to support processes, for example, algorithms, planning, and inductive logic programming. These methods allow for a unified and principal learning process among smart objects [3].

AI objects can process data in real time and interact with limited human involvement [22,26]. If a single agent is, or multiple agents are, not acting alone within an AI the desired process assumes that the assigned agent can reason through a degree of ambiguity. According to Wigmore [32], IoT objects in late years have developed significantly. In essence, such disruptive technology scale is due to a rapid increase in AI, ML, DL, commodity sensors, and others [32]. The demand for innovative design methods to power the IoT devices and organizations is paramount [33].

In 1982 a group of researchers discussed a revolutionized IoT technology at Carnegie Mellon University. This treatise was part of a technology concept on a "coke vending machine." It involved proof of a concept entitled "Internet-connected appliance."

2.11.1 INTERNET OF HEALTHCARE THINGS

The emerging of mobile technologies in the digital era has increased considerably. This rise is due to the constant demand for patient care, data provisioning, and the modernization of the electronic health record (EHR) system. Tech companies and governments are on the verge of modernizing and deploying their capabilities to meet patient's needs better. Most of these revolutionary transformations in public, private, and government sectors are due to the increase of patient data, for example, private EHRs [34].

The Internet of healthcare things (IoHT) is a concept that involves mobile technologies, ranging from intelligent systems, sensors and wearable devices. This method includes smartwatches that can be deployed in the distributed or decentralized IoT networks in support of the patient's EHR system. As a result of the disruptive and ubiquitous challenges that the governments and industry have dealt with in the past decade, data provisioning, visualization, parsing, and allocation, and the need to modernize legacy healthcare infrastructure are paramount [34].

IoHT involves intelligent devices: mobile technologies and smart machines. Vendors are still grappling with developing a new analytic baseline that supports the EHR modernization and data provisioning via integrated cloud solutions. In the IoHT environment, intelligent devices and sensors are deployed to collect and process data with less time than the traditional healthcare systems. The need for a flexible and integrated healthcare system will ensure that providers and patients have a continuous interaction and the ability to share information in real time. This interactive capability can only be possible if a robust healthcare infrastructure is built to support such a patient's need. Through biometric technology, patients and providers would be able to share private data and ensure for its protection [34].

2.11.2 REAL-TIME HEALTH SYSTEMS

Real-time health systems (RTHS) are convergence and integration, data collection, and intelligent sensor-based systems that can be deployed to collect patient information via IoT platforms. These devices comprise mobile technologies such as RTHS. Mobile devices collect, analyze, and process data between RHTS and IoHT devices. Data sharing is processed via the RTHS-integrated portal. In the clinical data ecosystem, EHR does not only address patient-based situational awareness [34]. It collects a patient's health information besides the care that each patient can receive at any medical facility, like hospitals or clinics. This application, captures patient's critical data. RTHS is responsible for gathering, analyzing, and processing the patient information and make it available to the assigned clinicians, who are accountable for the patient's continued care [34].

RHTS collects IoHT patient information, parses and provides actionable clinically based results. The data offer relevant indicators and trends concerning the current patient's health status as well as continuing treatment [34]. Within an integrated EHR/RTHS environment, patients and providers can exchange data securely and in real time. Researchers note that this revolutionized the healthcare capability industry predicting for many years [34]. Vendors argue that more work needs to be done to ensure that the EHR system continues to meet patients and health providers' needs. With the support of RTHS, health providers are now able to collect, process, and analyze patients' confidential data faster. This process encompasses identifying clinically relevant datasets, indicators, and forecast that provider use to predict future illnesses. Establishing an integrated and friendly solution that providers need to monitor data and alert the patient of potential or catastrophic diseases, some of which might lead to death. These health IT solutions include mobile devices that are paramount capabilities of a patient's continued healthcare [34]. With the new and integrated health, IT solutions will be able to provide the patient with real-time notifications of probable illnesses and ensure that such diseases are treated immediately before reaching an irretrievable stage [34].

2.12 NATURAL-LANGUAGE UNDERSTANDING AND INTERPRETATION

Natural-language understanding (NLU), is a subclass of NLP. In AI, NLU/NLI is a technology area that focuses on the study of machine reading comprehension [35]. It addresses AI-complex issues [36].

In 1964 Daniel Bobrow from Massachusetts Institute of Technology made the first computer, try to address NLU [36]. Aside from Bobrow's ambitious attempt presented in the dissertation entitled "natural language input focused on the computer problem-solving system," nearly 8 years later, a proven AI researcher, John McCarthy invented the term AI. NLU focuses on global marketable relevance, straddling AIAs, to systematized analysis and reasoning [36].

The NLI method involves the following AI areas [36]:

- Machine translation
- Question answering
- Newsgathering
- Text categorization
- Voice activation function
- Archiving
- Large-scale content analysis

These significant areas incorporate text postprocessing, which is central to NLP algorithms stage-processing along with some parts of speech identification using context from other recognizable devices such as automatic speech recognition and sight recognition [36]. NLU is a term that can broadly be working in AI, robotics, and other complex software engineering fields. Researchers suggest that this applied method can be used in computational applications involving small, medium, to large-scale tasks that are assigned to robotic systems [36]. It can support diverse computer applications, for instance, text classification needed for email automatic analysis, and others. NLU focuses on promoting and sharing pieces of standard algorithms elements, explicitly language lexicon, parser, and grammatical rules, needed to divide or to structure sentences as well as core illustration. Semantic theories must complement AIAs. These theories are developed as interpretative capabilities to translate applications in the language-understanding system.

In contrast, semantic methods consist of [36]: naive semantics, stochastic semantic analysis, pragmatics, and semantic parsers. This concept spans technologically advanced applications aimed at integrating logical inferences into the framework. In NLU, there are two types of logic, that is, "predicate logic and logical deduction." These logics, when simulated or applied to NLU, aid in reaching logical conclusions [36].

2.12.1 CLASSIC ARTIFICIAL INTELLIGENCE METHOD

Many researchers describe the classic artificial intelligence method (CAIM) as an early version of AI. Researchers invented CAIM to process and support extensive computer programs and network applications. It focuses on translation of complex mathematics, algorithms, and computer problems, preventing human brains from performing any activities. The need for user-friendly computer applications to interpret text messages and related software features is paramount to CAIM's functions. These applications traverse the human's ability to recognize AI objects in an image. Researchers argue that more research findings will improve AI solutions.

At present, there are millions to billions of IoT objects on the planet earth. These intelligent devices and systems are generally deployed to several areas to perform a straightforward or complex activity or mission. IoT devices are ubiquitous objects that can be found almost everywhere around the world. Researchers predict that in the coming years, there will be trillions of IoT objects around the world. These objects will be deployed to support various activities and missions [3,19]. The need for user-friendly computer applications to interpret text messages and related software features is paramount to CAIM's functions. These applications intersect the human's ability to recognize AI objects in an image. Researchers argue that more investigative findings will improve AI solutions [3,19].

As many researchers would antedate, natural intelligence (NI) is not a subset of AI; instead, it incorporates systems of control, which are not artifacts. NI involves the functioning of animals and human brains [37]. AI objects comprise neuroscience, researchers believe that NI continues to play an integral role in the medical field [37]. With the advent of AI, KRR, and OCR, many researchers suggest that "physicalism and functionalism," can be the two breakthrough assumption that often gives an insight into how the human's mind functionalities.

Identicality in human reasoning is documented as a reductive method that associates one's intellectual faculties with other human phenomena, like neuronal activities. The mind can trigger responsiveness and intentionality toward an environment. Such action leads to a perceived, responsiveness, and actable stimuli within the brain. This involuntary activity in the human's mind, prompt the human brain to begin thinking, while generating perceptive feelings toward others or an environment [38].

2.13 LEARNING USING PRIVILEGED INFORMATION

Learning using privileged information (LUPI) has been used in academia, industry, and other business sectors. LUPI is a process that transfers knowledge such as privileged information. In the new learning developmental paradigm, LUPI constitutes a part of the training phase that includes multidimensional methods needed for delivering tangible results. Vapnik and Vashist adopted and popularized LUPI as a transformational concept. In academia and industry, LUPI is used to compare data, provide a consistent level of reasoning. This approach involves logic, emotion, or metaphorical reasoning. LUPI privileged information may involve confidential communication, nondisclosure agreement, and need to know.

Privileged data is a process that handles confidential data, such as patient records. How data are handled, processed, and shared is essential to the provider and the patient. In ML, LUPI denotes a new paradigm that balances the digital era of information, technology, innovation, and processing digitized resources over the enterprise. Such concepts may be conditions for consistency and how machines may learn from their environments. LUPI focuses on practical algorithms, for example, support vector machines. In each environment, machines are programmed to execute diverse functions. These roles include processing or ruling potential algorithmic outliers. LUPI is embedded in technology and science. This method focuses on "human and classical ML" systems.

2.14 PICTURE ARCHIVING AND COMMUNICATION SYSTEM

Picture archiving and communication systems (PACS) are used in the healthcare industry. PACS is a medical imaging solution that ensures clinicians, have the permission to store the patient's confidential and nonconfidential information. The solution gives healthcare providers access to cost-effective storage capabilities. This repository retrieves patient's data, such as archived dental and medical records.

Using the AI and ML technological capabilities, medical providers are now able to view images that are daily received and processed from multiple sources of intelligent devices and sensors. In the PACS' healthcare ecosystem, electronic images and other data can be uploaded, processed, and parsed digitally. This healthcare e-capability focuses on streamlining the processing time that providers often take to send and share a patient's information to selected entities within the medical community. This technology has given medical practitioners and providers the capability they need to conveniently access, retrieve, parse view, and processing patient's data in real time or with a limited human error. Due to this critical process, today, practitioners can conduct computed tomography, magnetic resonance imaging via a dedicated, distributed, or secure network system. The process ensures that a patient's data can be transmitted and viewed through a secure portal. This capability protects and secures patient's confidential medical records. The technology streamlines and minimizes physical and time-barrier-based accessibility to confidential patient data, which unauthorized users may obtain. The data are stored on the provider's medical database systems. With the disruptive and ubiquitous IoT intelligent devices and sensors, providers can process patient's images or other confidential data with minimum/without disruption to the provisioning system. The PACS consists of four distinctive domains. Each of these domains can provide remote access, electronic image integration platform, and radiology workflow management. PACS has a robust capability to process data between local and wide area network distributed nodes. These capabilities can be deployed via a virtual private network and a secure socket layer. It supports interactive applications, like ActiveX, Javascript, and Java Applet.

2.15 INFRASTRUCTURE-BASED MOBILE NETWORKS

The industry continues to research and develop proven infrastructure-based mobile capabilities, like ad hoc networks, to fulfill customer's urgent and long-term needs. These decentralized capabilities come in many flavors, such as those equipped with intelligent sensors and devices that can provide quick and responsive results. The network infrastructure consists of a hardware and software solution that is implemented to support an IT environment [39]. These IT resources are used to provide seamless connectivity, communication, operations, and management solutions for small and large networks. Through these capabilities, users can share services and data through decentralized/distributed, unified, and interactive IT networks. Wireless and mobile ad hoc networks are vital to day-to-day customer's IT requirements. In the IoT, ad hoc systems are deployed to support other hardware and software solutions [39].

2.15.1 IoT CONSUMER APPLICATIONS

IoT consumer applications continue to play a crucial role in the integration of smart objects needed to support/interact with MI, ML, DL, and KRR applications [27]. IoT spans a total of six consumer applications. These applications comprise the following IoT consumer applications [27,39].

- ***Home security and smart domestic***: This type of IoT consumer application includes domestic safety (DS). DS is a critical area that has promoted the IoT services. Despite this progress, researchers concluded that in IoT, most devices and systems continue to experience network and security vulnerabilities. Such breaches are due to a series of weaknesses and threats affecting IT assets. Vendors have been working tirelessly to develop new software applications to avert or minimize some of the weaknesses and threats against the IoT devices and systems. Home security often ranges from a method that is designed to provide home appliances the level of protection needed to protect them from unauthorized users, for example, hackers and intruders. These devices are acquired and deployed in homes, business offices, and related facilities. Most of these objects can be attained in the market at an equitable economical price.
- ***Private healthcare, healthcare carriers, and healthcare players***: In healthcare, IoT devices and systems have been deployed to provide governments and enterprises with proven capabilities needed to support mission-critical solutions. These solutions can be deployed in smart homes and smart cities. These capabilities are designed to provide the ecosystem with a blend of continuous monitoring capabilities needed to protect data and patient records. Similar IoT devices are deployed or used as a wearable device that clinicians can use to diagnose patients. These wearable devices are ubiquitous in the sports industry. Athletes use these types of tools to monitor activities on the field, for instance, following heart rates, steps, and among other things. Such devices have the capabilities of monitoring, collecting, uploading, and storing data to a remote repository for future processing.
- ***Wearable technology***: In this context, are used in either individual or multiple persons' activities. These devices have been used in sports or in healthcare, but also for one's daily use. Some people use wearable devices instead of watches or other types of tools designed to serve a particular individual's purpose. In heavy industries, these smart objects are used as smartwatches and health monitoring devices. Vendors continue to develop new methods that can better meet the consumer's day-to-day requirements.
- ***Asset tracking or tracking valuable assets***: Wearable devices that can be employed as technology assets, which serve the consumer's needs. The global positioning system is a perfect example of intelligent machines that monitor or track an individual's requirements. IoT consumer applications have evolved over the years. This rapid evolution in IoT is due to the advent of advanced applications that vendors have built in recent years.
- ***Workplace***: Having IoT devices and objects deployed in the workplace can improve the level of work while providing the level of security that decision makers and employees might need.

Today, most of these devices and objects can track, store, collect, and disseminate actionable data to relevant sources within the workplace, by providing continuous monitoring within a physical security perimeter for the ability to function as sensors. Most sensors are designed to adjust the room temperature in residential and commercial buildings.

- ***Play***: IoT devices play an essential role in everyone's individual life and in the workplace. These devices and objects can perform activities beyond the human imagination. Some of the devices and objects can be found in luxurious resorts where people often plan annual vacations.

There are millions to billions of IoT objects on the planet earth. These devices and objects are generally deployed to several areas to perform complex activity or mission. IoT devices are identified as ubiquitous objects and can be deployed almost anywhere around the world. Researchers predict that in the coming years, there will be trillions of interconnected IoT objects in the world. These objects are deployed to serve activities and missions [27].

2.16 DYNAMICALLY CREATED MOBILE AD HOC NETWORKS

Dynamically created mobile ad hoc networks (DCMANs) consist of autonomous and dynamic network systems. In a distributed network, DCMANs are connected by nodes that can be deployed or redeployed to the decentralized networks. Despite this progress, DCMANs lack the infrastructure complexities and setup. Dynamic wireless networks are administered via the process of enabling smart objects [37]. These devices can be configured and deployed in assigned networks at any given time.

2.17 INTELLIGENT AGENTS, CONVERSATIONAL AND NATURAL INTELLIGENCE

IAs are AI-independent objects or software applications designed to analyze, retrieved, or display information collated from the Internet. IAs extract data from the Internet for immediate or future use [37]. These intelligent objects consist of computational activities ranging from macros in the Excel spreadsheet and Word pages [40]. Such objects process or monitor data flow through sensory devices and act within the environment as believed necessary. In an IA, sensing devices can interact with actuators known as agents [37]. This process allows actuators to autonomously send data to intended nodes or systems to meet a business goal.

In various business sectors, decision makers often rely on IAs to study or gather data needed to be used toward a business goal [40]. IA includes devices and objects, for instance, a thermostat, complex systems, and other reflex machines [37]. In AI, human's traditional perception of conversational intelligence (CI) means different things. Researchers define CI as a method that describes a human's most authoritative and hardwired ability to be able to engage with other humans through conversation mechanisms [37]. When decision makers talk to each other, whether by using verbal or gestural language and being able to translate whatever message they have into real action that can be called a part of conversational intelligence known as CI process [37]. The following are three dimensions humans rely on to communicate in an office setting or the public domain [37]:

- ***Biochemical***: Gives decision makers the ability to gain control of independent methods of thinking or cognition. If a manager in the workplace is either unfulfilled or exhausted, this type of dimensional way can display such reaction. Decision makers must be cognizant of their surroundings, when a similar response occurs in the presence of others, for example, employees, peers, and others [37].
- ***Relational***: Every human has ambitions and needs in life. When interacting with others, we tend to develop a last-long affinity and bond, which helps shape our relationships [37]. The relational dimension is a critical factor in the workplace. Managers should be aware of these steps when interacting with employees, whether on a project or other forms of engagements [37].
- ***Cocreational***: How to cocreate conversations among humans in the workplace? This type of discussion ensures people have the ability to interact and be able to show or create new conversational forums, which often lead to positive results. Concreating conversations is a concept that allows people to forgive, learn from past mistakes, and be able to refocus attention on more positive future interactions. The ability to help or work with others to achieve a common goal part of the cocreative conversation. For example, being humble, respect others, and is also listen to others' views among other things is key to proactive and/or healthy dialogues [37].

CI performs as a platform to connect these three-dimensional processes in the workplace and in our livelihoods [37]. Three-dimensional conversations focus on developing creativity among cultures and people while giving them the opportunity they need to thrive in society [37]. In the context of NI, organizations and decision makers should be more aware of how this process is accepted in the workplace. As many researchers would antedate, NI is not a subset of AI. Instead, it includes systems of control, which are not objects. NI involves the functioning of animals and human brains [37]. AI objects span neuroscience, researchers suggest that NI continues to play an integral role in the medical field [37].

2.18 ADVANCED METERING INFRASTRUCTURE

Advanced metering infrastructure (AMI) was coined and popularized to support the infrastructure and smart meter solutions. AMI involves two-way communication networks that businesses use as the metering capability in support of the enterprise utilities and consumers' needs. It provides automatic and embedded functions, in which legacy systems did not have in the past, to sustain the holistic efficiency or customer networks. AMI monitors the consumption of electricity that data centers and the comprehensive infrastructure need to power devices or systems. It is a time-based rate and incentives, which boosts a client's interest when deciding on the best cloud services that fit their immediate and long-term business needs. The solution reduces electricity cost, during a peak time, and manage the foreseeable consumption.

AMI provides global customers and businesses with the ability to monitor operations and maintenance cost savings. This process involves remote billing services designed to provide them with services and reduce the electricity cost. Its intelligent devices and sensors benefit from the adoption of AMI capabilities. The services provide a real-time metering capability that customers need to reduce energy consumption while boosting productivity. The service is designed to monitor, read the meter, remotely diagnose any issues concerning intelligent devices and sensors within the ecosystem. AMI capabilities are deployed in public, private, and government sector datacenter environments. These services provide clients with decentralized regional distribution of services, which includes top energy cost-saving results. The capabilities have a remote service connection and disconnection methods. These methods allow providers to connect or disconnect any customer service that has a past-due billing or payment balance.

2.18.1 CONTENT DELIVERY NETWORKS

Content delivery networks (CDNs) are decentralized proxy servers or data centers. In the digital network, CDNs deliver ease of use and "high performance" capabilities [41]. These technological resources provide the end users to distribute IT-server capability needed for businesses and consumers [41]. CDNs deliver large-scale Internet content services to many consumers around the world. CDNs' capabilities range from web objects, like text messaging, visuals, and scripts. These types of enterprise networks have empowered the end users with supreme capabilities needed to download small, medium, and large-scale files, applications, and documents [42].

Over the years, end users have taken advantage of these enterprise Internet services to video/live substantial streaming content of data, media materials, like on-demand streaming of substantial content of the information [41]. In IoT devices and machines are deployed to supply a range of content delivery services. These decentralized capabilities include [41,42]:

- Video streaming
- Software downloads
- Web/mobile content acceleration
- Licensed and managed CDN
- Transparent caching
- Internet services

These objects are capable to power devices and systems across multiple domains. Such interoperability ensures that there is a redundancy between objects, to reducing the cost of bandwidth, speed up the uploading and downloading time. In CDN, end users can access data in real time without experiencing a single point of failure. Devices and sensors benefit from continuous data synchronization, which can be surviving through a multitude of nodes [42]. With the rapid increase or the omnipresence of AI web analytics, nodes can provision data in real-time, augmenting for performance, and delegation of routes that these systems believe that is more convenient for transmitting data between nodes. Edge servers are the most excellent example of the CDN architectural environment. Edge servers can be deployed between the upper layer of the cloud infrastructure and the end-user environment.

These classes of servers give businesses and end users the advantage needed to readily access or provisioning data without having to access the upper cloud layer [42]. Edge servers have a unique advantage of performing within a point of presence. The nodes provide the end user with the capability needed to access data within reach, without having to deploy more solutions. Within CDNs, there is large-scale end-to-end leverage provided by the transport layer of the Open Systems

Interconnection (OSI) model. It allows for data distribution and provisioning and use of smart objects and applications deployed on the Internet. Several types of CDNs include but are not limited to peer-to-peer, provide CDNs and others [42].

2.19 DISTRIBUTED AUTOMATION NETWORKS

Distributed automation networks (DANs) are innovative technology-based end-to-end solutions. The DANs deliver intelligent automation cloud capabilities to a multitude of users worldwide. Today, the industry continues to reinvent its technology capabilities to serve its consumer's cloud requirements best. The DANs are designed to boost system dependability, for example, reliable or solid-state transformers. Researchers argue that in the future, DANs capabilities may transform into the next generation of advanced distribution automation (ADA). The ADA provides advanced functionality of hybrid cloud solutions within a single or distributed networks.

Future cloud capabilities will be built with embedded solutions. These capabilities will prevent an outage and allow for immediate datacenter self-healing. The reliable state transformer is designed to ensure that when deployed to the cloud environment, they can minimize power losses, also known as power failures. In the IoT ecosystem, the distributed automation network interactive technical approach brings many advantages to intelligent devices and sensors. These systems are deployed to distributed networks providing intelligent devices and sensors with the operational efficiency and minimization of potentially incurred operating costs. In data centers, distributed automation services deliver reliable electric capabilities that help to power machines and applications. The use of arc sense technology gives the detection capability of any faults within the network. In AI, distributed automation systems can be miniaturized into intelligent devices and sensors. These capabilities are deployed to plug-in electric vehicles or autonomous cars. The service ensures that vendors and consumers can save costs and minimize autonomous-vehicle-associated system losses.

2.20 OPTICAL CHARACTER RECOGNITION AND HUMAN MINDS

In the digital information era, many terminologies like optical character recognition (OCR) are often misconstrued. This dilemma is due to many in the technology space, not having a clear grasp or understanding of the term OCR and the context of usability [38]. Similarly, OCR is a term that was devised in concert with the software development landscape [43]. It is the process of scanning documents while trying to understand any designs or patterns that are exhibited on a computer screen. These patterns often are shown in the form of letters, characters, or even calligraphies. When arranged properly, these items can show results in the method of "pictures of letters" that can then be deciphered into texts [38]. Researchers have studied the difference between letters or texts displayed in word or picture format for many decades [43]. OCR technology still is a complex and dubious science. Infinite variations and screen displays have contributed to the continuing improvements of OCR software and underlying technology specifications [44]. This technological progress, in AI, KRR, and the genealogical fields, OCR are still an essential software that allows for various analytical studies. It involves the reviews on the descent of persons and family trees, and others [43]. The OCR's evolution has been useful, whether in reading human handwriting, letter varies, and the complexity of word text and word picture design/patterns [38]. The digitalization of calligraphies and document scripts has given a new view of the OCR evolution [43]. The birth of AI, KRR, ML, and DL has streamlined the time that professionals or consumers need to work on word text or word pictures [44]. OCR technology still uses today—OCR has taken on a pivotal part in criminal investigation and court proceedings [38].

In social science, perception of a collection of reasoning abilities includes, but is not restricted to [38,44]:

- Awareness
- Sensing
- Reckoning
- Mind
- Speech
- Storage

They may be labeled as a reasoning function that includes a human's brainwave and feelings. The mind can process, imagine, make out, and appreciate any human activities [43]. In scientific terms, the mind deals and processes a person who is an emotional state of being, reactions, opinions, and related activities [38]. In the ancient era, the phrenological mapping of the brain was among the prime efforts to associate cerebral functions with specific fields of HI [38]. The research involved other intelligence analogous methods, such as "dualism and idealism." Researchers continue to contend with the underlying nature of the cognizance [43]. With the advent of AI, KRR, and OCR "physicalism and functionalism" might be the two breakthrough assumptions that would yield some insights into how the human brain uses. Identicality in human reasoning is a reductive method that associates one's intellectual faculties with other human phenomena, like neuronal activities. The mind may trigger a reaction and intentionality toward an environment. Such activity leads to a perceived, responsiveness, and actable stimuli within the psyche. This involuntary activity in the human's mind, prompt the human psyche to begin thinking, while generating perceptive feelings toward others or an environment [38].

2.21 SIMPLE NEURAL AND BIOLOGICAL NEURAL NETWORKS

Many bay windows around the globe, such as research centers namely the International Business Machines or IBM Corporation at Thomas J. Watson Research Center is one of the global AI leading research and development institutions [45,46].

In recent decades, IBM researchers refocused their attention on parallel technological areas, for example, the simple neural and biological networks, and "biological neural networks," commonly known as BNNs. Such an advance is attributed to researchers, who devoted countless analytical and exploratory hours on critical technological domains—classic AI, simple neural networks, and biological neural nets. At the starting time, IBM did not have unique technological capabilities to solve AI problems with its early research inventions. In late years, IBM has had a competitive advantage over other significant players worldwide [45,46].

2.21.1 ARTIFICIAL GENERAL INTELLIGENCE

Artificial general intelligence (AGI) is an intelligent process that includes smart systems that may successfully execute intellectual activities, for example, tasks than a human [47]. The AGI supports complex tasks that do not humans' involvement [47]. AI has played indispensable parts in several technical areas: research & development, robotics, defense, healthcare, aeronautics, entertainment, social media, marketing, maritime, climatology/meteorology, and others [48]. Many industries define AI as "hard AI" or "full AI" [48]. These terms show that a machine could execute a general intelligent action [47]. AI is a technology that includes "strong and full" functions [48].

In AI, the terms "strong AI and full AI" give machines the capability required to do chores, like the general intelligence action [47]. In a hypothesis, "strong AI" gives machines the ability to officiate or perform the desired task [47,48]. This concept is reliably functional to complicated and specific research, problem-solving, and cognitive outcomes. There is a difference between "weak and strong AI solutions," like variances. Any inconsistencies may be well defined by random patterns that machines run through in the ecosystem. These results do not include human reasoning, though there might be a parallel between these capabilities [47,48].

2.22 MACHINE INTELLIGENCE LEARNING AND DEEP LEARNING

In machine intelligence learning (MIL), smart objects mimic activities that human intelligent performs [49,50]. Deep or hierarchical learning covers great ML methods. The world of advanced AI-based solutions and IoT systems/applications aims to give technological resources or technology assets the desired communication leverage. This stage of computing autonomy ensures that devices or objects can interact when connected to a physical or virtual network. The ability of these generation of tools and computers will ensure these solutions provide the services that consumers need to execute tasks. These devices/objects can be embedded in capabilities such as intelligent automation.

ML and DL capabilities continue to supply an acceptable level of computing automation that is required to scale the operation and optimization of IoT devices/objects [4,49]. These processes are developed to extract datasets and process data representations, aside from task-specific algorithms. In contrast, MIL is a supervised or an unsupervised method that involves subclass of AI and ML solutions [20,49,50]. In this setting, smart objects mimic activities that are done by HI [49,50].

Deep and hierarchical learning covers extensive ML processes and mining. ML and DL capabilities are the principal ingredients in processing data images as an alternative to task-specific algorithms. MLI functions involves supervised and unsupervised intelligent systems [20,50]. In the recent decade, technology companies developed DL and deep network learning solutions to solve AI problems. AI issues stem from language translation, email spam identification, the ability to arrange images, and others. SNNs are sophisticated mathematical methods developed to identify large volume patterns, notably a static set of data. Researchers predict that in the future, CAIM and SNN will be fitted with embedded with innovative software solutions capable of processing complex tasks analogous to machine intelligence systems. CAIM and SNNs continue to affect AI, especially the machine intelligence system performance. Despite these challenges, researchers have produced a new genesis of intelligent machine systems known as systems of intelligence [49].

2.23 UPPER ONTOLOGY AND MACHINE TRANSLATION

In 1993 ontologies were built to defend the "PANGLOSS," the accretion of facts or information for the machine translation system [51]. Ontology is a formal KR concept that spans theories or perceptions, that is, devices, processes, and objectives. These concepts stems from various fields and relations. In information science, ontology spans upper, top level, or foundation ontology. The terms range from objects, attributes, and relations [51]. These three elected attributes involve domain system structure. Upper ontology focuses on comprehensive semantic interoperability, notably domain-specific ontologies.

In a "natural philosophy computational implementation" method, these concepts continue to play a substantial part, whereas in "physicalism," physical ontology is a metaphysical thesis. This hypothesis means that everything under the sun is physical. It involves monism, substance, dualism, materialism, and others [51]. Large-scale ontologies have been produced to sustain the machine translation system [15,51]. The WordNet hierarchies, LDOCE aim to complement the ontology's upper region. The PANGLOSS MT system constitutes a pragmatic model of engineering innovations, like ontology concepts. These methods supplement knowledge base functions, that is, the generation element(s).

In PANGLOSS, there are active 50,000 nodes—these computer-based solutions are merged into manually built and smaller upper ontologies. Corresponding elements include definition match and hierarchy match algorithms [15,51]. A large-scale ontology is required for analyzing activities that would happen within the machine translation system's active modules. In PANGLOSS, each client can perform autonomous self-conception. PANGLOSS merges with LDOCE online and WORDNET applications. The objective is to syndicate definitions of Longman and semantic relations—this attack calls for an ontology semiautomatic taxonomization concept using a WordNet [15,51].

2.24 FRAME PROBLEM AND CYCL PROJECTS AND SEMANTIC WEB OF THINGS

Build or frame problem (FP) and CycL projects have been about since the advent of AI capabilities. FP is a process that defines a problem [52]. It functions using first-order logic (FOL). FOL determines how robots may interact with each other when deployed to the AI ecosystem [52]. In AI, the robotic state is represented by traditional FOL settings—the approach comprises several computational requirements.

In AI, the use of axioms ensures that there is no single change among objects—whereas, in centralistic terms, the FOL system grants more axioms. This concept also gives axioms the ability to make inferences about AI to which are deployed [53]. FP is a method that concentrates on identifying desired assemblages of axioms that may be applied to support any descriptive method that requires a robotic network. Despite any complexity involved in the research, some of the AI researchers at Stanford University suggest that setting up a problem or frame problem often involves an in-depth analytical investigation [52]. Further, they concluded that framing a problem would be a drawn-out process. It might include multiple or challenging methods, most of which might wander from the influence of action. If large numbers cannot be presented, there will be no effect of activity. Some researchers believed that framing a problem might be a suggestive procedure, which takes a broader epistemological series of issues [52,53].

Semantic web of things (SWoT) comprises technological semantic web processes and interfaces that can be utilized to support IoT solutions [54]. SWoT consist of architectural and programming design patterns [54]. These unified architectural-software capabilities are planned to interact with IoT and Swat intelligent devices and sensors [54].

In the next 4–5 years, Microsoft corporation plans on investing ***$5B*** in IoT [54]. This effort, will assure that customers have the first-hand capabilities they need to connect applications, devices, and systems to the digital environment [54,55]. It focuses on ensuring that billions of its intelligent devices, sensors, and other objects can interconnect around the universe; this includes interacting and provisioning data in real time [54]. Such a conversion ensures that many customers worldwide stay interconnected and can interact in real time [54]. Given this technological revolution, customers shall be able to reduce costs and bolster productivity [54]. The web of things consists of five interdependent unified pillars, notably social network, semantic network, programmable web, physical web, and real-time web [54,55]. Microsoft predicts that in the future, the interoperability between its and the semantic web/network of things (SWoT) will be defined by how these applications exhibit continuous interaction once deployed in the IoT environment [54,55]. Some researchers believe that more studies will be warranted for these unified domains [55]. Integrating legacy applications: JSON-LD, HTTP, JSON, REST, and Microdata interfaces are key examples of revolutionized technology trends. As these objects replicate by day, researchers predict that in the future, marketers will develop enhanced applications to confirm the workload delivering to support the daily operationalization of interconnected intelligent devices and sensors [54,55]. If properly integrated into KRR and AI, SWoT will support the infrastructure data provisioning. In some way, this process aids the collection and data processing through linked open data application [55]. It ensures that a unified capability will offer asynchronous device-to-device data processing. The approach spans continuous abstraction of intelligence devices, detectors, and other services [55]. For example, distributing generic nodes and ensuring that its applications may balance and processing power or communicate with objects deployed to a decentralized digital environment without a single spot of failure [54,55].

2.25 PRESENTING, REASONING, AND PROBLEM SOLVING

Early studies in AI and KRR focused on developing algorithms to mimic a consistent human performance and reasoning [2,4]. This theory involves techniques that people use to unravel puzzles and draw reasonable inferences. In the 1980s and 1990s, AI was described as an advanced scientific method that gathered, analyzed, and processed data. As a domain, AI involves mathematical probabilities and cost-effective results [56]. Its concept includes applying combinatorial analysis to process memory and computer activities. In a joint AI/KRR setting, humans might generate intuitive decisions—despite the systematic presumptions that early AI research relied on events.

Similarly, these systems are designed to perform tasks to support humans or animals' necessities—like problem solving, which is an area that has played a critical role for many years [56]. The use of conventional, ad hoc, and transformative analytical or mathematical methods has resulted in the unearthing of complex AI/KRR issues [56]. For example, problem-solving methods might be given to routine personal and job activities. These methods are not limited to interdisciplinary areas— AI, computer science, systems technology, math, medicine, project, and program management. As a result, these technologies might spin fields of exploratory or empirical scientific research [56]. The global research community defines reasoning as a procedure by which researchers, applied scientists, practitioners, and sellers may use or be able to measure, identify and find resolutions to issues are affecting enterprise business and professional settings [56,57]. This process stems from logical facts to adapting and defending sound practices. These methods describe institutional and moral principles. In general, reasoning applies to rational, perception, and intellectual capacity [56,57].

2.26 SIMPLE NEURAL NETWORKS, ARTIFICIAL NEURAL NETWORKS, AND NATURAL LANGUAGE PROCESSING

In the 1960s researchers developed ANNs to aid in solving global issues affecting science and technology sectors. On that occasion, researchers did not know much about ANN capabilities. ANNs changed how researchers view ANN technology capabilities—for example, the adoption of biological realism has substituted for early editions of similar technology solutions proven to be performing at a special rate. It changed the way ML and other smart objects performed mathematical calculations and statistical processes—ML supports large data workloads. It concentrates on extracting statistical information that researchers require to work out tasks, like turnouts. Researchers rely on AI solutions to support ML capabilities. These solutions include DL and deep network learning. ANNs are central systems that promote DL and deep network learning methods [22,23]. Such processes involve DL and related network learning solutions: ML algorithms intended for processing datasets, to selected multitier environments—have built in graphics processing units (GPUs). GPUs are embedded systems built for managing thousands of core processes. These operations can share large data workload transactions without interruption [19,22,23].

2.26.1 NATURAL LANGUAGE PROCESSING

NLP is a subcategory of computer science, AI, and computational linguistics. NLP centers on the interaction of information processing systems and natural human languages when deployed in the IoT environment [58]. Computer programming returns massive quantities of natural language data process [58]. This concept includes "machine-readable, logical forms, connecting words, machine perception, dialog systems, speech understanding, and natural language generation" [58,59].

In 1950 Alan Turing first coined the term "NLP." In the same year, Turing published a journal article entitled "Computing Machinery and Intelligence." Today, Alan's research body of work is known as the Turing test. Turing's research describes the criterion of intelligence, which remains relevant in today's scientific research community [58,59].

In 1954 the Georgetown scientific community held an experiment. The research includes a programmed conversion of sentences from Russian to English language [58,59]. Many scientists predicted that in the future, machine translation would solve AI, IoT, ML, and DL issues [58,59]. Fig. 2.2 illustrates a unified neural network structure and its relevant procedures.

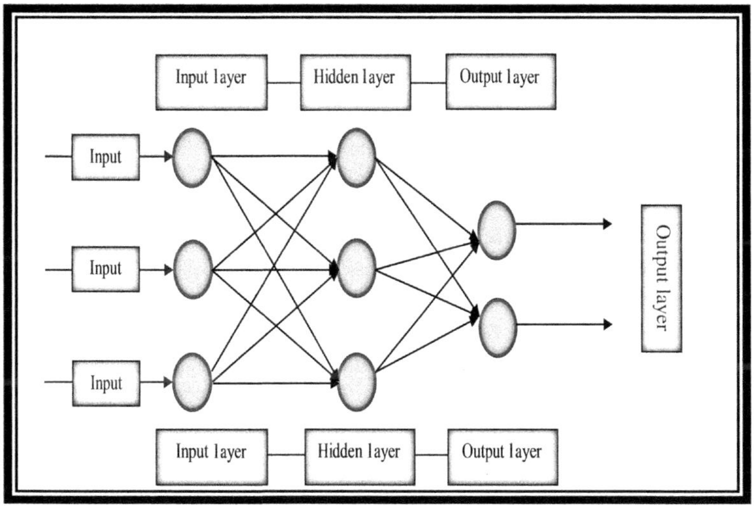

FIGURE 2.2

Unified neural network infrastructure.

2.26.2 UNIFIED NEURAL NETWORK INFRASTRUCTURE

Neural nets are a nonlearning examination of ANNs—Walter Pitts and Warren McCulloch developed ANNs to solve analytical and computational problems [45].

Frank Rosenblatt defines "perceptron," an algorithm-based method for supervising the learning of binary classifiers [45,46]. However, perceptron processes date back to the era of such algorithm-based research projects, which took place in the late 1950s [45]. These algorithms were shown to support the functioning of ANN custom hardware. These functions ensure that a statistical vector gives details on each input, output, and hidden entries [45]. The unified neural network infrastructure process consists of data, protected, and output layers. There is a synergistic interaction between information, protected, and output layers [23,60].

2.27 CONTEXTUAL ARTIFICIAL INTELLIGENCE PERSPECTIVES

In 1956 AI founders/researchers, like Newell et al. developed another computer application that was called "logic theorist." This application was built to solve issues associated with the English language [5,6]. Later, these successful inventions, Simon predicted that "machines will be capable, within twenty years, of doing any work a man may serve" [7,8]. While Minsky had a different prediction that summarizes "within a generation, the problem of creating 'artificial intelligence' is considerably solved" [11,61]. Despite little progress that was made in AI research tasks, some of the researchers who came before us failed to complete other inventions, due to the lack of research funding [5]. Counting how many times AI was redefined, in the 1959s, Samuel describes the term "machine learning" as a logical concept to support human interaction with smart objects [62]. Samuel invented another application called computer checkers [3,62]. During his research findings, Samuel made some corrections to his previous research theories, some of which are discussed in this chapter [3]. Preceding this scientific research, Samuel failed with some of his first innovations. A few months after, Samuel decided to pursue other inventions in data mining solutions, which were brought out in the 1990s. In the same year, Samuel documented his designs in data mining as a proven technology solution that focused on the role of algorithmic theories [4,62]. At the starting time, Samuel tried to assess dataset patterns, even with other researchers continued investigative efforts, which were more centered on data mining and ML solutions. These advances served as a benchmark for the future development of AI solutions and KRR applications [3,62].

In 1974 the US Congress insisted that Sir James Lighthill identified other feasible research projects. Hence, the United States and the British governments decided to stop funding future "exploratory AI research projects" [5]. Several years later, the project was called "AI winter." The AI winter was the name that was passed on to the AI project, which would not qualify for extra funding [5].

In the early 1980s commercial research experts launched another project known as "expert systems." This project was analogous to the one that the United States and the British governments decided not to carry on funding. This commercially funded program "expert systems" was later named the "AI program." Therefore the intent was to simulate human-machine cognitive and analytical sciences. This task was established to carry on scientific studies on simulated human experts [5].

In 1985 the AI market and research projects increased over a billion US dollars [5]. This progress allowed scientists to go on more innovative AI-based analytics solutions [6,62]. Researchers predict that in the future, AI and KRR capabilities will be more technologically advanced [1,5]. This prophecy is accredited to the presentation of advanced technologies, that is, AI, ML, DL, and IoT [1,2]. Many researchers predicted that the disruption of AI would put in new technological areas of meaning. Humans, thinking systems, and smart objects will rely on more or less of these emerging technologies, namely ML, DL, and IoT, focused on processing small- and large-scale datasets [3]. Whereas, technical and nontechnical users will equally rely on these innovative AI/KRR solutions to simplify the time asked to execute, implement, check, and deliver projects within schedule and under budget [1,3,62].

In 1959 Arthur Samuel coined the term "machine learning" or "machine intelligence concept" [6,62]. These similar nomenclatures date back to the 20th century when many scientists were focusing the efforts on parallel scientific research projects. In the same decade, Samuel's late and successful innovations set a new era of computer systems [3,5] Therefore Smith describes KR:

"Any mechanically embodied intelligent process will be made up of structural ingredients that a) we as external observers naturally take to represent a propositional account of the knowledge that the overall process exhibits, and b) independent of such external semantic attribution, play a formal but causal and essential role in generating the behavior that manifests that knowledge..." (as cited in ref. [63]).

2.27.1 USE CASE

Strong AI focuses on an intelligent machine's ability to undertake tasks and being aware of the respective environs [64,65]. In AI, "belief" may be described as nature and quality of perception. It illustrates humans' ability to pay attention to external objects within environs [64,65]. Strong AI was first cited in Latin and English literature interpretations, for example, sentience, qualia, awareness, subjectivity, feel, wakefulness, selfhood, and soul [64,65].

2.28 COMPLEX ARTIFICIAL INTELLIGENCE SYSTEMS AND SWARM INTELLIGENCE

Many researchers suggest that there is a parallel between AI and sophisticated AI systems. However, these domains might differ in how solutions may be exhibited to the IoT environment. In the last decade, vendors developed CS to complement AI's research activities (at the outset, 2006). The relationship between objects and attributes may further note the distinction between AI and complex artificial intelligence systems (CAISs). For instance, most attributes in AI and CAISs may be separated by nonlinear behavior. Having a better understanding of CS and how these attributes may be tested and associated with each is vital to the overall relationship [66]. In AI, the integration of these systems/objects determines how CAISs can be deployed with AI/MI, KRR, DL, and ML to solve pervasive issues affecting the IoT environment. Researchers suggest that there is a parallel or shared value between complex biological networks and AI. For example, the human's immune system, social insects, cellular metabolism, and SI may be a perfect case scenario of such a relationship [66]. SI is an interdisciplinary area that involves both artificial and natural systems. It consists of smart objects, human artifacts, or animals, notably ants, termites, fish, and groups of birds with similar collective behavior [67]. These colonies of ants and termites, as well as herds of land animals, communicated in a decentralized control environment [67]. SI research makes up four distinctive domains, markedly artificial, natural, scientific, and engineering.

Natural swarm intelligence (NSI) research concentrate on the study of biological systems. However, artificial swarm intelligence (ASI) addresses the study of human artifacts, particularly individual characteristics classify ASI and NSI. Such characteristics include the study of the nature of which dictates the objects being examined.

In scientific and engineering environments, SI may be categorized into alternative areas, which is defined through the informative arrangement to the specification of the nature of the system or object that is being evaluated [66,67]. Researchers conducted many experiments using a swarm of robots in recent years [66]. These studies span the regular stream of the SI domain. In the 1990s, however, Deneubourg et al. proposed a distributed probabilistic model. This model focused on clustering behavior—as part of lab experimentation, these researchers developed a model that allowed ants to pick up and drop items. Probabilities were the underlying factors of such a study. It relied on corpse density that they made it available for the ants [68].

In an ideal world, SI systems focus on scientific variables, whether those involving a little understanding of each object's natural behavior or the specific mechanism. This method, however, aims to ensure that systems may interact among themselves, or with the environment to which they are deployed [66]. Engineering stream is a field of science that focuses on exploiting the understanding of how scientific stream concepts and techniques may be shown. It allows for the designing of systems capable of solving problems of a practical significance. There are two contrasts in natural versus artificial and engineering versus science methods. Researchers suggest that regardless of any existing dichotomies between these scientific domains, SI has been used in the robotic scientific research sphere [66,68].

2.29 THE IMPACT OF SMART DUST ON IoT TECHNOLOGY

Many researchers have suggested wide-scale concerns that smartdust or smart dust (SD) poses to global consumers. These issues involve the adoption of technology, regulatory compliance for data protection, security, and privacy. Given the miniaturization of these devices, many consumers are worried about how the safety of data, considering that these sensors may be embedded in other units [69]. These devices have the ability to record data in real-time. The cost of intelligent sensors can be inflated, given that sensors are connected via satellite technology [69]. SD is an evolving wireless technology that consists of small devices that the world has ever experienced. It is also a newly unveiled technology that involves microelectromechanical devices [70]. SD is built on a sensory model concept to be embedded into miniaturized devices with sensors [69]. This advanced technology is found on CCTV cameras and similar devices that are built with an ability to send the data. They are useful technology assets that may be used in data collection, parsing, and storing [69].

Global tech and production companies namely General Electric, IBM, Cisco Systems, Cargill, and many others have led the charge in the development of SD devices. These technologically advanced companies continue to invest millions of United States dollars toward research and development of these miniaturized SD devices. Researchers predict that SD devices will disrupt the global economies in the upcoming years or so [70]. Such disruption will continue due to new applications that are being developed to support the adoption, deployment, and functioning of these devices [69]. In real-time brain function monitoring, SD has contributed immensely to the constructs of leaps or bounds. This technology may be used on humans to monitor and measure behaviors. Such activities involving single or multiple neurons, that is, uses magnetic resonance imaging. This technology may is used in interdisciplinary areas, like those involving the brain-machine interfaces, where humans could control machines by thoughts [70]. Aside from these technological advances, which might have led to more meaningful results, researchers suggest that the limited spatial resolution of these devices may yield a lack of movability and risky invasiveness [69]. These intelligent devices and sensors may generate ideas that may be used to sprinkle miniaturized electronic sensors.

Neural dust (ND) has been around for many years. Unlike SD, ND is equipped with embedded sensors that may generate ultrasound that may power ND sensors. These intelligent systems, when deployed into the environment, may listen to potential messages or responses and may have a similar effect in terms of capability as that of the RFID system. Partly, the system is tetherless, which means the data may be collected, parsed, and stored on the cloud for immediate or future use [70]. These types of systems often run in lower power capacity, despite the special high resolution, and the ability to be transported easily. Brain-machine interfaces, also known as BMIs, do not have an embedded implantable neural interface system. It makes it difficult for them to remain operational or functional for a long-time [69]. Now, a technology, such as microelectromechanical devices known as MEMS or motes may be rated by its hi-tech and agile platforms that are designed to interact with the SD operations.

In AI analytics capabilities, MEMS/Motes capabilities may be embedded into smart objects, to support the decentralized interoperability of SD decides and systems deployed to the IoT environment [70]. These devices are proven to be equipped with sensors, computing, and wireless capabilities that enable them to communicate seamlessly. Researchers suggest that the devices may collect data involving acceleration, stress, pressure, humidity, sound, and other relevant items within the environment [69]. Once the data are stowed into the sensors, these devices could process the information through an onboard computer system. Each sensor can store the data in internal memories for immediate or future accessibility [70]. Through embedded wireless capabilities, miniaturized devices may remotely upload data onto the cloud or even share with other MEMS/Motos within the IoT environment [69].

With the disruptive gain of AI, KRR, and IoT, SD technology has played an instrumental role in the aid of data collection, physical security, infrastructure monitoring, and more. In recent years, however, millions to billions of IoT intelligent systems have gained a footprint in the SD technology landscape. Most of these devices were used in various industries, like in the agricultural fields, to monitor crops on an unparalleled scale. The technology is deployed to determine sprinkling, fertilization, and pest-control requirements. This technology is implemented with monitoring capabilities to facilitate proper maintenance and harvest forecasts [69].

2.29.1 INTELLIGENT DEVICE MANAGEMENT

Intelligent device management (IDM) is a term that has been used in software engineering for many decades. IDM is a field of engineering science or scientific engineering software that involves products and services [3,19]. These areas of engineering science aim to support industry leaders, that is, managers, during the software development lifecycle management. This process focuses on hardware and software areas that every engineering company in the industry would need to monitor services and product manufacturing in an advanced engineering environment. IDM is built-in software that is deployed to ensure that managers have the capability necessary to get the job done [3,19]. This method stems from involving technical teams while ensuring they have the tools needed to effortlessly coordinate or monitor different layers of internal communication channels among various production teams. This process may take place within the same geographical area or deal with multifaceted approaches. The process spans the virtual networks or through production settings.

In a traditional engineering network, IDM spans a multitude of service management, to managing on-site technological resources. It supports day-to-day production or standard service processes [3,19]. This approach includes the following industrial production areas: incident, change, configuration management, and problem solving. In the IoT environment, IDM delivers a rapid solution deployment amid decentralized objects, mostly intelligent devices and sensors. It guarantees that if these objects are connected to AI, they interconnect with other peer smart objects without any downtime. IDM continues to play a vital role in the industrial production sector, the deployment of IoT applications, and the object [3,19].

2.29.2 COGNITIVE SIMULATION AND ANTILOGIC OR NEAT AND SCRUFFY

Several decades ago, Herbert Simon and Allen Newell conducted scientific research on social problem-solving skills. These researchers' scientific investigative efforts aimed to confirm the research hypothesis. These researchers' investigative findings led to the practicalities of what is now known as AI or CS. The research led to the fundamentals of other interdisciplinary areas, like operations research and management science [3,19]. The scientific investigation was founded on psychological experiments that they needed to develop simulated programs. The simulation of such programs further involves simulated techniques that are used on humans to solve complex issues. Most of this critical research work was conducted at Carnegie Mellon University in the 1980s. Thus CS applies to cross-vertical and horizontal scientific research areas. CS is functional in the medical, engineering, and stimulating fields. For instance, in the medical field, it is applied to therapeutic programs, such as rehabilitation techniques. When applied to therapeutic programs, CS solutions focus on cognitive reserve and neuroplasticity. In this context, the program intends to improve a patient's mobility and performance [3,19]. CS is used in medical or scientific laboratories, where experts are heavily engaged in complex brain simulation research and findings.

CS programs are designed to humans from existing lives, like adults, children, elderly/seniors, infants, teens, and others. In the clinical network environment, CS may be used to performs, implanted tasks associated with the human brain. It includes the repetitive projection of myelinated neural circuits into the mind, to restore its cognitive activities or functions [3,19]. This method essentials to restore human brain functions if concoction, nerve disorders, or even brain damage, which may eventually affect cognitive functions. However, antilogic is a branch of science that tends to lean toward the central difference involving humans, for example, men and women. This variance might range from how men and women interact with one another to identify at birth. According to researchers, these variances are unassailable and unchangeable among humans of different gender identity.

Despite these differences, researchers suggest that the existence of antilogic may develop personal tendency among men and women [71]. Researchers believe that in AI, there is a distinct difference between neat and scruffy. Neats are labels that determine elegance, clearness, and correctness. While scruffies view intelligence as much of a complex concept or computational obturation, such apparent differences can be determined for a similar system. Newell defines neat and scruffy, as new logical approaches that could lead to successful scientific breakthroughs [71]. In cognitive models, such as personal psychological data "neat and scruffy" may have a single system execution and representation. However, the rules are introduced to each ad hoc system. Neat solutions focus on logic or formal methods [71]. These concepts consist of pure and applied statistics. Thus scruffy are methods that hackers use to attack or break into AI systems. Scruffy hackers can assemble a team that focuses on conducting malicious attacks on AI systems, like ML [71].

Researchers argue that the difference between these two methods, for instance, neat and scruffy, is that the latter can perform diverse and successful activities based on randomized results [71]. If providential, the directed AI system produces intelligence insights that hackers often need to break into the ML system. Neat is a method that is founded on formalism. Despite its functions, it can appear to be sluggish, breakable, and boring, mainly when deployed to real systems. Researchers are still investigating whether "neat and scruffy" methods could mimic HI [71].

2.29.3 COGNITIVE INFORMATICS AND COGNITIVE COMPUTING

Cognitive informatics (CI) is a maturing scientific field that focuses on NI and core data processing methods. It involves computational-reasoning subcategories, such as brain, processes, sense, and cognition [72,73]. As an interdisciplinary area, CI provides researchers with proven capabilities needed for conducting scientific studies. Applying a sequence of mathematical functions helps advance analytical research [72,73]. These results range from research and engineering areas, primarily CS, cybernetics, systems science, software engineering, neuropsychology, KRR, and software engineering concepts [72,73].

CI comprise two distinct critical areas [73]: information technology and human information interaction (HII). HII is a process, which involves the advance of computational methods, to increased human activity, or performance of human-machine systems [72,73]. Learning and skill development are areas that focus on developing modernized system applications needed to interact with parallel technology solutions. These technological features have embedded functions built to support learning and skill development practices [72,73]. Cognitive systems include analytical processes—adaptive, interactive, iterative/frequentative, state of interaction/stateful, and contextual methods [4].

Researchers have argued that both "CI and cloud computing (CC)" are modern scientific disciplines consist of human reasoning, computational intelligence practices, and procedures, or the denotational of large-scale mathematical concepts [72,73]. The advance of innovative engineering program applications needs deep thinking and learning or reasoning [72,73].

2.30 THE NEXT GENERATION OF COMPUTERS AND FUNCTIONAL TRENDS

For many years, humans have performed some of the computational and functional mathematical tasks. Such responsibilities incorporate but are not limited to new standards, functions, procedures, management/security policies, regulations, and protocols. As these objects continue to acquire more intelligence, there will be an even-steven/shared level of collaboration between public and private sector organizations and the research community. The author predicts the next generation of computers will consist of highly developed smart devices and intelligent systems. The integration of these super objects with cognitive sensors will continue to mimic human perception, cognitive functions, and environmental awareness. It will help hyperscale sophisticated IoT devices and objects with processing data faster than today's machines. The author envisages that the next generation of computers will be smaller and equipped with powerful processors and sensors. The devices and objects will be embedded with hyperperformance computerized functions.

From an ecosystem perspective, the next-gen's intelligent devices and smart objects will be able to adapt, interact, process, and share data. The data will be provisioned through a decentralized physical or virtual computing environment. These faster AI chatbot functions are integrated with super objects, like central processing units or GPUs capabilities. The goal is to process a large amount of data faster than humans. This convergence ensures smart devices and intelligent machines have a unified autonomous-robotic capability to interact effortlessly by analyzing, exchanging, monitoring, managing, supervising, and predicting daily events with limited human's participation. The integration of these smart objects into the next generation of super machines will ensure that such purposes have the independent ability and influence to perform a collection of tasks with accuracy and faster than humans. The author predicts that future AI and IoT capabilities will be deployed through a decentralized node environment. The process of delivering tasks to each device or machine will be faster with a shorter timeframe for executing each instruction. These objects, analyze, predict task/project implementation and execution, receive/provision data, process/monitor, or interact with each other independently or without human participation. In the future, machines and systems will be performing various tasks within an unlimited boundary and unified ecosystem, commonly known as the artificial intelligence of things (AIoT). The author predicts that

AIoT will be an integrated technology with unified capabilities to support the next generation of computers. As these intelligent devices and systems continue to mature, they will be capable of performing other complex tasks.

2.31 CONCLUSION AND FUTURE READING

The author emphasizes how researchers and practitioners can solve issues affecting AI-based solutions, KRR, and the IoT applications. The researched process stems from human perceptions and the ability to warrant the functioning of smart objects and determine faculty to interact. AI has reshaped the global technology landscape through incorporated analytical capabilities.

AI-based solutions have been used to solve KRR and IoT complex problems around the world [1]. Checkers' strategies were a computer-based game that these researchers designed to address many issues involving the algebraic "proving logical theorems" [5]. The term "reasoning" derives from the fields of computer science, sociology, and psychology. Both BC and FC are concepts that can be implemented in AI, game theories, and other ES environmental prototypes [12]. DO and DSO are methods that describe things that can be observed throughout the world. There are performance limitations between AI-based system functions. Such technical constraints include "strong and weak" AI capabilities. These limitations are due to the ability these systems have for autonomous interaction [3].

In the beginning, IBM did not have unique technical capabilities to solve AI problems through its new research inventions. In recent years, IBM had a competitive advantage over other significant players worldwide [45,46]. While supporting AIAs, limited capability, like CAIM and SNNs, have proven can be the main factor affecting machine intelligence system performance. PANGLOSS was designed to help merge LDOCE online and WordNet applications. The intent was too concise syndicate definitions of Longman and semantic relations. The goal, however, was to give a semiautomatic taxonomization for other ontologies using WordNet [15]. There is a difference between "weak and strong AI solutions," for example, variances. Any inconsistencies can be defined by random patterns that machines go through in the ecosystem. These solutions do not include human reasoning, though there might be a parallel between these capabilities [47,48].

Humans, intelligent systems, and smart objects will rely on some of these emerging technologies, that is, ML, DL, and IoT focused on processing small- and large-scale datasets [74]. Technical and nontechnical users will equally rely on these innovative AI/KRR solutions to simplify the time needed to execute, implement, check, and deliver projects/tasks within schedule and under budget [1,3,62]. In the future, the advance of innovative engineering program applications will require in-depth thinking/analysis as well as learning or reasoning [72,73]. The author prophesies that AIoT is an integrated technology that will be the next generation of computers. These devices and sensors have become more intelligent to be able to execute similar tasks that are being performed by humans. For many years, humans have implemented some of these tasks—the robotic systems are beginning to take over. Some new standards, functions, procedures, management/security policies, regulations, and protocols are discussed in this chapter. As smart devices, sensors, or objects continue to acquire more intelligence, there will be an even-steven/shared level of collaboration between public and private sector as well as the research community.

REFERENCES

[1] V. Moret-Bonillo, Can artificial intelligence benefit from quantum computing? Prog. Artif. Intell. 3 (2) (2015) 89–105. Available from: <https://link.springer.com/article/10.1007/s13748-014-0059-0>. (accessed 06.01.19).

[2] G.P. Zarri, 2014 Special issue: sentiments analysis at conceptual level making use of the Narrative Knowledge Representation Language, Neural Netw. 58 (2014) 82–97. Available from: <https://science-direct.com/science/article/pii/s0893608014001099> (accessed 06.01.19).

[3] M. Purdy, P. Daugherty, Why artificial intelligence is the future of growth, Accenture. Available from: <https://www.accenture.com/us-en/_acnmedia/PDF-33/Accenture-Why-AI-is-the-Future-of-Growth.pdf>, 2016 (accessed 06.01.19).

[4] R. MacGregor, Using a description classifier to enhance knowledge representation, IEEE Expert. 6(3) (2017) 41–46. Available from: <http://ieeexplore.ieee.org/xpl/login.jsp?tp = &arnumber = 87683&url = http%3A%2F%2Fieeexplore.ieee.org%2Fxpls%2Fabs_all.jsp%3Farnumber%3D87683> (accessed 05.01.19).

[5] Artificial Intelligence, Wikipedia: the free encyclopedia. Available from: <http://en.wikipedia.org/wiki/Artificial_intelligence> (accessed 06.01.19).

[6] A. Kaplan, M. Haenlein, Siri, Siri my hand: who's fairest land? On the interpretations, illustrations, implications of artificial intelligence, Bus. Horiz. 62 (1) (2018) 15–25.

[7] H. Simon, Herbert Simon remembers Allen Newell, Ai Mag. 13 (4) (1992) 30. Retrieved 06.01.19. Available from: <https://dl.acm.org/citation.cfm?id = 170730> (accessed 06.01.19).

[8] H.A. Simon, Allen Newell: 1927–1992, IEEE Ann. Hist. Comput. 20 (2) (1998) 63–76. Retrieved 06.01.19. Available from: <http://dl.acm.org/citation.cfm?id = 2237917> (accessed 06.01.19).

[9] J. Malpas, Donald Davidson (Winter 2012 Edition), The Stanford Encyclopedia of Philosophy, Edward N. Zalta (ed.). Available from: <https://plato.stanford.edu/archives/win2012/entries/davidson/> (accessed 23.08.20).

[10] J. McCarthy, Inversion of functions defined by turning machines, Automata Studies, Annals of Mathematical Study No. 34, Princeton University Press, 1956.

[11] M.L. Minsky, Heuristic Aspects of the Artificial Intelligence Problem: Lincoln Laboratory Report, 1956, pp. 34–55.

[12] A.T. Schreiber, H. Akkermans, A. Anjewierden, R. Dehoog, N, Shadbolt, W. Vandevelde, et al., Knowledge Engineering and Management: The CommonKADS Methodology, first edition, The MIT Press, Cambridge, MA, 2000. ISBN 978-0-262–19300-9.

[13] J.F. Sowa, Semantic networks, in: S.C. Shapiro (Ed.), Encyclopedia of Artificial Intelligence, 1987.

[14] W.A. Woods, Theoretical Studies in Natural Language Understanding, and Annual Report, Bolt Beranek and Newman, Cambridge, MA, 1979 (Technical Report No. 4332).

[15] R.J. Brachman, J.G. Schmolze, An overview of the KL-ONE knowledge representation system. Schlumberger Palo Alto Research/BBN Laboratories Inc. Available from: <http://csjarchive.cogsci.rpi.edu/1985v09/i02/p0171p0216/MAIN.PDF>, 1985 (accessed 20.01.19).

[16] W.A. Woods, What is important about knowledge representation? IEEE Comput. 16 (10) (1983) 22–27.

[17] R.J. Brachman, What's in a concept: structural foundations for semantic nets, International Journal of Man-Machine Studies 9 (1977) 127–152.

[18] W.A. Woods, What's in a link: Foundations for semantic networks, in: D.G. Bobrow, A.M. Collins (Eds.), Representation and Understanding: Studies in Cognitive Science, Academic Press, New York, 1975, pp. 35–82.

[19] G.A. Miller, The cognitive revolution: a historical perspective, Trends Cognit. Sci. 7 (2003) 141–144. PMID 12639696. Doi:10.1016/S1364–6613(03)00029-9.

[20] Y. Bengio, A. Courville, P. Vincent, Representation learning: a review and new perspectives, IEEE Trans. Pattern Anal. Mach. Intell. 35 (8) (2013) 1798–1828. arXiv:1206.5538. http://doi.org/10.1109/tpa-mi.2013.50 (accessed 06.01.19).

[21] Y. LeCun, Y. Bengio, G. Hinton, Deep Learning, Nature 521 (2015) 436–444. Available from: http://doi.org/10.1038/nature14539.

[22] G. Luger, W. Stubblefield, Artificial Intelligence: Structures and Strategies for Complex Problem Solving (fifth edition), Benjamin Cummings, 2004. ISBN 0-8053-4780-1.

[23] S.J. Russell, P. Norvig, Artificial Intelligence: A Modern Approach, 2nd ed., Prentice Hall, Upper Saddle River, New Jersey, 2003. ISBN 0-13-790395-2.

[24] M. Lungarella, G. Metta, R. Pfeifer, G. Sandini, Developmental robotics: a survey, Connect. Science. 15 (2003) 151–190. Available from: https://doi.org/10.1080/09540090310001655110. CiteSeerX 10.1.1.83.7615.

[25] M. Asada, K. Hosada, Y. Kuniyoshi, H. Ishiguro, T. Inui, Y. Yoshikawa, et al., Cognitive developmental robotics: a survey, IEEE Trans. Autonomous Ment. Dev. 1 (1) (2009) 12–34. Available from: http://dx.doi.org/10.1109/tamd.2009.2021702. Archived from the original.

[26] P.Y. Oudeyer, On the impact of robotics in behavioral and cognitive sciences: from insect navigation to human cognitive development, IEEE Trans. Autonomous Ment. Dev. 2 (1) (2010) 2–16. Available from: https://doi.org/10.1109/tamd.2009.2039057.

[27] E. Brown, Who needs the Internet of Things?. Linux.com, 2016.

[28] K. Ashton, That 'Internet of Things' thing, 2009.

[29] P. Magrassi, T. Berg, A world of smart objects, Gartner research report R-17-2243, 2002.

[30] Y. Ruan, A. Durresi, L. Alfantoukh, Trust management framework for the Internet of Things, in: IEEE 30th International Conference on Advanced Information Network and Applications (AINA), 2016, pp. 1013–1019.

[31] H. Dong, G. Singh, Attri, A. El Saddik, Open data-set of seven Canadian cities, Access. IEEE 5 (2017) 529–543.

[32] I. Wigmore, Internet of Things (IoT), TechTarget, 2014.

[33] Ahmed, K.Z., Kar, M., Mukhopadhyay, S. (2016). (Invited paper) Energy delivery for self-powered IoT devices, in: 21st Asia and South Pacific Design Automation Conference (ASP-DAC) 2016, pp. 302–307.

[34] K. Madden, Optimal timing thrombolytic therapy in acute ischemic stroke, CNS Drug 16 (4) (2002) 213–218.

[35] P. Semaan, Natural language generation: an overview, J. Comput. Sci. Res. (JCSCR) (2012) 50–57.

[36] R.V. Yampolskiy, Turing test as a defining feature of AI-completeness, in: Xin-She Yang (Ed.), Artificial Intelligence, Evolutionary Computation and Metaheuristics (AIECM)—In the footsteps of Alan Turing, Springer, London, 2013, pp. 3–17 (Chapter 1).

[37] S. Hirokazu, N.A. Christakis, Locally noisy autonomous agents improve global human coordination in network experiments, Nature 545 (7654) (2017) 370–374. Available from: https://doi.org/10.1038/nature22332. PMC 5912653. PMID 28516927.

[38] J.J.C. Smart, in: Edward N. Zalta (Ed.), The Mind/Brain Identity Theory, The Stanford Encyclopedia of Philosophy, 2011.

[39] M.M. Zanjireh, H. Larijani, A survey on centralised and distributed clustering routing algorithms for WSNs, in: IEEE 81st Vehicular Technology Conference, Glasgow, Scotland, 2015. doi:10.1109/VTCSpring.2015.7145650.

[40] S.V. Albrecht, P. Stone, Autonomous agents modelling other agents: a comprehensive survey and open problems, Artif. Intell. 258 (2018) 66-95. Available from: <https://doi.org/10.1016/j.artint.2018.01.002> (accessed 10.01.19).

[41] E. Nemeth, Chapter 19, Web hosting, content delivery networks, UNIX and Linux System Administration Handbook, fifth edition, Pearson Education, Boston, 2018, p. 690. ISB 9780134277554.

[42] M. Noormohammadpour, QuickCast: fast and efficient inter-datacenter transfer using forwarding tree cohorts, in: IEEE INFOCOM 2018—IEEE Conference on Computer Communications, 2018.

[43] N. Karunamuni, R. Weerasekera, Theoretical foundations to guide mindfulness meditation: a path to wisdom, Curr. Psychol. (Submitted Manuscr.) (2017). Available from: https://doi.org/10.1007/s12144-017-9631-7.
[44] N.D. Karunamuni, The five-aggregate model of the mind, SAGE Open. 5 (2) (2015). Available from: https://doi.org/10.1177/2158244015583860. 215824401558386.
[45] Y. Freund, R.E. Schapire, Large margin classification using the perceptron algorithm, Mach. Learning. 37 (3) (1999) 277–296. Available from: https://doi.org/10.1023/A:1007662407062.
[46] N. Nilson, Artificial Intelligence: A New Synthesis, Morgan Kaufmann, 1998. ISBN 978-1-55860-476-4.
[47] A. Newell, H.A. Simon, Computer science as empirical inquiry: Symbols and search, Communications of the ACM 19 (3) (1976) 113–126.
[48] Kurzweil, R. (2005). Advanced human intelligence, where he defines strong AI as "machine intelligence with the full range of human intelligence," p. 260.
[49] Y. Bengio, Y. LeCun, G. Hinton, Deep learning, Nature. 521 (2015) 436–444. Available from: https://doi.org/10.1038/nature14539. PMID.
[50] J. Schmidhuber, Deep learning in neural networks: an overview, Neural Netw. 61 (2015) 85–117. Available from: https://doi.org/10.1016/j.neunet.2014.09.003. PMID 25462637. arXiv:1404.7828.
[51] D. Stoljar, in: E.N. Zalta (Ed.), Physicalism, The Stanford Encyclopedia of Philosophy, 2009.
[52] P. Carruthers, On Fodor's problem, Mind Lang. 18 (5) (2003) 502–523. The Architecture of the Mind, Oxford University Press.
[53] S.J. Chow, What's the problem with the frame problem? Rev. Philos. Psychol. 4 (2013) 309–331.
[54] V. Trifa, Building blocks for a participatory web of things: devices, infrastructures, and programming frameworks (Ph.D. thesis), ETH Zurich, 2011.
[55] D. Guinard, V. Trifa, E. Wilde, A resource oriented architecture for the web of things, in: Internet of Things 2010 International Conference, 2010.
[56] D.L. Schacter, et al., Psychology, second edition, Worth Publishers, New York, 2009, p. 376.
[57] A. MacIntyre, Dependent Rational Animals: Why Human Beings Need the Virtues (The Paul Carus Lectures), Open Court, 2013. ISBN 9780812697056. Retrieved 12.01.14. [...] the exercise of independent practical reasoning is one essential constituent to full human flourishing.
[58] J. Hutchins, The history of machine translation briefly, 2005. [self-published source].
[59] P.M. Nadkarni, L. Ohno-Machado, W.W. Chapman, Natural language processing: an introduction, Journal of the American Medical Informatics Association, 18, (5), 544–551. Available from: https://doi.org/10.1136/amiajnl-2011-000464.
[60] D. Poole, A. Mckworth, R. Goebel, Computational Intelligence: A Logical Approach, Oxford University Press, New York, 1998. ISBN 0–19-510270-3.
[61] M.L. Minsky, Semantic Information Processing, MIT Press, Cambridge, MA, 1968.
[62] A.L. Samuel, Some studies in machine learning using the game of checkers, IBM J. Res. Dev. 3 (3) (1959) 210–229. Available from: https://doi.org/10.1147/rd.33.0210.
[63] B.C. Smith, Prologue to reflections and semantics in a procedural language, in: R. Brachman, H.J. Levesque (Eds.), Readings in Knowledge Representation, Morgan Kaufmann, 1985, pp. 31–40. ISBN 0–934613-01-X.
[64] G. Farthing, The Psychology of Consciousness, Prentice-Hall, 1992. ISBN 978-0-13-728668-3.
[65] R.V. Gulick, Consciousness, Stanford Encyclopedia of Philosophy, 2004.
[66] M. Mitchell, Complex systems: network thinking, in: P. Norvig, D. Perlis (Eds.), Artificial Intelligence., vol. 170, Elsevier B.V, 2006, pp. 1194–1212.
[67] M. Dorigo, T. Stützle, Ant Colony Optimization, MIT Press, Cambridge, MA, 2004.
[68] L.J. Deneubourg, S. Aron, S. Gross, J.M. Pasteels, The self-organizing exploratory pattern of the Argentine ant, J. Insect Behav. 3 (159) (1990) 1990.

[69] K.U. Römer, Tracking real-world phenomena with smart dust, Wireless Sensor Networks, Lecture Notes in Computer Science Series, vol. 2920, Springer, 2004, pp. 28–43. DOI: 10.1007/978-3-540–24606-0_3.

[70] F. Zhao, L. Guibas, Wireless Sensor Networks: An information Processing Approach, Elsevier Science & Technology, 2004, p. 376. ISBN 10:1-55860-914. Available from: <https://textbooks.elsevier.com/web/product_details.aspx?isbn = 9781558609143> (accessed 19.02.19).

[71] J. Kolodner, The "neat" and the "scruffy" in promoting learning from analogy: we need to pay attention to both, J. Learn. Sci. 11 (1) (2002) 139–152. Available from: <http://www.jstor.org/stable/1466725> (accessed 21.02.19).

[72] J. Kelly III, Computing, cognition and the future of knowing, IBM Research: Cognitive Computing, IBM Corporation, 2015.

[73] W. Kinsner, D. Zhang, Y. Wang, J. Tsai, Cognitive informatics, in: Proceedings of the Fourth IEEE International Conference (ICCI'05), IEEE CS Press, Irvine, CA, 2005.

[74] Purdy, M., & Davarzani, L. (2015). The Growth Game-Changer: The Growth Game-Changer: How the Industry Internet of Things can drive progress and prosperity".

FURTHER READING

D. Amodei, C. Olah, J. Steinhardt, P. Christiano, J. Schulman, D. Mané, Concrete problems in AI safety. Available from: <https://arxiv.org/abs/1606.06565>, 2016 (accessed 06.01.19).

A. Bahrammirzaee, A comparative survey of artificial intelligence applications in finance: artificial neural networks, expert systems, and hybrid intelligent systems. Neural Comput. Appl. 19(8) (2010) 1165–1195. Retrieved 05.01.19. Available from: <https://link.springer.com/article/10.1007/s00521-010-0362-z> (accessed 06.01.19).

R. Barcia, T. Berardi, A. Kak, H. Kreger, K. Schalk, Cloud customer architecture for mobile, Cloud Standards Customer Council. Available from <http://www.cloud-council.org/deliverables/CSCC-Cloud-Customer-Architecture-for-Mobile.pdf>, 2015 (accessed 06.01.19).

L. Bonin, Boy wonder. Available from: <http://www.ew.com/ew/article/0,165660,00.html>, 2017 (accessed 06.01.19).

J. Clark, Artificial intelligence has a 'sea of dudes' problem, Bloomberg, 2016. Available From: <https://www.bloomberg.com/news/articles/2016-06-23/artificial-intelligence-has-a-sea-of-dudes-problem> (accessed 06.01.19).

C. Corbett, C. Hill, Solving the Equation: The Variables for Women's Success in Engineering and Computing, The American Association of University Women, 2015. Available from: <http://www.aauw.org/files/2015/03/Solving-the-Equation-report-nsa.pdf> (accessed 06.01.19).

M.A. Campbell, J. Hoane Jr., F. Hsu, Deep blue, Artif. Intell. 134 (2002) (2002) 57–83. nos. 1 and 2.

K. Crawford, Artificial intelligence's white guy problem. The New York Times. Available from: <http://www.nytimes.com/2016/06/26/opinion/sunday/artificial-intelligence-white-guy-problem.html>, 2016 (accessed 06.01.19).

P. Domingos, The Master Algorithm: How the Quest for the Ultimate Learning Machine Will Remake Our World, Basic Books, New York, 2015.

C. Frank, AI, deep kearning, and machine learning: a primer, Andreessen Horowitz, 2016. Retrieved from: <http://a16z.com/2016/06/10/ai-deep-learning-machines>.

E. Felten, T. Lyons, Public input and next steps on the future of artificial intelligence, Medium. Available from: <https://medium.com/@USCTO/public-input-and-next-steps-on-the-future-of-artificial-intelligence-458b82059fc3>, 2016 (accessed 06.01.19).

J. Furman, Is this time different? The opportunities and challenges of artificial intelligence, (presentation, AI Now: The Social and Economic Implications of Artificial Intelligence Technologies in the Near Term),

New York, NY. Available from: <https://www.whitehouse.gov/sites/default/files/page/files/20160707_cea_ai_furman.pdf>, 2016 (accessed 06.01.19).

G. Graetz, G. Michaels, Robots at work, CEPR Discussion Paper No. DP10477. Available from: <http://papers.ssrn.com/sol3/papers.cfm?abstract_id = 2575781>, 2015 (accessed 06.01.19).

N. Jean, M. Burke, M.W. Xie, M. Davis, D.B. Lobell, E. Stefano, Combining satellite imagery and machine learning to predict poverty, Science 353 (6301) (2016) 790–794.

C. Liersch, Vehicle technology timeline: from automated to driverless, Robert Bosch (Australia) Pty. Ltd., 2014. Available from: <http://dpti.sa.gov.au/__data/assets/pdf_file/0009/246807/Carl_Liersch_Presentation.pdf>, 2016 (accessed 06.01.19).

P. McCorduck, in: M.A. Natick, A.K. Peters (Eds.), Machines Who Think: A Personal Inquiry into the History and Prospects of Artificial Intelligence, second edition, W.H. Freeman, San Francisco, 2004. 1979.

IBM, Native, web or hybrid mobile-app development, Somers, New York, USA. Available from: <ftp://public.dhe.ibm.com/software/pdf/mobile-enterprise/WSW14182USEN.pdf>, 2012 (accessed 06.01.19).

S. Kenaw, Hubert L. Dreyfus's critique of classical AI and its rationalist assumptions, Minds Mach. 18 (2) (2008) 227–238. Available from: <https://link.springer.com/article/10.1007/s11023-008-9093-7> (accessed 06.01.19).

J. Kingston, Expert systems with applications 2001. High performance knowledge bases: four approaches to knowledge acquisition, representation and reasoning for workaround planning, Expert. Syst. Appl. 21 (4) (2001) 181–190. Available from: <http://dblp.uni-trier.de/db/journals/eswa/eswa21.html> (accessed 06.01.19).

M. Malkawi, O. Murad, Artificial neuro fuzzy logic system for detecting human emotions, Human-centric Comput. Inf. Sci. 3 (1) (2013) 3. Available from: <https://hcis-journal.springeropen.com/articles/10.1186/2192-1962-3-3> (accessed 06.01.19).

J.D. Meier, A. Homer, D. Hill, J. Taylor, P. Bansode, L. Wall, et al., Mobile application architecture guide. Retrieved from: <http://robtiffany.com/wp-content/uploads/2012/08/Mobile_Architecture_Guide_v1.1.pdf>, 2008 (accessed 06.01.19).

V. Müller, N. Bostrom, Future progress in artificial intelligence: a survey of expert opinion, Fundamental Issues of Artificial Intelligence, Springer, 2014.

J.N. Nilsson, The Quest for Artificial Intelligence: A History of Ideas and Achievements, Cambridge University Press, Cambridge, 2010.

C. Twardy, R. Hanson, K. Laskey, S.T. Levitt, B. Goldfedder, A. Siegel, et al., SciCast: collective forecasting of innovation, Collective intelligence, 2014.

D. Wang, A. Khosla, R. Gargeya, H. Irshad, A.H. Beck, Deep learning for identifying metastatic breast cancer. Available from: <https://arxiv.org/pdf/1606.05718v1.pdf>, 2016 (accessed 06.01.19).

J. Weng, McClelland, A. Pentland, O. Sporns, I. Stickman, M. Sur, et al., Autonomous mental development by robots and animals, Science 291 (2001) 599–600. Available from: http://dx.doi.org/10.1126/science.291.5504.599-via. msu.edu.

M.M. Zanjureh, A. Shahrabi, H. Larijani, ANCH: a new clustering algorithm for wireless sensor networks, in: 27th International Conference on Advanced Information Networking and Applications Workshops, 2013. Available from: <WAINA 2013. http://dx.doi.org/10.1109/WAINA.2013.242> (accessed 17.05.19).

CHAPTER

3 ARTIFICIAL INTELLIGENCE, INTERNET OF THINGS, AND COMMUNICATION NETWORKS

Garima Singh and Gurjit Kaur

Department of Electronics and Communication Engineering, Delhi Technological University, Delhi, India

3.1 INTRODUCTION

Artificial intelligence (AI) and machine learning (ML) are coming up as the two most diverse technologies creeping into every industry and influencing the business world with their features. The data shows that 81% of industries were affected positively by AI in 2017 and that has jumped to 54% this year [1–3].

To reveal the hidden potential of consumer data businesses are progressively are using AI and ML.

With the help of AI and ML in networks, one can think of a world having a universal network, which will have minimum issues on connectivity and data availability with no lags for surely. One can think of it as all the connections are pooled by a single dynamic automated network that is capable of analyzing the user behavior and capable enough to continuously switch the low-speed connection with a fast one to ensure uninterrupted services. This will leave individuals free from switching the network from mobile data to WiFi for a better connection.

The merger of AI and ML seems like IoT in nature because as the Internet of Things (IoT) means all things connected through the Internet the same is in AI. AI and ML are capable enough to automatically connect, switch with the better network providing uninterrupted services that result in increased workload over the network. AI is the best solution to deal with the exponential increase of data as AI can analyze and computes at a much faster rate than humans. Additionally, there are other ways also in which AI and ML affect the network.

3.1.1 TECHNOLOGY COMBINED HUMAN GENIUS

Human inventiveness and brightness joined with the gigantic learning approach of AI and ML will help networks to handle the burden of countless management techniques and new designs. None of this would be conceivable without each other, that is, AI depends on humans and humans on AI would change the systems drastically. In any case, there is a requirement of human insight in building the actual network infrastructure.

Artificial Intelligence to Solve Pervasive Internet of Things Issues. DOI: https://doi.org/10.1016/B978-0-12-818576-6.00003-4

3.1.2 USING METRICS TO FIX NETWORK ISSUES

Traditional networks have a dependency on deep-packet inspection (DPI) techniques to creep into one's networks to gather the information required to trace and fix the network problem. While this method is much time consuming and due to an authentication problem, these technologies have limited access to one's data and will be unable to trace and fix the network problem. This problem will be removed by incorporating AI and ML. AI with ML competencies can be skilled to diagnose the network problems by using the data gathered from various channels. By using the existing metrics, AI can diagnose the problems and gives 80% accurate practical solutions.

The use of AI and ML in networking implies that some fields like analysis and detection can be automated to find a solution automatically. This enables IT, professionals, to shift their attention to other issues that cannot be resolved without human intervention. In the coming future, AI and ML will become that much smart, that they will not require the metrics and DPI also.

3.2 MACHINE LEARNING/ARTIFICIAL INTELLIGENCE-ASSISTED NETWORKING

The volume of data crossing over the networks is so astronomical that it has become unmanageable for humans to analyze, act, and process. By exploiting the benefits of ML and AI in harnessing for this data and process in an efficient, faster, and smarter way. As the generation is moving toward autonomous era, ML and AI-based solutions can automate the operations, improve the security, and enhance the user experience as mentioned in the following text.

3.2.1 PROACTIVE OPERATIONS

Through automation, proactive actions, and self-healing capabilities the IT operations have been revolutionized by extreme networks. The engine based on ML provides predictive analytical technology that is capable enough to automatically understand the nature of wireless networks and heals it. It offers proactive resolutions to IT despite traditional reactive mode, making IT free for other business-related activities.

To improve the security of network ML and AI are used. Unsupervised form of ML is used in the networks to study the estimated behavior of IoT network endpoints. If IoT endpoint behaves unusually then an alert will be raised automatically. Therefore with the help of AI, threats for IoT devices can be detected and mitigated without any human intervention.

3.2.2 ENHANCES USER EXPERIENCE

AI-based systems can detect and correct the problem before they appear to the end user. This will reduce the maintenance burden on IT and will enrich the end-user experience.

3.3 ARTITICIAL INTELLIGENCE IN COMMUNICATION NETWORKS

AI can be used to enhance the operation and configuration of devices used in the network, optical performance monitoring devices, fiber nonlinearities mitigation, modulation format recognition, and estimation of the quality of transmission (QoT) that upgrades the overall performance of the network. The use of AI in optical networks will make it *cognitive* [4] and cognitive optical networks are those who are supported by an intelligent "brain" or an intelligent control functions implemented in software defined networking framework. This cognitive controller uses the data gathered from the network monitors and additional data sources also and will use the AI techniques to enhance the performance of the network.

All together with benefits, AI possesses challenges along with opportunities in decision making and automating operations because of the huge amount of data collected through heterogeneous sensors and network devices, and time-varying parameters. AI is particularly valuable for performance prediction and optimization of complex optical networks. In this viewpoint, traditional algorithms used for signal processing are not as productive as AI techniques. Nowadays ML, a subpart of AI, is getting popularly used in the optics field, covering all its areas like optical networks and communication along with Nanophotonics and quantum mechanics.

A couple of simple ML algorithms have been used in the optical area by delivering favorable outcomes. For example, upgradation in the limit of transmission was seen without adjusting the infrastructure of the network. This is considered that ML devices will gain much more usage beyond usual systems like for diversity of fiber plant, applications in a meshed network, and adaptable network. Since it has been observed that ML can recuperate random noise from noisy signals, this will be conceivably advantageous for different applications, for example, networks for 5G, visible-light-based communication, satellite-based communication, and even for optical sensing also. AI can be yoked to catch laser dynamics and parameters that are hard to model by standard approaches.

3.3.1 THE MOTIVATION FOR USING MACHINE LEARNING IN COMMUNICATION NETWORKS

From the last few years, researchers have tried to practice mathematical approaches in optical communications and networks and ML is one of the examples of that effort. The overall motivations for the use of ML in optical networks can be identified as follows:

1. **Increased system complexity:** The implementation of innovative coherent technology-enabled transmission practices [5], along with exceptionally flexible networking principles, for example, the elastic optical networks technique. These new technological advancements require a high number of tunable parameters like symbol rates, adaptive coding rates, modulation formats, adaptive channel bandwidth, and so on for their implantation that has ended up with an extremely complex design and operation of optical networks. Modeling of such systems with the help of closed-form formulas is next to impossible and in fact "margins" are normally assumed in the analytical models, resulting in the underutilization of resources that lands up with increased system cost. To resolve these issues ML methods are gaining attention as they can handle these complex and nonlinear systems with comparatively simple training of supervised and/or unsupervised algorithms. These algorithms are capable of solving typical cross-layer problems of optical network fields through exploiting the historical network data information [6].
2. **Increased data availability:** Current optical networks have several monitors that are used to give different forms of information about the whole system like monitoring of traffic, indicators

of signal quality bit error rate (BER), system failure notifications, and so on. The development brought by ML comprises simultaneously leveraging the plethora of gathered information and discover hidden relations among numerous information types.

3.3.2 MACHINE LEARNING IN OPTICAL NETWORKS

Nowadays optical networks are getting themselves ready for future applications through making their availability at edge level networks. Which requires smart and intelligent working to handle high-speed data with minimum latency, for example, 5G. The greatest obstruction in using software controls in optical networks is its analog nature of optical transmission, which increases the operational and managerial complexity of the networks. Moreover, there is a threat to traditional capacity-based applications because of spectral density limits on optical systems. There is a need for new efficient scaling methods to further enhance the cost/bit/s despite depending on capacity improvements alone.

AI with ML algorithms provides a novel approach having the potential to for wider use of software controls in optical systems and also optimizes their efficiency across all dimensions. Reference data sets for ML would advance the operability and functionality across the industry, which further enables scaling and efficiency. For the enhancement of scalability and performance, software-driven networking is required in optical systems. For optical communication systems, there is a need for reference training data sets of ML including needs for new or different methodologies to be analyzed to make optimum use of ML in optical networks. There is a need to understand the basic models of ML to utilize it fully, that is, neural network and genetic algorithm.

Neural networks are motivated by the conduct of biological neurons leads to the development of artificial neurons. This is a software-based module that is connected in layer format. Each neuron can transmit signals to the neuron of the next layer along with connections that have weights according to input importance from a previous layer. After receiving the desired strength signals a neuron will send its signals. The sent signal is further tuned by ML algorithm and connection weights are decided with the help of proper training process.

Genetic algorithms are also a nature-inspired technique. Multiple methods of detecting the right output based on input have been developed by the developers. Then ML is used to mimic the behavior of nature, by them. Weed out the least fit options, mix and mutate the survivors, and repeat the cycle to improve results over time. This is all done in a genetic algorithm to solve the network issues [7].

3.4 TRANSFORMING OPTICAL INDUSTRIES BY ARTIFICIAL INTELLIGENCE

The introduction of AI in networking gives an abundance of opportunities for connectivity to the most emerging systems like IoT-enabled autonomous vehicles and always-connected smart-city systems. AI has revolutionized the present business model by combining the advancements of ML and data mining making it feasible to analyze a huge amount of data gathered from numerous sources, to recognize the patterns, give interactive understanding and to make intelligent predictions.

One of the most emerging technologies of AI is software-defined networking that can dynamically schedule the traffic between the internet and private network. This helps in ensuring the seamless, secure, and super-fast access of the network for all applications and data, wherever in the world they are. Through this technological advancement, networks have gone through more profound evolutions across all industries.

If we talk about the retail sector, AI has vast applications as it can provide accurate and faster detection products for the customer that increases the order value with the conversion rate for their company's shopping portals. Similarly in the healthcare industry, it is very easy to analyze thousands of documents in a minute through a network built on AI that helps doctors to make well-informed decisions about patient care. For the moment, AI is also a big support for call center business because AI-supported network can quickly and accurately route and service customer inquiries.

The aforementioned are very few industrial examples where AI has gained popularity and served their requirements. All the available expertise along with data and research are combined into an AI algorithm. Moreover, this algorithm can be further developed and augmented, so that people can use it except restricted to few. Undeniably, this alteration will need a lot of work from the human's side first. This will require a summary of all the continued research and data form all networks at a global level to make a reliable AI-based system to meet the growing digital demand for businesses.

After understanding the use of AI in networking for business purposes, now it is a turn of data that how it is acquitted. The desired raw data is gathered by using axial motion sensors like gyroscopes and accelerometers installed in wearable devices and portable devices like smartphones. This motion data is obtained along the three axes (x, y, z) in an entirely unobtrusive way, that is, movements are continuously tracked and evaluated in a very user-friendly manner and then processed through supervised learning approaches to AI. For better and accurate results it is advisable to gather several sets of samples from limited users than smaller sets of the sample from a large number of users. Only getting raw sensor data is not sufficient enough. There is a need for a highly precise organization of that data to carefully define certain features because AI is an iterative process. Thus there will be some guesstimates also based on domain knowledge.

3.5 ARTIFICIAL INTELLIGENCE IN OPTICAL TRANSMISSION

The table shows the use of AI techniques in the field of optical transmission that has been used until now and is available in the literature [8].

Application in Optical Transmission	AI Technique Used
Transmitters	1. Bayesian filtering and expectation-maximization [9] 2. Simulated annealing [10] 3. Machine learning and genetic algorithms [11]
Optical amplification control	1. Kernelized linear regression [12] 2. Linear/logistic regression [13] 3. Multilayer perception neural network [14]
Linear impairment identification	1. Kalman filter [15] 2. Neural networks [16] 3. Principal component analysis [17]

OSNR monitoring	1. Deep neural network [18] 2. Neural networks [19]
Modulation format recognition	1. Principal component analysis [17] 2. Support vector machines (SVM) [19] 3. Clustering k-means [20]
Receiver nonlinearity mitigation	1. Maximum a posteriori [21] 2. Maximum-likelihood [22,23] 3. Maximum-likelihood and maximum a posteriori [24] 4. Bayesian filtering and expectation-maximization [25] 5. Nonlinear support vector machines [26] 6. K-nearest neighbors [27] 7. Clustering k-means [28] 8. Nonlinear support vector machines and newton method [29]
QoT estimation	1. Case-based reasoning (CBR) [30] 2. CBR with learning/forgetting [30–31] 3. Random forests classifier [33] 4. Linear regression [34] 5. Support vector machines [35]

3.6 ARTIFICIAL INTELLIGENCE IN OPTICAL NETWORKING

Below given table gives the use of AI in optical networking that has been used till now and is available in the literature.

Optical Networking Application	AI Technique Used
Survivable optical networks	1. Genetic algorithms [36] 2. Ant colony optimization [37]
Regenerator placement	3. Genetic algorithms [38,39] 4. Ant colony optimization [37]
Resource allocation	5. Genetic algorithms [38,39] 6. Particle swarm optimization [40] 7. Ant colony optimization [41] 8. K-means clustering [42] 9. Markov decision processes [43]
Connection establishment	10. Swarm intelligence [44] 11. Genetic algorithms [38,45–49] 12. Ant colony optimization [50–51] 13. Case-based reasoning [52] 14. Simulated annealing [53,54] 15. Tabu search [55,56] 16. Backpropagation neural network [57] 17. Q-learning [58] 18. Game theory [59] 19. Neural networks and principal component analysis [60] 20. Kalman filters [61] 21. Markov decision processes [62,63]

Continued

Optical Networking Application	AI Technique Used
Network reconfiguration	1. Genetic algorithms and colony optimization [64]
	2. Genetic algorithms [65,66]
	3. Genetic algorithms and cognition [67]
	4. Neural networks [68]
Failure/fault detection	1. Bayesian networks, clustering [69]
	2. Cognition-based methods [70]
	3. Bayesian inference networks [71,72]
Software-defined networking	1. Methods based on Cognition [73,74]
	2. Neural networks [75]
Reduction/estimation of burst loss	1. Learning automata [76]
	2. Q-learning [58,77,78]
	3. expectation-maximization and hidden Markov model [79]
	4. Bayesian networks [80]
	5. Feed-forward neural network and Q-learning [81]
	6. Extreme learning machine [82]
	7. Ant colony optimization [83]
Statistical solution for prediction	1. Hidden Markov model (HMM) [84]
	2. Bayesian methods and game theory [85]
Intelligent ROADM	1. Linear Regression [34]

3.7 ADVANTAGES OF MACHINE LEARNING IN NETWORKING

ML-driven analytics tools are best in the learning process where traditional networks fail. There are three main reasons for which ML is used in networking and that are: managing the performance, managing the health, and security.

3.7.1 MANAGING THE PERFORMANCE

ML tools can handle traffic management moment-by-moment having larger planning for range capacity and management. These ML tools can automate the traffic if it is spiking in some areas to another area. Management tools of ML may move half of the traffic set out toward a back-end framework starting with one server farm then onto the next dependent on traffic conditions. ML tools additionally can extend patterns of traffic for the future decision-making process [86].

3.7.2 MANAGING THE HEALTH

Initial failure stages of the network can be detected through ML-driven analytics and also predict about their appearance as a healthy node for those initial stages. Vendors of network equipment are increasingly using these types of analytics in management tools, particularly for SaaS offerings.

3.7.3 SECURITY

Detecting irregularities in the behavior of the network can support the cybersecurity group in finding all the information whether it is about a hardware node that has been compromised or an employee getting rouge on the company's network. ML-based techniques have immensely enhanced the analytics space for behavioral threat along with distributed denial-of-service detection and remediation.

3.8 OPTICAL TECHNOLOGIES TO SUPPORT INTERNET OF THINGS

Optical technologies, in particular, fiber optic technologies, contribute to a loT of networks and applications in various aspects, ranging from data transportation, networking, and sensing and imaging.

3.8.1 TRANSMISSION AND SWITCHING

loT generates a large amount of data. Even though some of the data are processed locally at the fog layer and do not require further analysis or storage, a large portion of data requires processing in the cloud. Therefore transporting a large amount of data between the devices and local fogs and the cloud is an important part of the loT network. Besides high transmission bandwidth for large data volume, the data transmission in loT has other requirements like low latency, long distance, security, and flexibility. Fiber-optic communication network provides the most suitable platform, due to its stable channel with low attenuation, high transmission speed, and bandwidth along with multidimensional multiplexing capability. Optical circuit switching is by nature bit-date and protocol-independent and can support heterogeneous signal formats simultaneously. Flexibility can be further enhanced through utilizing CDC ROADMs, flexible grid WDM networks, and variable rate transponders, with the intelligent centralized control through transport SON. Optical layer encryption adds physical level security to the existing security measures at the higher layers [87].

For the last segment of the network, that is, between the devices and the rest of the network, wireless technologies are mostly used due to the mobility advantage. However, for devices that generate a large amount of data, especially in industrial loT applications, optical transmission is still a good solution. Fiber-optic communication is also advantageous in locations where the RF interference is an issue or RF channels are unavailable. The current passive optical network and FTTx can be expanded to support more loT of data and applications. Free space optical (FSO) communication provides another alternative to connect an IoT of devices to the network. This includes laser-based high-speed, long-range FSO systems, and LED-based low-data rate, short-distance indoor visible light communication systems.

3.8.2 DATA CENTER NETWORKING

Cloud computing, a major part of IoT, is driving up the traffic volumes in the data center. It is forecasted that 83% of global data center traffic will come from cloud services and applications by 2019, with a total data center traffic volume of 10.4 zettabytes per year [88]. Optical technologies have been used in the data center networks (DCNs) for high-speed point-to-point links. The developments of complementary metal oxide semiconductor (CMOS) technologies, photonic integrated circuit technologies, and digital signal processing (DSP) technologies enable optical transceiver module hardware with a higher data rate, longer transmission reach, and lower cost [89]. Spatial division multiplexing technologies will further increase the bandwidth capacity per fiber without increasing the fiber count [90].

There is also an increasing need to achieve geo-distribution for the cloud service data centers, due to the requirements of low latency, redundancy, computation power, scalability, and location restriction. Inter-DCNs are almost entirely based on fiber optic technology [91,93].

3.8.3 OPTICAL SENSING AND IMAGING

Besides transporting and routing data, optical technologies can be used to generate data in the IoT network, especially in terms of optical sensing and imaging. Optical sensors measure various physical phenomena, like temperature, pressure, displacement, vibration, acceleration, electrical field, chemical, and so on, by observing the optical property change in the light beam caused by the phenomena. The optical properties to be monitored include intensity, phase, wavelength/frequency, polarization, spectral distribution, and so on. Optical sensing offers high sensitivity, low latency, and long sensing distance. It is also immune to electromagnetic interference and can be implemented in a harsh environment.

Based on how the medium that the light travels, optical sensing technologies can be divided into free space optical sensing and fiber optic sensing (including optical waveguide sensing). Fiber optic sensing does not require line-of-sight and can perform distributed or quasi-distributed sensing over a long distance with remote monitoring capability. Distributed fiber optic sensing utilizes the backscatter of light pulses directed down a fiber optic cable. Because the backscatter occurs down the entire length of the cable, every single part of the optical fiber acts as a monitoring device. Common distributed fiber optic sensors include Rayleigh scattering-based vibration and acoustic sensors, Brillouin scattering-based temperature and strain sensors, and Raman scattering-based temperature sensors. There are also single point sensors that use sensing elements like fiber Bragg grating to conduct sensing at a targeted location. Quasi- distributed sensors contain arrays of multiple sensing elements along with the optical fiber. FSO sensing and imaging does not require fiber installation and is thus more flexible and nonintrusive. The sensing and imaging distance can be range from less than a millimeter (like optical coherent tomography) to hundreds of kilometers (like space lidar). The spectrum range in free space optical sensors is also broad. The optical source and receiver can be located at the same end (like in most standoff detection) or different ends of the light path.

3.9 APPLICATIONS OF INTERNET OF THINGS WITH OPTICAL TECHNOLOGIES

IoT is a ubiquitous network that connects a huge amount of devices, aiming to cover every aspect of daily life and every business sector. Therefore it has a wide range of applications. Here are a few examples where optical technologies play a key role.

3.9.1 UTILITY NETWORK

The fiber optic broadband connections to homes and other buildings are used with wireless networks to connect individual smart meters to the utility company's network, allowing the utility operators to monitor the usage in real-time or near real-time, pinpoint outrages and reduce restoration time, automate billing validation, and better manage and balance energy load during different usage periods. The fiber optic support network also ensures always-on, gigabit per second speed Internet service required for supercomputers that constantly monitor the power grid throughout the city.

3.9.2 DIGITAL OIL FIELD

Digital oil field (DOF) is a direct example of industrial loT. It consists of both the tools and the processes surrounding data and information management across the entire suite of oil/gas exploration and production activities. The combination of advanced fiber optic sensing technologies with integrated networking and big data analytic technologies allows more accurate underground resource exploration and smart good monitoring and management, realizing DOF with improved production and optimize facility performance. For example, distributed fiber sensors provide the operators with high-resolution 3D or 4D vertical seismic profile images for accurate underground reservoir characterization.

3.9.3 AUTOMATIC TOLL BOOTH

All toll booths equipped with emission cameras and optical gas sensors of multiheight multispecies to do the real-time scanning of the vehicles passing through to check the vehicles with failed emission regulations without disturbing the traffic. Aided by automatic plate recognition, the vehicles' information, including manufacturer, model and year can be linked with the emission screening results. Other information, like vehicle occupancy can also be detected. The toll booths can also exchange information with the passing by vehicles, like extracting vehicle running status data and sending test results and repair instructions.

3.10 CONCLUSION

To attain a viable improvement in the network edge, there is a need to improve the services for better customer experience. This requires technologies that are capable of handling such a massive amount of user data daily. These requirements have created pressure on present optical networks for performance, security, and bandwidth availability and to handle this pressure there is a need for implementing AI and ML in available networks.

ML is a part of AI that mainly focusses on using the computer to solve the problems despite human guiding it on how to solve the issue. When it comes to networking part ML can be used to improve management, analytics, and security. However, to fully understand this there is a need for understanding the ML models, for example, neural networks and genetic algorithms.

REFERENCES

[1] S.J. Russell, P. Norvig, Artificial Intelligence: A Modern Approach, Pearson Education Limited, Malaysia, 2016.

[2] J. Ferber, G. Weiss, Multi-agent Systems: An Introduction to Distributed Artificial Intelligence, Vol. 1, Addison-Wesley, Reading, 1999.

[3] I. Arel, D.C. Rose, T.P. Karnowski, Deep machine learning-a new frontier in artificial intelligence research, IEEE Comput. Intell. Mag. 5 (4) (2010) 13−18.

[4] E. Wenger, Artificial Intelligence and Tutoring Systems: Computational and Cognitive Approaches to the Communication of Knowledge, Morgan Kaufmann, 2014.

[5] R. Proietti, X. Chen, K. Zhang, G. Liu, M. Shamsabardeh, A. Castro, et al., Experimental demonstration of machine-learning-aided QoT estimation in multi-domain elastic optical networks with alien wavelengths, IEEE/OSA J. Optical Commun. Netw. 11 (1) (2019) A1−A10.

[6] D. Kedia, G. Kaur. Examining different applications of cloud-based IoT, in: Examining Cloud Computing Technologies Through the Internet of Things, IGI Global, 2018, pp. 125−146.

[7] G. Singh, G. Kaur. Role of communication technologies for smart applications in IoT, in: Handbook of Research on Big Data and the IoT, IGI Global, 2019, pp. 300−313.

[8] J. Mata, I. De Miguel, R.J. Duran, N. Merayo, S.K. Singh, A. Jukan, et al., Artificial intelligence (AI) methods in optical networks: a comprehensive survey, Optical switching Netw. 28 (2018) 43−57.

[9] D. Zibar, L.H.H. de Carvalho, M. Piels, A. Doberstein, J. Diniz, B. Nebendahl, et al., Application of machine learning techniques for amplitude and phase noise characterization, J. Lightwave Technol. 33 (7) (2015) 1333−1343.

[10] A. Hraghi, M.E. Chaibi, M. Menif, D. Erasme, Demonstration of 16QAM-OFDM UDWDM transmission using a tunable optical flat comb source, J. Lightwave Technol. 35 (2) (2016) 238−245.

[11] S.L. Brunton, X. Fu, J.N. Kutz, Self-tuning fiber lasers, IEEE J. Sel. Top. Quantum Electron. 20 (5) (2014) 464−471.

[12] Y. Huang, W. Samoud, C.L. Gutterman, C. Ware, M. Lourdiane, G. Zussman, et al., A machine learning approach for dynamic optical channel add/drop strategies that minimize EDFA power excursions, in: ECOC 2016 42nd European Conference on Optical Communication, VDE, 2016, pp. 1−3.

[13] Y. Huang, P.B. Cho, P. Samadi, K. Bergman, Dynamic power pre-adjustments with machine learning that mitigate EDFA excursions during defragmentation, in: 2017 Optical Fiber Communications Conference and Exhibition (OFC), IEEE, 2017, pp. 1−3.

[14] E.D.A. Barboza, C.J. Bastos-Filho, J.F. Martins-Filho, U.C. de Moura, J.R. de Oliveira, Self-adaptive erbium-doped fiber amplifiers using machine learning, in: 2013 SBMO/IEEE MTT-S International Microwave & Optoelectronics Conference (IMOC), IEEE, 2013, pp. 1−5.

[15] B. Szafraniec, T.S. Marshall, B. Nebendahl, Performance monitoring and measurement techniques for coherent optical systems, J. Lightwave Technol. 31 (4) (2012) 648−663.

[16] X. Wu, J.A. Jargon, R.A. Skoog, L. Paraschis, A.E. Willner, Applications of artificial neural networks in optical performance monitoring, J. Lightwave Technol. 27 (16) (2009) 3580−3589.

[17] M.C. Tan, F.N. Khan, W.H. Al-Arashi, Y. Zhou, A.P.T. Lau, Simultaneous optical performance monitoring and modulation format/bit-rate identification using principal component analysis, J. Optical Commun. Netw. 6 (5) (2014) 441−448.

[18] T. Tanimura, T. Hoshida, J.C. Rasmussen, M. Suzuki, H. Morikawa, OSNR monitoring by deep neural networks trained with asynchronously sampled data, in: 2016 21st OptoElectronics and Communications Conference (OECC) held jointly with 2016 International Conference on Photonics in Switching (PS), IEEE, 2016, pp. 1−3.

[19] J. Thrane, J. Wass, M. Piels, J.C. Diniz, R. Jones, D. Zibar, Machine learning techniques for optical performance monitoring from directly detected PDM-QAM signals, J. Lightwave Technol. 35 (4) (2016) 868–875.
[20] N.G. Gonzalez, D. Zibar, I.T. Monroy, Cognitive digital receiver for burst mode phase modulated radio over fiber links, in: 36th European Conference and Exhibition on Optical Communication, IEEE, 2010, pp. 1–3.
[21] M.G. Taylor, Phase estimation methods for optical coherent detection using digital signal processing, J. Lightwave Technol. 27 (7) (2009) 901–914.
[22] F. Karinou, N. Stojanovic, C. Prodaniuc, M. Agustin, J. Kropp, N.N. Ledentsov, Solutions for 100/400-Gb/s ethernet systems based on multimode photonic technologies, J. Lightwave Technol. 35 (15) (2016) 3214–3222.
[23] A.P.T. Lau, J.M. Kahn, Signal design and detection in presence of nonlinear phase noise, J. Lightwave Technol. 25 (10) (2007) 3008–3016.
[24] F. Rottenberg, T.H. Nguyen, S.P. Gorza, F. Horlin, J. Louveaux, ML and MAP phase noise estimators for optical fiber FBMC-OQAM systems, in: 2017 IEEE International Conference on Communications (ICC), IEEE, 2017, pp. 1–6.
[25] D. Zibar, M. Piels, R. Jones, C.G. Schäeffer, Machine learning techniques in optical communication, J. Lightwave Technol. 34 (6) (2015) 1442–1452.
[26] D. Wang, M. Zhang, Z. Li, Y. Cui, J. Liu, Y. Yang, et al., Nonlinear decision boundary created by a machine learning-based classifier to mitigate nonlinear phase noise, in: 2015 European Conference on Optical Communication (ECOC), IEEE, 2015, pp. 1–3.
[27] D. Wang, M. Zhang, M. Fu, Z. Cai, Z. Li, Y. Cui, et al., KNN-based detector for coherent optical systems in presence of nonlinear phase noise, in: 2016 21st OptoElectronics and Communications Conference (OECC) held jointly with 2016 International Conference on Photonics in Switching (PS), IEEE, 2016, pp. 1–3.
[28] J.J.G. Torres, A. Chiuchiarelli, V.A. Thomas, S.E. Ralph, A.M.C. Soto, N.G. González, Adaptive nonsymmetrical demodulation based on machine learning to mitigate time-varying impairments, in: 2016 IEEE Avionics and Vehicle Fiber-Optics and Photonics Conference (AVFOP), IEEE, 2016, pp. 289–290.
[29] E. Giacoumidis, J. Wei, S. Mhatli, M.F. Stephens, N.J. Doran, A.D. Ellis, et al., Nonlinear intersubcarrier intermixing reduction in coherent optical OFDM using fast machine learning equalization, in: 2017 Optical Fiber Communications Conference and Exhibition (OFC), IEEE, 2017, pp. 1–3.
[30] T. Jiménez, J.C. Aguado, I. de Miguel, R.J. Durán, M. Angelou, N. Merayo, et al., A cognitive quality of transmission estimator for core optical networks, J. Lightwave Technol. 31 (6) (2013) 942–951.
[31] T. Jiménez, J.C. Aguado, I. de Miguel, R.J. Durán, N. Fernandez, M. Angelou, R.M. Lorenzo, A cognitive system for fast quality of transmission estimation in core optical networksMarch OFC/NFOEC (2012) 1–3. IEEE.
[32] A. Caballero, J.C. Aguado, R. Borkowski, S. Saldaña, T. Jiménez, I. de Miguel, et al., Experimental demonstration of a cognitive quality of transmission estimator for optical communication systems, Opt. Express 20 (26) (2012) B64–B70.
[33] L. Barletta, A. Giusti, C. Rottondi, M. Tornatore, March). QoT estimation for unestablished lighpaths using machine learning, Optical Fiber Communications Confererence, Optical Society of America, 2017, pp. Th1J–Th11.
[34] S. Oda, M. Miyabe, S. Yoshida, T. Katagiri, Y. Aoki, T. Hoshida, et al., A learning living network with open ROADMs, J. Lightwave Technol. 35 (8) (2017) 1350–1356.
[35] J. Mata, I. de Miguel, R.J. Durán, J.C. Aguado, N. Merayo, L. Ruiz, et al., A SVM approach for lightpath QoT estimation in optical transport networks, in: 2017 IEEE International Conference on Big Data (Big Data), IEEE, 2017, pp. 4795–4797.

[36] R. Morais, C. Pavan, A. Pinto, C. Requejo, Genetic algorithm for the topological design of survivable optical transport networks, IEEE/OSA J. Optical Commun. Netw. 3 (1) (2010) 17–26.

[37] M. Morgan, An ant colony approach to regenerator placement with fault tolerance in optical networks, in: 2015 Seventh International Workshop on Reliable Networks Design and Modeling (RNDM), IEEE, 2015, pp. 85–91.

[38] F. Martinelli, N. Andriolli, P. Castoldi, I. Cerutti, Genetic approach for optimizing the placement of all-optical regenerators in WSON, J. Optical Commun. Netw. 6 (11) (2014) 1028–1037.

[39] I. Cerutti, F. Martinelli, N. Sambo, F. Cugini, P. Castoldi, Trading regeneration and spectrum utilization in code-rate adaptive flexi-grid networks, J. Lightwave Technol. 32 (23) (2014) 3894–3901.

[40] L. Velasco, P. Wright, A. Lord, G. Junyent, Saving CAPEX by extending flexgrid-based core optical networks toward the edges, J. Optical Commun. Netw. 5 (10) (2013) A171–A183.

[41] I. de Miguel, R. Vallejos, A. Beghelli, R.J. Durán, Genetic algorithm for joint routing and dimensioning of dynamic WDM networks, J. Optical Commun. Netw. 1 (7) (2009) 608–621.

[42] F.R. Durand, T. Abrão, Energy-efficient power allocation for WDM/OCDM networks with particle swarm optimization, J. Optical Commun. Netw. 5 (5) (2013) 512–523.

[43] M. de Paula Marques, F.R. Durand, T. Abrão, WDM/OCDM energy-efficient networks based on heuristic ant colony optimization, IEEE Syst. J. 10 (4) (2014) 1482–1493.

[44] Á. Rubio-Largo, M.A. Vega-Rodriguez, J.A. Gomez-Pulido, J.M. Sanchez-Pérez, A comparative study on multiobjective swarm intelligence for the routing and wavelength assignment problem, IEEE Trans. Syst. Man Cybern. C Appl. Rev. 42 (6) (2012) 1644–1655.

[45] L. Gong, X. Zhou, W. Lu, Z. Zhu, A two-population based evolutionary approach for optimizing routing, modulation and spectrum assignments (RMSA) in O-OFDM networks, IEEE Commun. Lett. 16 (9) (2012) 1520–1523.

[46] S.G. Petridou, P.G. Sarigiannidis, G.I. Papadimitriou, A.S. Pomportsis, On the use of clustering algorithms for message scheduling in WDM star networks, J. Lightwave Technol. 26 (17) (2008) 2999–3010.

[47] R.R. Reyes, T. Bauschert, Adaptive and state-dependent online resource allocation in dynamic optical networks, IEEE/OSA J. Optical Commun. Netw. 9 (3) (2017) B64–B77.

[48] A. Rubio-Largo, M.A. Vega-Rodriguez, J.A. Gomez-Pulido, J.M. Sanchez-Perez, A comparative study on multiobjective swarm intelligence for the routing and wavelength assignment problem, IEEE Trans. Syst. Man. Cybernet. Part C (Appl. Rev.) 42 (6) (2012) 1644–1655.

[49] D. Monoyios, K. Vlachos, Multiobjective genetic algorithms for solving the impairment-aware routing and wavelength assignment problem, IEEE/OSA J. Optical Commun. Netw. 3 (1) (2010) 40–47.

[50] C.A. Kyriakopoulos, G.I. Papadimitriou, P. Nicopolitidis, E. Varvarigos, Energy-efficient lightpath establishment in backbone optical networks based on ant colony optimization, J. Lightwave Technol. 34 (23) (2016) 5534–5541.

[51] Y. Sun, J. Yuan, M. Zhai, An optimized RWA method with services selection based on ACO in optical network, in: 2016 15th International Conference on Optical Communications and Networks (ICOCN), IEEE, 2016, pp. 1–3.

[52] Z. Chen, S. Wang, H. Zhang, Y. Liu, Y. Peng, Cognitive routing and wavelength assignment algorithm for dynamic optical networks, in: 2014 12th International Conference on Optical Internet 2014 (COIN), IEEE, 2014, pp. 1–3.

[53] K. Christodoulopoulos, I. Tomkos, E.A. Varvarigos, Elastic bandwidth allocation in flexible OFDM-based optical networks, J. Lightwave Technol. 29 (9) (2011) 1354–1366.

[54] J. Perelló, J.M. Gené, A. Pagès, J.A. Lazaro, S. Spadaro, Flex-grid/SDM backbone network design with inter-core XT-limited transmission reach, J. Optical Commun. Netw. 8 (8) (2016) 540–552.

[55] K. Walkowiak, M. Klinkowski, B. Rabiega, R. Goścień, Routing and spectrum allocation algorithms for elastic optical networks with dedicated path protection, Optical Switching Netw. 13 (2014) 63–75.

[56] R. Goścień, K. Walkowiak, M. Klinkowski, Tabu search algorithm for routing, modulation and spectrum allocation in elastic optical network with anycast and unicast traffic, Computer Netw. 79 (2015) 148–165.

[57] W.B. Jia, Z.Q. Xu, Z. Ding, K. Wang, An efficient routing and spectrum assignment algorithm using prediction for elastic optical networks, in: 2016 International Conference on Information System and Artificial Intelligence (ISAI), IEEE, 2016, pp. 89–93.

[58] Y.V. Kiran, T. Venkatesh, C.S.R. Murthy, A reinforcement learning framework for path selection and wavelength selection in optical burst switched networks, IEEE J. Sel. Areas Commun. 25 (9) (2007) 18–26.

[59] J. Zhu, B. Zhao, Z. Zhu, Leveraging game theory to achieve efficient attack-aware service provisioning in EONs, J. Lightwave Technol. 35 (10) (2017) 1785–1796.

[60] D.R. de Araújo, C.J. Bastos-Filho, J.F. Martins-Filho, Methodology to obtain a fast and accurate estimator for blocking probability of optical networks, IEEE/OSA J. Optical Commun. Netw. 7 (5) (2015) 380–391.

[61] M. Yannuzzi, X. Masip-Bruin, S. Sánchez-López, E.M. Tordera, J. Solé-Pareta, J. Domingo-Pascual, OPN08-01: Interdomain RWA based on stochastic estimation methods and adaptive filtering for optical networks, IEEE Globecom 2006 (2006) 1–6.

[62] T. Tachibana, I. Koyanagi, MDP-based lightpath establishment for service differentiation in all-optical WDM networks with wavelength conversion capability, Photonic Netw. Commun. 20 (2) (2010) 183–192.

[63] C. Sue, Y. Hsu, P. Ho, Dynamic preemption call admission control scheme based on Markov decision process in traffic groomed optical networks, IEEE/OSA J. Optical Commun. Netw. 3 (4) (2011) 300–311.

[64] F.C. Ergin, E. Kaldırım, A. Yayımlı, A.Ş. Uyar, Ensuring resilience in optical WDM networks with nature-inspired heuristics, J. Optical Commun. Netw. 2 (8) (2010) 642–652.

[65] N. Fernández, R.J. Durán, I. de Miguel, N. Merayo, J.C. Aguado, P. Fernández, et al., Survivable and impairment-aware virtual topologies for reconfigurable optical networks: a cognitive approach, in: 2012 Fourth International Congress on Ultra Modern Telecommunications and Control Systems, IEEE, 2012, pp. 793–799.

[66] X. Gao, Z. Ye, J. Fan, W. Zhong, Y. Zhao, X. Cao, et al., Virtual network mapping for multicast services with max–min fairness of reliability, J. Optical Commun. Netw. 7 (9) (2015) 942–951.

[67] N. Fernández, R.J.D. Barroso, D. Siracusa, A. Francescon, I. de Miguel, E. Salvadori, et al., Virtual topology reconfiguration in optical networks by means of cognition: evaluation and experimental validation, J. Optical Commun. Netw. 7 (1) (2015) A162–A173.

[68] F. Morales, M. Ruiz, L. Gifre, L.M. Contreras, V. López, L. Velasco, Virtual network topology adaptability based on data analytics for traffic prediction, IEEE/OSA J. Optical Commun. Netw. 9 (1) (2017) A35–A45.

[69] M. Ruiz, F. Fresi, A.P. Vela, G. Meloni, N. Sambo, F. Cugini, et al., Service-triggered failure identification/localization through monitoring of multiple parameters, in: ECOC 2016 42nd European Conference on Optical Communication, VDE, 2016, pp. 1–3.

[70] X. Zhang, W. Hou, L. Guo, S. Wang, Y. Sun, X. Yang, Failure recovery solutions using cognitive mechanisms for software defined optical networks, in: 2016 15th International Conference on Optical Communications and Networks (ICOCN), IEEE, 2016, pp. 1–3.

[71] S.R. Tembo, S. Vaton, J.L. Courant, S. Gosselin, A tutorial on the EM algorithm for Bayesian networks: application to self-diagnosis of GPON-FTTH networks, in: 2016 International Wireless Communications and Mobile Computing Conference (IWCMC), IEEE, 2016, pp. 369–376.

[72] R. Montero, F. Agraz, A. Pagès, J. Perelló, S. Spadaro, Dynamic topology discovery in SDN-enabled transparent optical networks, in: 2017 International Conference on Optical Network Design and Modeling (ONDM), IEEE, 2017, pp. 1–6.

[73] J. Oliveira, J. Oliveira, E. Magalhães, J. Januário, M. Siqueira, R. Scaraficci, et al., Toward terabit autonomic optical networks based on a software defined adaptive/cognitive approach, J. Optical Commun. Netw. 7 (3) (2015) A421–A431.

[74] S.J.B. Yoo, Multi-domain cognitive optical software defined networks with market-driven brokers, in: 2014 The European Conference on Optical Communication (ECOC), IEEE, 2014, pp. 1–3.

[75] S. Yan, F.N. Khan, A. Mavromatis, D. Gkounis, Q. Fan, F. Ntavou, et al., Field trial of machine-learning-assisted and SDN-based optical network planning with network-scale monitoring database, in: 2017 European Conference on Optical Communication (ECOC), IEEE, 2017, pp. 1–3.

[76] J. Praveen, B. Praveen, T. Venkatesh, Y.V. Kiran, C.S.R. Murthy, A first step toward autonomic optical burst switched networks, IEEE J. Sel. Areas Commun. 24 (12) (2006) 94–105.

[77] T. Venkatesh, A. Sankar, A. Jayaraj, C.S.R. Murthy, A complete framework to support controlled burst retransmission in optical burst switching networks, IEEE J. Sel. Areas Commun. 26 (3) (2008) 65–73.

[78] S. Haeri, W.W.K. Thong, G. Chen, L. Trajković, A reinforcement learning-based algorithm for deflection routing in optical burst-switched networks, in: 2013 IEEE 14th International Conference on Information Reuse & Integration (IRI), IEEE, 2013, pp. 474–481.

[79] A. Jayaraj, T. Venkatesh, C.S.R. Murthy, Loss classification in optical burst switching networks using machine learning techniques: improving the performance of TCP, IEEE J. Sel. Areas Commun. 26 (6) (2008) 45–54.

[80] H. Elbiaze, Cognitive mechanisms for providing QoS in OBS networks, in: 2011 13th International Conference on Transparent Optical Networks, IEEE, 2011, pp. 1–4.

[81] S. Haeri, L. Trajković, Intelligent deflection routing in buffer-less networks, IEEE Trans. Cybern. 45 (2) (2014) 316–327.

[82] H.C. Leung, C.S. Leung, E.W. Wong, S. Li, Extreme learning machine for estimating blocking probability of bufferless OBS/OPS networks, IEEE/OSA J. Optical Commun. Netw. 9 (8) (2017) 682–692.

[83] Y. Coulibaly, G. Rouskas, M.S. Abd Latiff, M.A. Razzaque, S. Mandala, QoS-aware ant-based route, wavelength and timeslot assignment algorithm for optical burst switched networks, Trans. Emerg. Telecommun. Technol. 26 (11) (2015) 1265–1277.

[84] K. Chitra, M.R. Senkumar, Hidden Markov model based lightpath establishment technique for improving QoS in optical WDM networks, in: Second International Conference on Current Trends In Engineering and Technology-ICCTET 2014, IEEE, 2014, pp. 53–62.

[85] O. Awwad, A. Al-Fuqaha, B. Khan, D. Benhaddou, M. Guizani, A. Rayes, Bayesian-based game theoretic model to guarantee cooperativeness in hybrid RF/FSO mesh networks, in: GLOBECOM 2009-2009 IEEE Global Telecommunications Conference, IEEE, 2009, pp. 1–7.

[86] G. Kaur, A.K. Yadav, V. Anand, Design and implementation of artificially intelligent microcontroller based chess opponent, in: Proceedings of the World Congress on Engineering, 2010, vol. 1.

[87] B.T. Intelligence, The Internet of Things: examining how the IoT will affect the world, market report, 2015.

[88] Cisco, Cisco Global Cloud Index: forecast and methodology, 2014-2019, White paper, 2015.

[89] P.N. Ii, Inter-datacenter optical networks, in: International Symposium on VICTORIES Project and Workshop, 2014.

[90] G. Milione, et al., Real-time bi-directional lOGbE transmission using MIMO-less space-division-multiplexing with spatial modes, OFC, WI.2, 2016.

[91] C. Kachris, T. Ioannis, A survey on optical interconnects for data centers, IEEE Commun. Surv. & Tutor. (2012) 1021–1036.

[92] P.N. i, Hybrid optical-electrical data center networks, OSA Advanced Photonics Congress, NeM3B.3, 2016.

CHAPTER

AI AND IOT CAPABILITIES: STANDARDS, PROCEDURES, APPLICATIONS, AND PROTOCOLS

4

Aditya Pratap Singh[1] and Pradeep Tomar[2]

[1]Ajay Kumar Garg Engineering College, Ghaziabad, India [2]Department of Computer Science and Engineering, University School of Information and Communication Technology, Gautam Buddha University, Greater Noida, India

4.1 INTRODUCTION

The wave of development in computing, pulling us outside the realm of the traditional desktop era. Now the computing world is not limited to conventional computers like desktop or laptop, instead it is expanded to many things being used in day-to-day life of human being like air conditioners, refrigerators, washing machines, cars, and many more. Many new entities are participating in the meaningful computation to make human life more organized and comfortable. The inclusion of these new entrants in the world of computing was made possible by a new computing paradigm popular as Internet of Things (IoT) paradigm. In the IoT paradigm, the gadgets and appliances around humans will become part of network and participate as a node. The data collected at different levels, transferred, stored, and processed for making decisions based on data. The devices and type of data ranges from normal CCTV cameras, sensors for different type of data like temperature, distance, elevation, and so on to small pill-shaped cameras in human digestive tract to collect several images for identification of illness. Now the traditional farming is also getting benefitted from IoT and intelligent information system through which farmers can get suggestions regarding farming conditions, irrigation need, and crop selection for particular areas [1]. The IoT technologies and its applications are creating a basic transformation in individual and society views toward collaboration of technology and business. The business models are also getting updated and sometimes dynamic models used like offering different discount to different customers based on buying preferences. The efficiency of these business models can be increased with introduction of various kinds of sensors in controlling different processes like manufacturing. The usage of IoT is widespread and getting support from underlying technologies. The improvement in wireless communication and standardized communication protocols make data collection from IoT nodes/sensors more efficient and fast. Now the collection of data is possible almost anytime, anywhere, and of any size. The support from cloud computing has drastically increased storage and computing power at a reasonable cost. With the excessive creation and flow of information, the world nowadays is becoming an information system itself.

The upcoming developments in IoT infrastructure and related services will interact with technologies responsible for creating autonomous systems in order to deliver more efficient advance

Artificial Intelligence to Solve Pervasive Internet of Things Issues. DOI: https://doi.org/10.1016/B978-0-12-818576-6.00004-6

functionality. This blend on technology will lead to development of new business models and open new funding opportunities for business organizations. The effective use of data analysis enhances decision-making capability. The business processes are governed or modified by actuators on the basis of feedback coming through network. This feedback is generated by the commands created by data analysis of data collected through IoT. In this way the data becomes root for automation and control in business processes. The use of artificial intelligence (AI) to make applications capable of adjusting itself for complex situations with limited or nil human intervention increases productivity. In this way the human reactions to specific situations can be imitated up to a satisfactory performance levels with AI decision-making.

The chapter will address all the procedures involved in IoT, standards for IoT and AI both, various protocols supporting IoT, and application domains where both technologies are being implemented.

4.2 INTERNET OF THINGS

Many appliances (electronic, electrical, and nonelectrical) and gadgets are being embedded with sensors and becoming capable to communicate. The IoT is created by using devices capable to be part of communication-actuating network. Such devices are exploding rapidly around us wherein sensors and actuators composition with the environment around us and the information shared across platforms creates a common operating picture [2]. IEEE described the phrase "Internet of Things" as [3]: *A network of items—each embedded with sensors—which are connected to the Internet.*

Another definition of IoT given by OASIS is [4]: "System where the Internet is connected to the physical world via ubiquitous sensors." The things in IoT refers to computers, sensors, people, actuators, refrigerators, TVs, vehicles, mobile phones, clothes, food, medicines, books, cameras, and so on.

The Internet Society describes the IoT as follows: "Internet of Things generally refers to scenarios where network connectivity and computing capability extends to objects, sensors and everyday items not normally considered computers, allowing these devices to generate, exchange and consume data with minimal human intervention" [5].

So it is easy to understand that the network of anything which is capable of communicating is simply IoT. The things participating in an IoT network have identities, physical attributes, and simulated personalities [6]. As per the report of Economist [7], the proposed smart cities projects will keep IoT at its base. The people and things work together digitally to achieve much greater efficiency.

The radiofrequency identification (RFID) and sensor network technologies made it possible to cater increasing number of things as node of networks. The interaction with physical environment is achieved using sensors and actuators. To get useful inferences form the data collected from sensors, it is important to collect, store, and process data intelligently.

Fig. 4.1 shows the relationship of things with respect to IoT, all the things equipped with compatible sensors to collect data and compatible communication device and medium (mostly wireless) to send and receive data. All of these things are connected to an IoT-enabled server. These IoT-

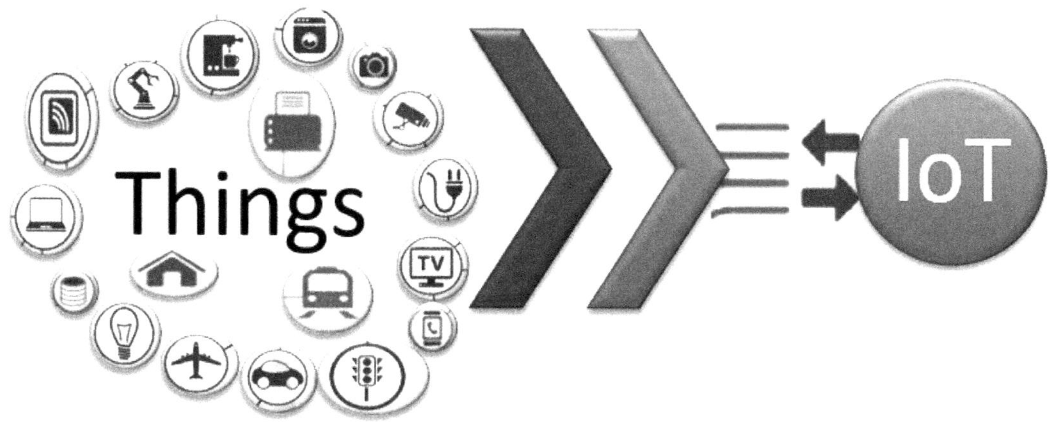

FIGURE 4.1

Internet of Things (IoT).

enabled things automate certain activities through machine-to-machine (M2M) interaction. The various communication standards suitable for IoT-compatible devices for M2M communication are Bluetooth, ZigBee, IPC Global standards, and low-power wireless fidelity (WiFi) [8].

The IoT has a ubiquitous applicability that makes it more beneficial. Many organizations and individuals find IoT useful as it utilizes different kind of physical objects (things) coordinate to help in decision-making and sharing information. The IoT applicability has three aspects: individual, business, and society. For instance, an individual may employ IoT devices to control household appliances remotely and can get alerts for healthcare based on vital heath monitoring. In business scenario, IoT can help in consignment tracking, inventory monitoring and ordering, automatic security monitoring and alerts, optimization of manufacturing process, and so on. At social front, IoT can be used as part of smart city project, for effective transportation planning, for reducing energy consumption, and so on.

To understand the IoT more closely, IoT architecture, standards, protocols, and application domains are need to be discussed.

4.2.1 ARCHITECTURE OF INTERNET OF THINGS

IoT is a vast concept already being implemented. Many architectures for IoT are proposed but there is no uniformly accepted architecture is available yet. The working of IoT includes a range of sensors, network, communication, and computing technologies. Some of the popular architectures are discussed here.

4.2.1.1 IEEE standard for an architectural framework for Internet of Things (P2413)

IEEE P2413 standard [9] considers IoT as a three layer architecture as depicted in Fig. 4.2. The different layers of this architecture address the similarities, interactions, and relationships among different domains and elements. This architecture considers different IoT domains, abstractions, and

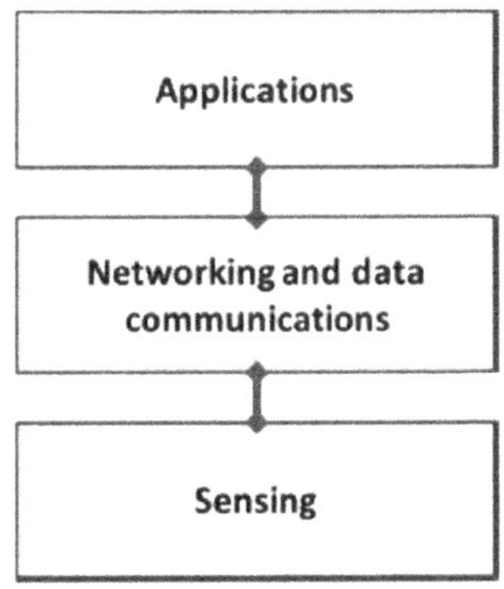

FIGURE 4.2

IEEE P2413 architecture.

commonalities among domains [10]. P2413 follows the architecture defined in ISO/IEC/IEEE 42010 [11]. The IEEE P2413 leverages existing standards for the specified architecture. The abstract "thing" object of this standard may incorporate apps, things, and services. In this standard the information can be shared horizontally or vertically or in both directions. This reference architecture incorporates the essential parts of basic architecture as well as its proficiency to become part of multitiered systems.

4.2.1.2 International Telecommunications Union reference model for Internet of Things

As discussed by Kafle et al. [12], the International Telecommunications Union (ITU) reference model is described as layered architecture. This ITU reference model is based on ITU-T Y.2060 [13]. This architecture includes the common understanding, functionalities, and capabilities of IoT. This architecture has four layers capped with management capabilities and security capabilities:

- Application layer
  - IoT applications
- Service support and application support layer
  - Generic support capabilities
  - Specific support capabilities
- Network layer
  - Networking capabilities
  - Transport capabilities
- Device layer
  - Device capabilities
  - Gateway capabilities

This layered architecture is capped with two capabilities:

- Management capabilities
  - Generic management capabilities
  - Specific management capabilities

- Security capabilities
    - Generic security capabilities
    - Specific security capabilities

The top most application layer represents different IoT applications like e-governance, e-health, smart education, smart traffic system, smart grid, and so on. The service support and application support layers are below application layer. This layer provides generic support capabilities with some application-specific support capabilities too. The next layer is network layer, which is responsible for networking of things and managing their connectivity. Various networking needs like routing, mobility, resource management, access control, and so on are fulfilled. The other responsibility of network layer is transportation facilities. Here the transportation of control instructions along with data generated from IoT applications is managed. At last the device layer is placed. This layer is responsible for physical connection between things hardware with network directly or through gateways. The devices are equipped with ubiquitous sensor networking functions. The gateway functionalities include privacy protection, security, and translation of protocols. These functionalities make the secure communication possibilities through heterogeneous technologies like ZigBee, Bluetooth, and WiFi.

4.2.1.3 Layered architecture of Internet of Things

A three-layered architecture as depicted in Fig. 4.3 is considered as basic architecture: perception layer, network layer, and application layer [14]. The perception layer sometimes treated as device layer is responsible for identifying objects and collects information like temperature, location, and so on. It is a physical layer consisting devices like sensors, actuators, controllers, RFID, barcode readers, and so on.

The network layer is responsible for establishing connection to other smart things, network devices, and servers. This layer is treated, as is the core of IoT architecture. The network layer addresses objects using a unique address respectively and securely transmit the information from the perception layer to the application layer and vice versa. The protocols like ZigBee, WiFi, and Bluetooth responsible for communication and transmission medium are part of this layer.

The application layer is on top of the architecture; it is responsible for delivering application-specific services to end user. This layer integrates the information received from lower layer and uses it in the management of applications implemented by IoT. These applications may be deployed for smart homes, smart cities, smart health, and many more.

With rapid growth in the field of IoT and to cover more functionality, three-layered architecture was not found sufficient. A five-layered architecture by adding two new layers of the processing layer and business layer proposed by Tan and Wang [15] as shown in Fig. 4.4.

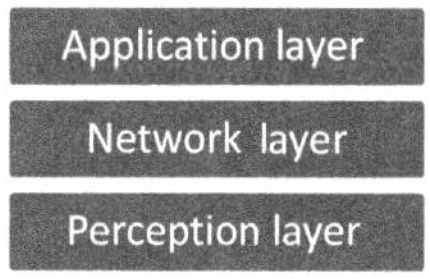

FIGURE 4.3

Three-layer architecture.

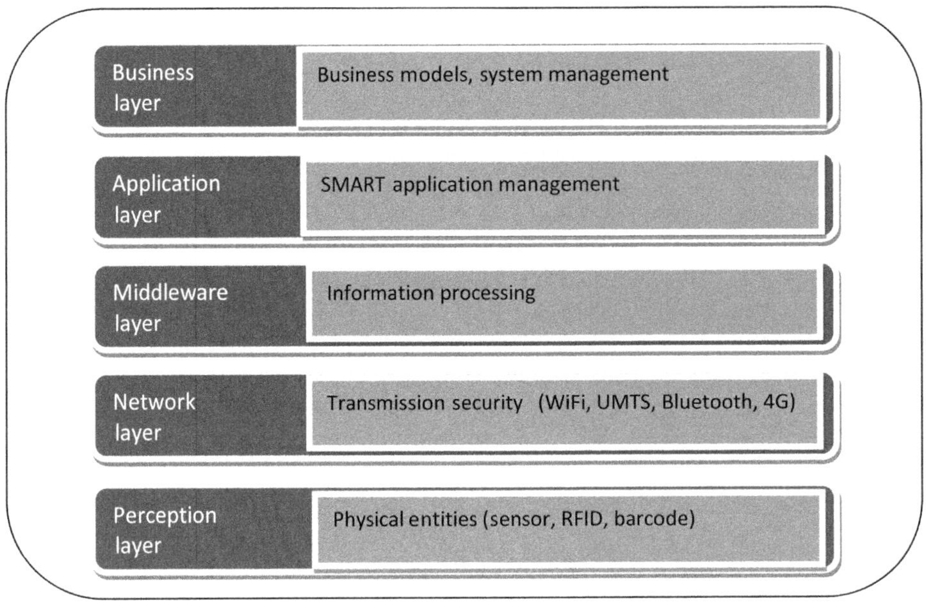

FIGURE 4.4

Five-layer architecture.

The five-layer architecture contains the following layers:

- business layer,
- application layer,
- middleware layer,
- network layer, and
- perception layer.

The business and middleware layers are the additional layers. The business layer is added to manage the IoT system for implementing business and profit model, at the same time user privacy is also handled by this layer. This layer manages the overall applications and services in line with business model. It may build models and graphs as per the data received from application layer. The success of any business depends on its business model and efficient implementation of that model. This layer is responsible for fulfilling the requirements of business model like analysis of information, presentation of important information to determine future actions, and strategies.

The other addition is the middleware layer, which is sometimes treated as processing layer also. The middle layer is responsible for storing and processing a bulk of data received from the transport layer. This layer not only manages a diverse set of services but also made them available to lower layers. These services may include various technologies like cloud services, database services, and big data processing services. This layer has link to database, where it stores the information received from network layer. The automated decisions are made after information processing. The various types of devices deployed in IoT implement compatible services and communicate with devices implementing similar service types.

The other layer is business layer that is responsible for managing business and profit model, user privacy through various application management. This layer is responsible for overall management of whole IoT system with applications and services. The business layer is responsible for building business model, graphs, flowcharts, and other tools helpful for managing business model. These business models are created on the basis of data coming from application layer. The analysis of result becomes a tool for laying out future strategies and action needed.

4.2.2 IOT STANDARDS AND PROTOCOLS

The standards are proposed to facilitate application and service providers. These standards ease the job of both application, service providers as a fixed standard can be followed by many IoT applications and services, these applications, and services can be utilized for several other IoT applications and services following same standard. Various groups were created to create protocols and standards for IoT domain. The organizations like Institute of Electrical and Electronics Engineers (IEEE), Internet Engineering Task Force (IETF), World Wide Web Consortium (W3C), European Telecommunications Standards Institute (ETSI), and more as all cannot be listed in this chapter.

The most popular among them is IEEE, which aims to nurture technological innovation as a professional engineering group created for the benefit of humanity in the field of engineering. Section 4.2.1.1 illustrates the IEEE standard P2413 for IoT.

4.2.2.1 ETSI Internet of Things standard

The ETSI has also contributed by providing standards applicable globally in the field of information and communications technologies (ICT). These standards ranges from fixed communication devices, mobile, radio, converged, broadcast, and Internet technologies. The European Union in the form of a European Standards Organization officially recognizes it. The IoT is not directly referred by ETSI; rather the label of "machine-to-machine (M2M) communication" [16] is applicable for IoT. ETSI defines M2M communication as:

> Machine-to-Machine (M2M) communications is the communication between two or more entities that do not necessarily need any direct human intervention. M2M services intend to automate decision and communication processes.

In this standard, the access network becomes the medium for devices to connected to the network domain.

The M2M devices perform the procedures like registration, authentication, authorization, management, and provisioning with the network domain. It may also provide its services to connected devices that are hidden from network. The M2M gateway works as proxy for the devices connected through it. The M2M devices and M2M gateway forms a M2M area network. This network provides IP connectivity, network control functions, services, interconnection with other network, and roaming.

4.2.2.2 ITU Internet of Things standard

The ITU supports ICT by taking responsibility of allocation global radio spectrum and specifications of satellite orbits. For seamless communication, ITU develops various standards and provides improved access to underserved communities globally. It is said to be a ubiquitous network as per the report by ITU for IoT [17]. In this ubiquitous network the availability and access of network is ensured everywhere at any time. The various roles of different technologies are also discussed by ITU. These technologies include RFID for labeling, sensor for feeling, and applications to smartly thinking and helping in decision-making. All these use nanotechnology for micro devices.

4.2.2.3 IETF Internet of Things standard

The IETF is an open community of network engineers along with operators, researchers, and vendors. All members work for smooth operation of Internet and responsible for evolution of Internet architecture. The IETF has also given standards for IoT, with definition [18]:

> The basic idea is that IoT will connect objects around us (electronic, electrical, non-electrical) to provide seamless communication and contextual services provided by them. Development of RFID tags, sensors, actuators, mobile phones make it possible to materialize IoT which interact and co-operate each other to make the service better and accessible anytime, from anywhere.

From the beginning, IPv6 emphasizes on wireless communication and created a Low-Power WPAN (6LowPAN) as a working group to work with sensor network having scarcity of resources.

This working group performed protocol optimization for working on IEEE 802.15.4 network as well as developed different standards for mesh routing and other routing protocols.

4.2.2.4 W3C Internet of Things standard

The development of web standards is a prime responsibility of W3C. The W3C has also recognizes IoT in the form of Web of Things (WoT) with a changed perception toward web technologies.

W3C defines "Web of Things" as follows [19]:

> The Web of Things is essentially about the role of Web technologies to facilitate the development of applications and services for the Internet of Things, i.e., physical objects and their virtual representation. This includes sensors and actuators, as well as physical objects tagged with a bar code or NFC. Some relevant Web technologies include HTTP for accessing RESTful services, and for naming objects as a basis for linked data and rich descriptions, and JavaScript APIs for virtual objects acting as proxies for real-world objects.

Everything required for IoT environment like systems, devices, and data inclusion in WoT establishes as a Plateform.

The charting of a W3C Interest Group for the WoT (www.w3.org/WoT/IG) was a output of a workshop on the WoT in June 2014 (www.w3.org/2014/02/wot). WoT interest group works toward creating open markets to cater upcoming applications and services. These services will handle interaction of IoT with web of data. Table 4.1 summarizes these standardization efforts done by various agencies [19].

4.2.2.5 Application protocols

The protocols are system of rules governing various activities happening in network transmission. Various activities like data format for exchange over network, addressing of network nodes, and routing of data packets between source and destination. The protocols for communication give strength to IoT systems and make data exchange with nodes possible. The constrained application protocol

Table 4.1 Internet of Things Standardization Efforts by Different Agencies.

Application Protocol		DDS	CoAP	AMQP	MQTT	Web Sockets	XMPP	HTTP REST
Infrastructure Protocol	Routing				RPL			
	Network layer	6LoWPAN				IPV4/IPV6		
	Link layer	IEEE 802.15.4						
	Physical/device layer	LTE-A		EPCglobal		IEEE 802.15.4		Z-Wave

(CoAP) and message queue telemetry transport (MQTT) protocol are lightweight application layer protocols to deal with IoT and M2M communication in a resource-constrained environment.

4.2.2.6 Constraint application protocol

The IETF CoRE is a working group, which developed CoAP. It is designed for a particular purpose of dealing with nodes having scarcity of resources and constrained networks. Here the constrained network refers to less power and lossy network. This protocol is specialized for M2M environment like various automation required for smart city projects. This is a web transfer protocol on top of HTTP and based on REpresentational State Transfer (REST) [20]. With stateless client server architecture, REST protocol becomes suitable for mobile and social network. Under CoAP servers hold resources and made available through URL. These resources can be accessed using methods like GET, POST. The CoAP bounds to UDP instead of TCP in contrast to REST. This difference makes it more compatible for IoT. CoAP uses modified HTTP functionalities to work with IoT like reduced power consumption, and support to lossy and noisy links. The IoT nodes may use 8-bit microcontrollers with very small primary memory (e.g., 10 KB RAM 100 KB code space).

4.2.2.7 Message queue telemetry transport

The MQTT is client server publish-subscribe protocol designed for IoT and M2M with support to lightweight and fast communication. It is an optimal connection protocol in which connection operations uses routing method (one to one, one-to-many, and many-to-many). The MQTT is compatible with TCP/Internet protocol (IP) or other protocols, which provides bidirectional, ordered, and lossless connection. It is best suited for IoT devices (resource constrained) that work in unreliable and low bandwidth connections. It follows subscriber, broker, and publisher model as described in Fig. 4.5.

It is easily understandable from the Fig. 4.5 that the publisher in this model creates the data of specific interest. The subscriber is consumer of data of interest. The subscriber would register for specific type of data so that it can get notification for that type of data. The publisher when generates the specific type of data and transmit it to registered subscribers through broker. The broker is responsible for security and checks the authorization of publisher and subscribers [21].

4.2.2.8 Extensible messaging and presence protocol

Extensible messaging and presence protocol (XMPP) is an IETF working group protocol designed for real-time communication. It supports instant messaging, collaboration, and presence services

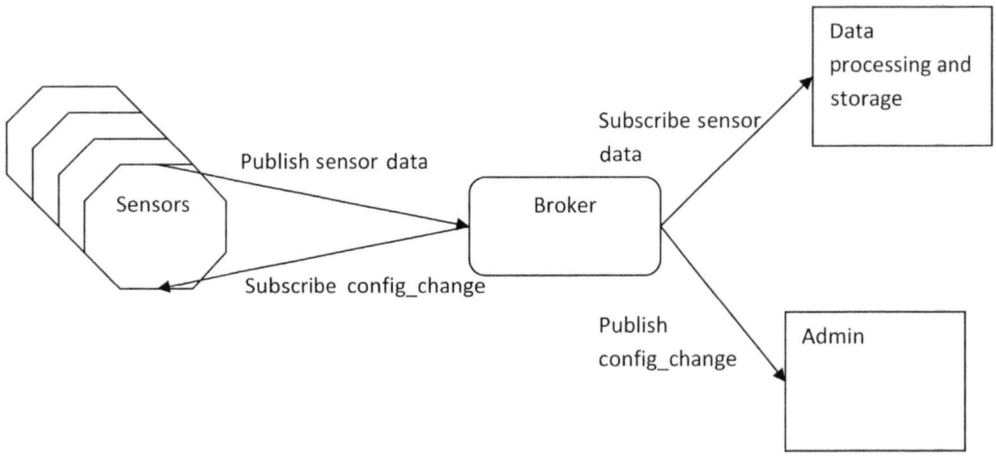

FIGURE 4.5

Message queue telemetry transport publish-subscribe model.

[22]. The XMPP draws attributes from XML to address voice and video streaming facilities, simultaneous chatting among multiple parties, and tele-presence. For this purpose, XMPP possess scalability addressing and security as its strength. Three basic roles: client, server, and gateway are performed to enable bidirectional communication between nodes. The object-to-object communication in IoT can be supported by XMPP with XML-based text messaging.

4.2.2.9 Advanced message queuing protocol

The application layer protocol advanced message queuing protocol (AMQP) is an asynchronous messaging service. The services like security, reliability with queuing and routing are included. The three type of message delivery guarantee (at-most-once, at-least-once, and exactly once) are used to provide reliable communication. With reliability as a feature it requires TCP like protocol. It enables communication between differently designed (heterogeneous programming languages) clients and middleware. The implementation of AMQP includes publish-subscribe, store-forward, message distribution, message queuing, and point-to-point [23]. It is freely available to be used by any organization.

4.2.2.10 Data distribution service

The data distribution service (DDS) also supports M2M communication using publish-subscribe protocol. It is designed by object management group. The key difference from AMQP and MQTT is broker-less communication model. The broker-less model makes it suitable for device-to-device communication and real-time IoT communication.

The other protocols like service discovery protocols, infrastructure protocols also have a significant role in IoT communication. The description of these is not in scope of this chapter.

4.2.3 INTERNET OF THINGS TECHNOLOGIES

4.2.3.1 Radiofrequency identification

RFID is a noncontact communication technology, and is used to identify and track objects without contact. It uses radio signals for data exchange over a short distance. Initially RFID was being referred for node to node communication in IoT, later on the practitioners and researchers found other technologies to be used with IoT like sensors, actuators, Mobile devices, and so on the RFID gives facility to automatic identification of objects. The system consists of multiple objects equipped with RFID tags and one reader. The tags have a small electronic circuit with antenna and possess a unique identity (ID). Whenever tag enters in area of a reader and receives an enquiry message from reader, it transmits its ID on a wireless medium [24]. These RFID tags can be classified in two types as active tag and passive tag based on their power source. These RFID tags are passive in nature (not always) and operate in different frequency bands (124–135 kHz) and operates up to 0.5 m. The active tags are less frequently used as they require a power source (e.g., a battery) for transmitting data covering higher range up to >100 m.

4.2.3.2 Electronic product code

Electronic product code (EPC) code represents a unique ID number used to allotted to items in supply chain management or other applications. It is kept on RFID tag. EPCglobal controls EPC, and related RFID standards apart from development of EPC. This architecture was found promising for IoT-based applications also. EPC and RFID work hand in hand to support IoT requirement like ID for objects and service discovery as well as it provides openness, scalability, interoperability, and reliability [20]. These IDs are 64–96 bits long.

4.2.3.3 Internet protocol

IP protocol is a primary and well-adopted standard protocol for network communication. Now two versions of IP, that is, IPv4 and IPv6 are being used. The IoT devices are connected to the Internet through IP stack. This connection requires significant amount of power and memory. IoT devices may also get connected using NON-IP communication channel like Bluetooth, RFID, and NFC. These NON-IP communication channels are only suitable for small area network (very limited range). The IP stack was needed to be modified to be used in low-power communication. One such example is 6LoWPAN. It uses IPv6 PAN with very less power consumption. Such protocols can be used within rage equivalent to LAN and power consumption as low as PAN [25].

4.2.3.4 Wireless fidelity

WiFi/802.11 is a communication technology supporting mobility within 100 m range. It can be used to communicate for ad hoc configurations without need of router. The 802.11 standard deliver data rates from 1 Mb/s to 6.75 Gb/s [26]. It functions in 2.4/5 GHz band. The IoT nodes can also communicate their data over WiFi. WiFi is meant as backbone to communicate higher bandwidth data like video or audio. To reduce power consumption, the WiFi alliance designed IEEE 802.11ahstandard named as “WiFi HaLow.” This 802.11ah adapts WiFi to the IoT. It works with low power consumption and provides better range. The WiFi HaLow has a wider range than WiFi [25]. 11ah has restricted access window as a main component for making it compatible with IoT by providing different access channel to IoT devices.

4.2.3.5 Bluetooth

IEEE 802.15.1 is a standard for Bluetooth communication. The Bluetooth is a popular communication medium for small wireless devices due to its low-power intake and lesser cost. It works well in a short range of 8–10 m. It works in 2.4 GHz band. It supports less data rate, that is, 1–24 Mb/s. To make it more suitable to IoT communication another version with ultra low power evolutes as Bluetooth low energy (BLE). BLE is also popular as Bluetooth Smart. This technology eliminates the requirement of cabling between objects like camera, speaker, printers, and so on. The BLE on the other hand uses minimal power and short-range radio, which can operate for a very long duration in comparison to its basic version. BLE's credit can be given to smart phone companies and now available in almost models.

The network stack of BLE consists of physical layer to transmit and receive bits, link layer for flow and error control with facilities like connection setup and medium access. The logical link layer and adaption protocol (L2AP) are responsible for multiplexing of data channels, fragmentation, and assembly of large packets. Next layer is generic attribute protocol; it is responsible for efficient data collection from sensors, or other small nodes. The generic access profile as top layer provides facility for advertising or scanning, and connection initiation and management modes apart from application configuration [20]. BLE is better in power consumption in comparison to ZigBee.

4.2.3.6 ZigBee

ZigBee is an answer for short-duration wireless communication requiring less energy consumption mainly in IoT for low-rate sensors. IEEE 802.15.4 is the base protocol for ZigBee protocol. IEEE 802.15.4 specifies low-rate wireless personal area networks (LR-WPAN) standards. ZigBee technology is a product of ZigBee Alliance.

The protocol stack for ZigBee has MAC, network, and application layer [27]. ZigBee supports three types of devices: ZigBee coordinator, router, and end device. The coordinator is fully functional device with functions for managing the whole network. The ZigBee router is also a fully functional device in tree and mesh networks only. These types of nodes are having routing functions like finding best path to forward packets. ZigBee end device is reduced-function device just to send and receive packets. Application object (APO) is used at application layer. It is a software responsible for controlling hardware units like transducer or switches on devices. These APOs have a unique id to enable them to communicate with other APOs. The ZigBee DeviceObject (ZDO) performs task like device discovery, security, managing requests, and so on. [28].

4.2.3.7 Z-Wave

Z-Wave supports low power consumption and applied as a wireless communication protocol. It is being used widely for home automation networks. It is best suitable for applications for remote control in small size commercial domain or smart homes. It was conceived by ZenSys (currently Sigma Designs) initially. The further improvement was done by Z-Wave Alliance [20]. It is a wireless, radiofrequency-based communication technology. It is used for control reading, monitoring appliances, and status reading. Z-Wave works with five layers: physical layer, MAC, transfer, routing, and application layer.

The key variance between ZigBee and Z-Wave is operating frequency. Z-Wave becomes superior in range then ZigBee as it operates in the sub 1 GHz band [27].

4.3 ARTIFICIAL INTELLIGENCE

AI is a field of mathematical engineering, which has potential to enhance many real-world application domains like healthcare, industrial automation, service sector, and so on, it is the science of implanting intelligence in nonintelligent systems (machine) to make these machines take decisions as human mind. The environment for AI is laid down with the popularity of IoT, universal existence of sensors, and development of E-Commerce. AI becomes the most suitable solution to manage large data flow and storage in IoT applications. The real-time decision-making (RTDM) is the heart of IoT because it is about decision informatics and it cuddles with advance sensing technology like Big Data, processing like real-time processing, and proceeding with RTDM. The AI is becoming an integral part of IoT like it improves its abilities like voice recognition and language understanding.

The machine learning (ML) empowers the AI. In its basic form, it practices algorithms to parse data, learns from this data to make predictions or decisions. These predictions and decisions provides base for implementation of AI. The IoT will be benefitted and become more useful with use of AI, ML, and robotics. These technologies will convert IoT in smart IoT. Let us imagine one air conditioning system that works as per ambient temperature. It also controls cooling as per the number of persons in that area, or based on habit of the person using it, and as per time constraints. Smart phones may learn about audio level for the particular user after few days of use. These features make system intelligent but it is not easy to implement such features without AI. Because the system has to learn the habit of person using it, and adapt accordingly (teaching). These features are not easy to implement and require complex algorithms to teach the system like AI methods.

The IoT causes unprecedented amount of data. This huge amount of data needs to be used intelligently for smart system. This may be achieved by centralized or decentralized algorithms of AI. The AI algorithms are implemented in business, application, and in information processing layer as described in Fig. 4.6.

Intelligence is required to take appropriate decision based on the data received. This is applicable in every domain like computer science, philosophy, social science, medical science, astrology, geography, and so on, this kind of decision-making and prediction can be done by human mind but it require more time and efforts, which is not possible all the time for humans. This requires some automated algorithms to simulate human like decisions based on data received. The data received also has some unwanted properties, which makes the process more complex. These properties of data may be like very high volume, unstructured form, different data source types, and need for real-time processing.

The growing research in AI-based algorithms is luring the organizations to use AI in their business strategies. Many organizations deal with huge amount of data and therefore require AI to make smart decisions for growth of business. Similarly, in case of IoT-based organizations, the small or large amount of data is integral part of huge set of connected devices. In IoT the smart objects also have some small-scale processing with some inherent intelligence. However, at larger scale great amount data is required to be processed and it does not happen at object level. If the decision is for these smart objects, the data is transmitted to distributed location where analysis is performed to take decision and the decision is sent back to smart object. The actuator present there will perform its job according to the decision [29]. The time required to for this process must be

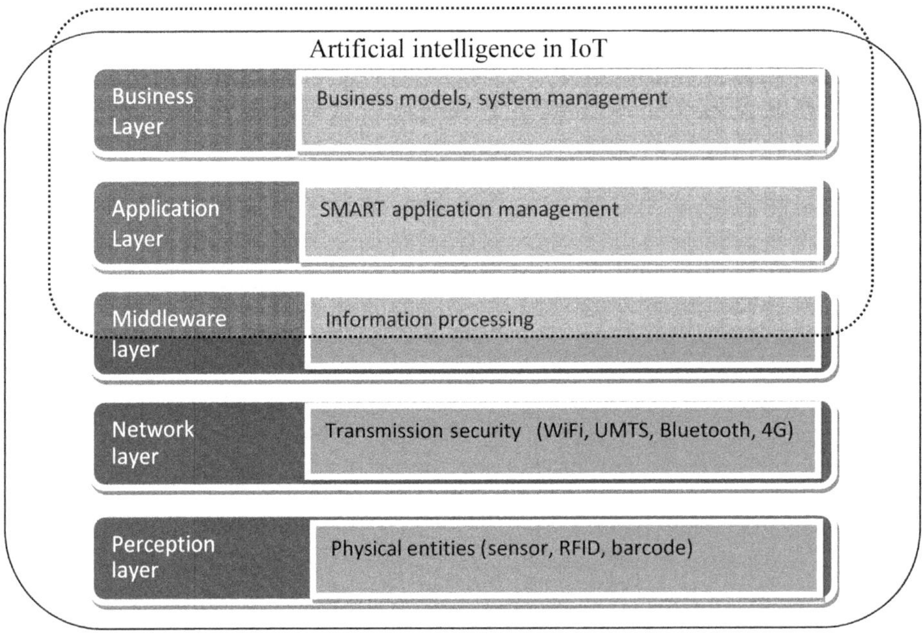

FIGURE 4.6

Role of artificial intelligence in Internet of Things architecture.

very less otherwise it will not be useful. Sometimes the data sent from each object is analyzed collectively at centralized location, which will be helpful for taking decision at greater level.

The conventional analytics tools are not capable of handling this enormous volume of data in real time. A good ML system will be required with capabilities like data preparation, advanced learning algorithms, automation and adaptive processes, scalability, and real-time decision-making.

4.3.1 MACHINE LEARNING

ML is a popular technique for AI. It provides skill to machine to learn without external programming. ML started with pattern recognition and computational learning theory. ML has a vast set of robust and workable algorithms now being applied to medical science, bioinformatics, intrusion detection, forecasting, and so on. The ML is a potential tool for data analytics in IoT. The ML algorithms being used frequently for IoT-based smart data analytics are given in Table 4.2 as discussed by Mahdavinejad et al. [30].

To get suitable decision by smart data analytics, it is mandatory to find out which task should be completed with the help of finding uncommon data points, structure discovery, prognosticate values/categories, and feature extraction.

The clustering algorithms become appropriate tool to find out structure of unlabeled data. For handling a huge data volume having sufficient number of data types, k-means is suitable and frequently applied clustering algorithm [30]. Principle component analysis (PCA) along with class

Table 4.2 Machine Learning Algorithms Application in Internet of Things (IoT) Use Case.

Machine Learning Algorithm	IoT, Smart City Example Use Cases	Metric to Optimize
Classification	Driverless car	Object detection, scene classification
Clustering	Netflix, medical	Recommendation, categorization
Anomaly detection	Traffic management	Trend in specific direction, dip in traffic
Support vector regression	Weather forecasting	Prediction
Linear regression	Disease epidemics, stock prices	Trend prediction
Classification and regression trees	GPS	Gini index
Support vector machine	Power systems, traffic management, handwriting recognition, and many more	Forecasting, variance, recognition, classification
K-nearest neighbors	E-commerce, smart citizen	Pattern prediction, efficiency of the learned metric
Naive Bayes	Stock trading, agriculture, medical	Future trends, heart disease prediction, location prediction
K-means	Air traffic control, smart home, eavesdropping detection	Data irregularities detection and pattern matching, grouping
Feed forward neural network	Medicine	Forecast, finding redundancy, tracking energy consumption
Principal component analysis	Security, spectral analysis	Fault detection, feature extraction

support vector machine (SVM) are the choice for finding uncommon data points and anomalies from data. The linear regression and support vector regression (SVR) is used for predicting values and classify data. The models in these processes are applied to execution and train high velocity data. The multiclass neural network is suitable for predicting data categories in function approximation problems with support to accuracy and long training requirement. The SVM as a well-known classification algorithm is able to handle huge amount of data and classify it in different types. Due to this ability smart data processing algorithms use SVM.

The feature extraction of data is commonly performed by PCA and canonical correlation analysis (CCA). Even CCA is capable of showing correlation between two categories of data. These two algorithms can also be used for discovering anomalies.

Mahdavinejad et al. [30] has identified some challenges for IoT to overcome in order to gain full potential of data analytics. These challenges are categorized in three different types: more accurate solution for data characteristics, privacy and security of IoT applications, and requirement of specialized data analytic algorithms.

4.4 CONCLUSION

The IoT is a source of huge unstructured data from a set of large number of connected objects of different types. The most popular application of IoT is smart city projects. In this IoT can be used to provide varied domain services like traffic, weather, energy, and mobility with planning for urban environment. The promise shown by IoT will become realistic only when technical challenges along with business policies undertook it as a key part. To get benefitted from the services of IoT, the smart data collected from smart objects must be analyzed properly. The chapter focuses on the standards and the protocols used for IoT along with technology being imparted into it. Many algorithms of AI are being used to extract knowledge from smart data collected by IoT means. In this chapter many such algorithms with their specialization are pointed out; however, the detail discussion of these algorithms is not in scope of this chapter.

REFERENCES

[1] C. Michael, L. Markus, R. Roger, The internet of things, McKinsey Q. 2 (2010) (2010) 1–9.

[2] J. Gubbi, R. Buyya, S. Marusic, M. Palaniswami, Internet of Things (IoT): a vision, architectural elements, and future directions, Future Gener. Comput. Syst. 29 (7) (2013) 1645–1660.

[3] IEEE, Special report: the Internet of Things, 2014.

[4] C. Cosgrove-Sacks, Open protocols for an open, interoperable Internet of Things, Org. Adv. Struct. Inf. Stand. (2014) 18.

[5] K. Rose, S. Eldridge, L. Chapin, The Internet of Things: an overview – understanding the issues and challenges of a more connected world, The Internet Society (ISOC), 2015. <https://www.internetsociety.org/iot>.

[6] European Research Cluster on the Internet of Things – IERC, The Internet of Things 2012 - New Horizons, Cluster Book 2012, 2012.

[7] Economist report. It's a smart world. Available from: <http://www.managementthinking.eiu.com/sites/default/files/downloads/Special%20report%20on%20smart%20systems.pdf>, 2010.

[8] M. Weyrich, C. Ebert, Reference architectures for the Internet of Things, IEEE Softw. 33 (1) (2016) 112–116. Available from: https://doi.org/10.1109/MS.2016.20.

[9] IEEE P2413, <http://grouper.ieee.org/groups/2413/>, IEEE Standards Association.

[10] H.P. Breivold, A survey and analysis of reference architectures for the Internet-of-things, ICSEA 2017 (2017) 143.

[11] ISO/IEC/IEEE 42010:2011 Systems and software engineering – architecture description. <https://www.iso.org/standard/50508.html>.

[12] V.P. Kafle, Y. Fukushima, H. Harai, Internet of things standardization in ITU and prospective networking technologies, IEEE Commun. Mag. 54 (9) (2016) 43–49.

[13] Recommendation ITU-T Y.2060, Overview of the Internet of Things, 2012.

[14] M. Yun, B. Yuxin, Research on the architecture and key technology of Internet of Things (IoT) applied on smart grid, in: 2010 International Conference on Advances in Energy Engineering, IEEE, June 2010, pp. 69–72.

[15] L. Tan, N. Wang, Future internet: the Internet of Things, in: 2010 Third International Conference on Advanced Computer Theory and Engineering (ICACTE), IEEE, August 2010, vol. 5, pp. V5–V376.

[16] ETSI technical specification, Machine-to-machine communications (M2M); M2M service requirements, Technical Specification, ETSI TS 102 689 V1.1.1, 2010.

[17] ITU, The Internet of Things executive summary, ITU Internet Reports, 2005. Available from: <http://www.itu.int/osg/spu/publications/internetofthings/InternetofThings_summary.pdf>.

[18] IETF, The Internet of Things - concept and problem statement, 2010. <http://tools.ietf.org/id/draft-lee-iot-problem-statement-00.txt>.

[19] W3C Web of Things (WoT). <https://www.w3.org/WoT/>.

[20] A. Al-Fuqaha, M. Guizani, M. Mohammadi, M. Aledhari, M. Ayyash, Internet of things: a survey on enabling technologies, protocols, and applications, IEEE Commun. Surv. Tutor. 17 (4) (2015) 2347–2376.

[21] MQTT. <www.mqtt.org>.

[22] P. Saint-Andre, Extensible messaging and presence protocol (XMPP): Core, Internet Eng. Task Force, Fremont, CA, RFC 6121, 2011.

[23] J. Lin, W. Yu, N. Zhang, X. Yang, H. Zhang, W. Zhao, A survey on internet of things: architecture, enabling technologies, security and privacy, and applications, IEEE Internet Things J. 4 (5) (2017) 1125–1142.

[24] E. Vahedi, R.K. Ward, I.F. Blake, Performance analysis of RFID protocols: CDMA versus the standard EPC Gen-2, IEEE Trans. Autom. Sci. Eng. 11 (4) (2014) 1250–1261.

[25] P. Sethi, S.R. Sarangi, Internet of things: architectures, protocols, and applications, *J. Electr. Comp. Eng.* **30**, 2017.

[26] P.P. Ray, A survey on Internet of Things architectures, J. King Saud. Univ. Comp. Inf. Sci. 30 (3) (2018) 291–319.

[27] J. Tan, S.G. Koo, A survey of technologies in internet of things, in: 2014 IEEE International Conference on Distributed Computing in Sensor Systems, IEEE, May 2014, pp. 269–274.

[28] P. Baronti, P. Pillai, V.W. Chook, S. Chessa, A. Gotta, Y.F. Hu, Wireless sensor networks: a survey on the state of the art and the 802.15.4 and ZigBee standards, Comp. Commun. 30 (7) (2007) 1655–1695.

[29] A. Ghosh, D. Chakraborty, A. Law, Artificial intelligence in Internet of things, CAAI Trans. Intell. Technol. 3 (4) (2018) 208–218.

[30] M.S. Mahdavinejad, M. Rezvan, M. Barekatain, P. Adibi, P. Barnaghi, A.P. Sheth, Machine learning for Internet of Things data analysis: a survey, Digital Commun. Netw. 4 (3) (2018) 161–175.

CHAPTER

INTERNET OF INTELLIGENT THINGS: INJECTION OF INTELLIGENCE INTO IOT DEVICES

5

Simar Preet Singh[1], Arun Solanki[2], Tarana Singh[2] and Akash Tayal[3]

[1]*Department of Computer Science and Engineering, Chandigarh Engineering College (CEC), Landran, Mohali, India*

[2]*School of Information and Communication Technology, Gautam Buddha University, Greater Noida, India*

[3]*Indira Gandhi Delhi Technological University for Women, Delhi, India*

5.1 INTRODUCTION

Internet of Things (IoT) and artificial intelligence (AI) are individually fascinating; their consolidated use cases hold all the incredible potential considerably, as indicated by analysts and industry specialists. When you think about the term IoT, most likely, a lot of gadgets that can see each other through availability, click in our mind [1–3]. IoT enables us to associate free devices together and benefit as much as possible from them by allowing them to transfer information and interface. Envision the intensity of information gathered through every one of these gadgets. To help understand this vision, AI comes into the image. AI has made a full-range application in various areas, including medicinal services, accounts, and our everyday lives with individual associates like Alexa and Siri [4,5]. Joining AI with IoT brings the guarantee of another future, as the two innovations impart and rotate around shared factor information. Information acquired from associated gadgets can be utilized further with the assistance of AI, and more brilliant experiences can be gained from it [6].

AI proves to be useful with another innovation under its umbrella: machine learning or ML [7]. The terms AI and ML work at the standard of creating software programs that know. This knowledge enables them to examine the information and settle on choices of how a human brain does likewise [8]. Since the embodiment of IoT gadgets is to accumulate information and utilize it, putting information obtained from physical devices through AI and ML enables us to develop those procedures [9,10]. The Internet of Intelligent Things (IIoT) uses automated reasoning to carry more an incentive to the IoT space by better deciphering information obtained from associated gadgets. At the point when a lot of related devices gather raw information and join it, the product projects empowered with machine insight abilities take this information and investigate them. After careful examination, the output contains essential data [11,12].

Intelligent IoT allows IoT applications to understand maximum capacity. ML and AI bring more detailed information to tables faster. Companies already expect the following rewards for using the IoT:

Artificial Intelligence to Solve Pervasive Internet of Things Issues. DOI: https://doi.org/10.1016/B978-0-12-818576-6.00005-8

- *Improving accuracy rate*: If you have ever attempted to examine the information from numerous sheets on your computer, you more likely than not understood that it is a hectic activity. Human brains are restricted from playing out specific assignments at a particular rate, and when the minds are exhausted, we have considerably inclined the frequency of making mistakes. The IIoTs can break-downvast amounts of information traveling every path through gadgets. The best part about this is since the entire procedure is machine and programming driven, it very well may be performed with no human attention, which makes it mistake-free and improves precision rates [13–15].
- *Predictive analysis and maintenance*: Predictive investigation indicates a part of the examination that takes the current information and dependent on the results; it predicts conceivable future occasions. Right now, IoT gadgets are being utilized by organizations to report any incidents or concerns, similar to hardware frailer, and so on., in a robotized way without human assistance [16–18].
- *Improved customer satisfaction*: The center of each business is consumer loyalty. As of now, organizations like Amazon have earned the identification of being the most client-driven organization by keeping the needs of their clients before everything else. Be that as it may, human-based client experience comes up short on specific occasions because of a few factors, for example, language boundaries, time imperatives, and so on. Organizations are perceiving the intensity of AI by empowering chatbots for associating with clients [19,20].
- *Expanded operational efficiency*: Predictions made through AI are exceptionally valuable as far as developing the operational effectiveness of the business. Joined inside and out bits of knowledge acquired through AI can be utilized to improve the general business forms from scratch, which can bring about expanded operational productivity and reduced expenses [21].

5.2 METHODOLOGY

Today we are witnessing the miracle of new technology, the formation of the IoT. This structure allows a particular operational entity to be combined into all areas around us. This chapter introduces the vision of the IoT world and the application of this methodology in various fields like healthcare, education, commerce, agriculture, and transportation. Learn more about the application of the IoT concept in different areas of medicinal cultivation. We also analyzed the significant impact of IoT on the physical world.

5.3 ARCHITECTURE OF INTERNET OF THINGS

In one article by Chang-le Zhong et al. in 2015, the author examined an IoT system architecture having five layers in it. These layers are the network access layer, perception layer, network layer, presentation layer, and application support layer [22]. All the different layers are having some related tasks with it. Later in 2017, Weigong et al. discussed an IoT framework engineering having seven-layer in its design, as appeared in Fig. 5.1.

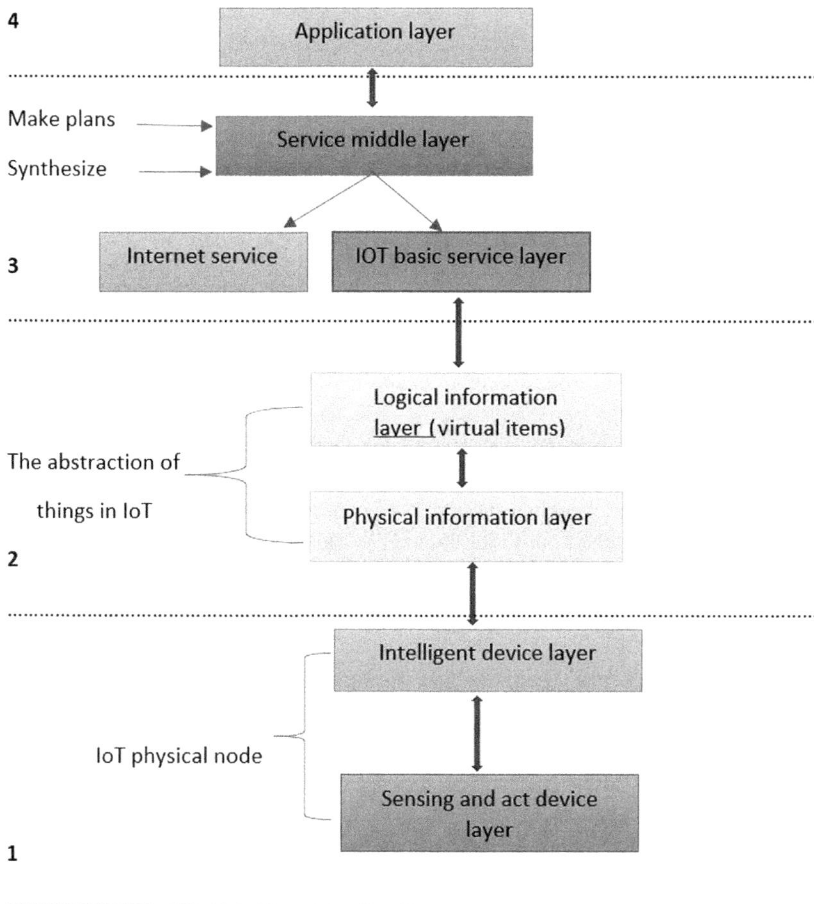

FIGURE 5.1

Seven-layer architecture IoT. *IoT*, Internet of Things.

Devices in the detection and execution device layer are divided into detection devices and execution devices. The sensing device may be a sensor that can reasonably reach the smart accessory and convert an external physical signal into a signal that the smart device can recognize. It is also test equipment that can convert physical data to digital data and transfer it to smart devices via the bus. The detection gadget can be a sensor network that cannot collect data exclusively, but it can not only move the data along a specific path and send it to the smart device, but also get the control flag when running the gadget smart devices can send data based on control signals and control the behavior of "things" [23].

The smart gadget layer contains smart gadgets that can collect data about objects and legally access the Internet. Smart gadgets are divided into IoT terminals, computers, and local area networks. IoT devices are embedded systems that include smart chips, because the two devices that enter the network are general-purpose computers, modern PCs, or mobile phones.

However, to provide transaction data, it must be associated with monitoring and control equipment. The device layer of the IoT provides the physical data of the element to higher layers. This hardware data contains the necessary information about the IoT. This information is reasonable after considering all physical hubs of the IoT, and this information usually determines byte groups. Used to build something, a physical data layer. In line with these ideas, IoT data configured in this way constitutes the true data layer of IoT. Starting from the IoT management layer, IoT reliable data describes computer-generated elements. The basic service layer is the most basic interface and Internet layer [24,25]. The basic management of the IoT can be divided into three types:

- *Things identification*: It is the frankest assistance that words can provide. To recognize the elements, initially, it is required to set authentic-identifiers for things. There is not a specific typical way for IoT things identifiers at current. The usual stuff id contains a standardized tag, radio-frequency identification (RFID), and two-dimensional code, and it was likewise recommended to see IPv6 addresses as an identifier [26].
- *Data collection of things*: In this stage, it starts to gather and get outer critical data by sensing gadgets, which are perhaps the most fundamental approaches to stay utilized, and the arenas comprise mechanical control, ecological checking, programmed pattern perusing, and so on. It is not significant for this information to be stretched out to the Internet because of information safety. However, as long as this stage utilizes a little piece of these, here can be a great deal of significant IoT administrations without considerable expense, for instance, in areas of ecological checking, horticulture, and so on. The information gathered from such administrations is exceptionally dynamic to the application, and many include industry benchmarks [27].
- *Things behavior control*: It is additionally perhaps the earliest approach to be utilized. This sort of power is functioning behavior, for the most part, for mechanical control, water, and different areas. If they are stretched out to the Internet, it is frequently identified with advanced security problems. There is another range, which is an "intelligent-home," and it is a decent method to regulate the behavior of household apparatuses through the IoT systems [28].

5.4 SECURITY AND PRIVACY

IoT presents several benefits to customers and can change the manners in which that purchaser associates with technology in critical methods. Further, IoT is going to merge the physical and virtual universes in the behaviors that are at present difficult to understand. From a safety and defense perspective, the expected unavoidable performance of gadgets and sensors, as of now, have limited spaces, for instance, the home, wearables, and the vehicles, even the body presents specific difficulties. As the physical training in our consistent day-to-day existence progressively identify and share insights about us, buyers will keep on needing protection. There are different may be various threats and attacks possible while the AI and IoT devices are being used [29,30]. Some of them are discussed as pursues:

- *Collaboration and more association among programmers and cybercriminals*: Hackers have been arranged into various gatherings, for example, traditional hackers, ideological hackers, state-supported attackers, and hackers-for-contract. Going ahead, we expect these gatherings will begin to cover, and in the end, team up for the simplicity of activity. Besides, we likewise hope to see some critical collusions among these gatherings of hackers, which will exploit each other's items and administrations [31].

- *Attack-as-an administration (AaaS)*: Malware-as-an administration and ransomware-as-an administration are not new ideas. Their appropriation was very special, however exceptionally useful. In 2019, we were expecting malware, explicitly ransomware, to progressively utilize the remote work area convention as a section point for the disease [32].
- *ML as the next weapon*: In a previous couple of years, we have seen malware utilizing avoidance procedures to bypass the ML engines. One of the ongoing models from 2018 was Plucky ransomware that used InnoSetup to bundle the malware and maintain a strategic distance from AI recognition. Thus bypassing the AI is now on the criminal plan for the day [33,34].
- *Data theft is the new money bovine for Hackers*: 2018 had milestone models for the most significant information break throughout the entire existence of humanity, for example, Facebook (87 + million), My Heritage (92 million), Under Armor (150 million), and supposedly 1.1 billon records from Aadhaar Program (India's extraordinary personality strategic). In a previous couple of years, both the advanced change and IoT have pushed increasingly corporate and individual information to the cloud. In 2019, we expected a noteworthy increment in information ruptures, particularly at the cloud level [35].
- *Smart home gadgets and edge gadgets will be vulnerable to attack in 2019*: Smart home gadgets are separate objectives to strike and send ransomware as they record and store individual information and have less protection. Moreover, edge gadgets are equipped with constrained resources, for the most part running on elementary OS. Consequently, these IoT edge gadgets cannot give any self-preservation highlights, for example, the production of a safe zone to ensure put away information and installed programming. Edge gadgets were seen as helpless against match up attacks, false information infusion, uninvolved attacks, and malicious hubs [36].

Some of the security solutions for securing IoT and AI framework are discussed below:

- *Collaboration and more organizations among cybersecurity arrangement suppliers*: Cyber threat alliance is perhaps the best case of these coordinated efforts that framed to improve the cybersecurity of the comprehensive computerized biological system. These joint efforts bring remarkable resources that pack the gifts and aptitudes of IoT security organizations to unite their best answers for make progressively voluntary contributions that can battle back against malware and botnets as well as even learn and advance [37,38].
- *Multifactor validation and gadget personality intelligence*: Identity is a crucial segment in verifying IoT. Secure recognizable proof between the gadget and human or the other way around was one of the past obstacles. Checking the character between gadget to-gadget collaborations and maintaining a strategic distance from harmful guile is the way to verifying IoT in 2019 [39,40].
- *ML as shield*: In the most recent year, the adoption of AI in IoT security has expanded fundamentally. At present, AI arrangements are regularly used to screen movement and act if strange practices are recognized. Besides, AI will not just process and examinations information a lot sharper than conventional apparatuses yet, will give a careful investigation of dangers and assaults. This implies break identification times can be diminished fundamentally, limiting the

potential disturbance. It likewise means that the data security group can organize work all the more successfully. Be that as it may, the degree for AI will go past observing client movement on the framework. AI, as an IoT security instrument, will not arrive at its maximum capacity in 2019. However, its utilization will quicken [41–43].

- *Increasing interest for the security workforce in governments and private division*: General data protection regulation (GDPR) guaranteed that all associations directly or by implication engaged with information the board concerning EU natives are obliged to conform to the guidelines, independent of where they are based. This has made a progressively outstretching influence of interest for the gifted security workforce among both government and the private part, which, thus has brought about expanded hierarchical planning for staff and preparing on information insurance. We anticipate that this pattern should increase in 2019 [44].

5.5 ARTIFICIAL INTELLIGENCE AND INTERNET OF THINGS

AI, the limit of an advanced computer-controlled robot to achieve assignments usually associated with humans. Since the improvement of the advanced computer during the 1940s, it has been demonstrated that Computers can be altered to finish confused tasks as, discovering proofs for measurable theories or playing chess with who is having extraordinary abilities. Regardless, paying little personality to continue with advances in PC planning rate and memory limit, there are so far no undertakings that can arrange human versatility over increasingly incredible spaces or in assignments requiring a great deal of regular learning. AI, on a fundamental level, demonstrates to computational gadgets that can substitute for human insight in the execution of a specific task. Computer-based intelligence causes it serviceable for machines to pick up change by new information sources and perform human-like tasks. AI lives in any place now, from labs to our homes. AI has started taking employments from people as of now and will continue taking occupations from people as it is widely adopted and received by the associations and organizations [45]. Every industry has fascinated by AI limits. AI is being used in:

- *Healthcare*: To give customized prescription and personal healthcare assistants, X-ray readings [46].
- *Retail*: To give virtual shopping capacities, that offers customized proposals as well as talks about various buying choices available with the shopper.
- *Manufacturing*: To divide processing plant information, as it streams from associated gear to measure the expected burden and for requesting utilizing intermittent systems.
- *Banking*: To distinguish which exchanges are probably going to be false, hold quick and exact credit scoring, and mechanize physically exceptional information, the executives' errands [47].

AI can reason without feelings, settling on intelligent choices with less or no mix-ups. It needs no rest, take breaks, or get engaged as AI does not get exhausted or tired [43,48–50].

The IoT depicts the system of physical items "things" that are installed with sensors, programming, and different advances to associate and trading information with various gadgets and frameworks over the web. These gadgets extend from standard family unit articles to advanced mechanical devices. From ordinary articles, for example, kitchen machines, vehicles, indoor regulators, infant screens to the web using inserted gadgets, everything is associated [51–53]. Same as the AI innovation, IoT is additionally boundlessly utilized in numerous associations.

- *Manufacturing*: By utilizing creation line checking to empower active support on equipment when sensors identify an impending failure.
- *Automotive*: Sensors can identify approaching hardware failure in vehicles already indicate caution to the driver with details and suggestions.
- *Retail*: To manage stock, improve consumer's experience, upgrade the inventory network, and decrease operational expenses.
- *Healthcare*: To track IoT asset-monitoring applications.

AI has a vital role in IoT applications and deployments [54,55]. There is reasonable conjunction between the IoT and AI. IoT is tied in with interfacing machines and utilizing the information produced from those machines. AI is tied in with reproducing intelligent behavior in devices of various types. As IoT gadgets will provide a large amount of data, at that point, AI will be practically essential to manage these vast volumes of the data in case we get an opportunity of understanding the data. Data may be valuable if it creates an activity. To make data significant, it should be enhanced with setting and inventiveness. IoT and AI together are "connected intelligence and not just connected devices" [56–58].

According to IoT Analytics, there were more than 17 billion connected gadgets worldwide in 2018, including more than 7 billion IoT devices. The IoT is a collection of these different sensors, gadgets, and different technologies, and their goal is not to interact directly with consumers (phones, computers, and so on). Instead, IoT devices provide information, control, and analytics to help connect the world's best hardware devices to the Internet. With the advent of cheap sensors and low-cost connections, IoT devices are growing. No wonder businesses are flooding devices with large amounts of data, managing devices using AI, and gaining more insight and intelligence from data posted by large numbers of users. Discussion system. However, managing and extracting valuable information from these systems is much more difficult than we think. Another layer of IoT complexity relates to the scale of functionality [59,60]. In general, creating sensors that can be accessed from smart devices is easy, but creating reliable remote controls, upgrades, and secure and cost-effective tools is much more complicated. By combining AI and the IoT, you can visualize and properly control various Internet-related gadgets and sensors.

The IoT has changed the business model by helping companies transition from purely manufactured products and services to companies that provide the desired results to their customers. By influencing an organization's business model, IoT-compatible devices, and sensors are combined with ML to create a collaborative and interconnected world consistent with results and innovation. The combination of IoT and AI is changing the relationships among many industries, businesses, and customers. Companies can now use the IoT to collect data and turn it into useful and valuable information [61]. As organizations apply the principles of digital transformation to operations, the combination of IoT and AI disrupts the entire industry. Will your organization use IoT and AI to attract customers, implement customer-facing chat agents, personalize user experiences, get insights, and optimize productivity through information and predictions? IoT and AI are how companies start out with a look, and the way they communicate is how employees, suppliers, and partners interact in different ways to get a high-quality information ecosystem with each data [62,63]. IoT devices can not only model software business processes in a way closer to the real world, but also provide systems with a real interface to the real world. Wherever you place sensors and methods to measure, interact, or analyze something, you can deploy AI-compatible cloud-connected IoT devices to add significant value.

Challenges organizations are facing today with AI and IoT include the application, accessibility, and analysis of IoT data. However, if you anticipate events and take appropriate follow-up actions, like replacing a drill bit or predicting a mechanical component failure, the company uses these techniques to apply them. You need to learn to identify it. Data and process type [64,65].

An organization that deploys large amounts of IoT data (especially sensors or beacons) at the level of a single unit. Managing growing information using traditional strategic observation and analysis tools is difficult. This is where AI comes in. ML systems use unsupervised learning and clustering techniques to automatically identify normal and abnormal patterns in the data and to signal deviations from observed standards when anomalies occur. You. No operator adjustment is required. Similarly, these AI-compatible IoT systems are almost invisible because they can automatically display relevant information that is not available from large amounts of data [66].

Companies implement AI-compatible IoT systems in several different ways. Solution companies create code and package models that include reliable models for specific application areas, like transportation and logistics, manufacturing, energy, environment, building operation and installation, and other models. Other companies use cloud providers to take advantage of external processor capabilities to create client solutions, architectures, and models [67,68]. Some solutions focus AI capabilities on local solutions or cloud-based products. However, other goals are to decentralize AI capabilities, push ML models to the limit, move data closer to the device, and improve performance. There are many ways to implement this technology, but the challenge is how to apply and access it properly.

Today we have witnessed the tremendous progress of AI and the IoT. When combined with these technologies, you can achieve higher levels of automation and productivity while reducing costs. As consumers, businesses, and governments begin to manage the IoT in a variety of environments, our world will change dramatically, and we will all be able to make better choices. Everything from retail to supply chain to healthcare is evolving rapidly. The AI-compatible IoT has transformed the energy industry with smart energy solutions. Cities and towns want to use solar panels to create a nonlocalized electricity business for their families [69,70].

With the creation of IoT devices, the future is closely linked to immediate access to the information world. AI must manage all these devices and make the most of their data. Thus AI and the IoT are highly symbiotic and will continue to maintain intertwined relationships in the future [71–73].

5.6 APPLICATIONS OF INTERNET OF THINGS

IoT acts as the backbone of every network nowadays. The various forms of the IoT are:

- *Smart home*: Why to get up from bed to switch on/off the lights, when you can do it while lying over the bed by just a single click. Home Automation is one of the critical applications of IoT, which makes your life more relaxed and makes you feel comfortable [74]. Using a processor to regulate the home activities, which will be connected to your phone through a network like

WiFi, Internet, Bluetooth, and so on. This application gave a boost to the industry and opened a new scope to earn a profit. Today nearly 250 companies and startups working over Home Automation projects to develop new products to make our home activities easily controllable [75]. Fig. 5.2 represents the smart home mechanism that communicates with various devices in the smart home.

- *Wearables*: Today, many of us using smartwatches to maintain the count of our steps, blood pressure, and many more [76]. Even many smartwatches are capable of regulating and controlling your mobile phone activities using Bluetooth network or many more. Many other projects are going on, like Smart Jacket, to maintain the metabolism of your body [77]. Many companies like Apple, Mi, and so on already launched a series of smart wearables and earning millions. Fig. 5.3 shows the rapid growth of these devices.
- *Smart city*: Many smart city projects are going on. If we talk about India, there are more than 50 smart city projects are in progress [78,79]. Everyone wants to live in a city where every public service is hassle-free to use and also accessible to all of the dwellers of that region, where everyone is connected to the internet. To make it possible, we need to understand the requirement of the smart city where we need to manage the traffic, transportation system, water management, waste management, crime detection, and many more using IoT [80,81]. Smart Dubai, Amsterdam Smart Project, and so on are the numerous projects. Fig. 5.4 represents a smart city.
- *Smart grid*: The idea of the smart grid is to control and manage the supply of electricity in a particular locality by obtaining the information about the supplier and customers for that

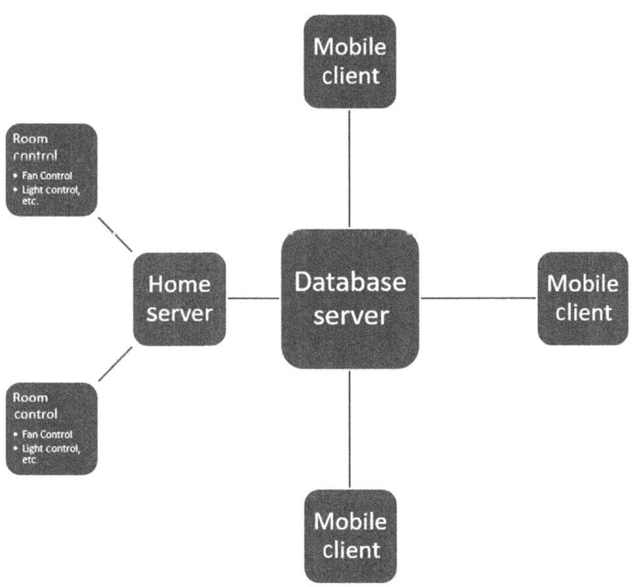

FIGURE 5.2

Smart home mechanism.

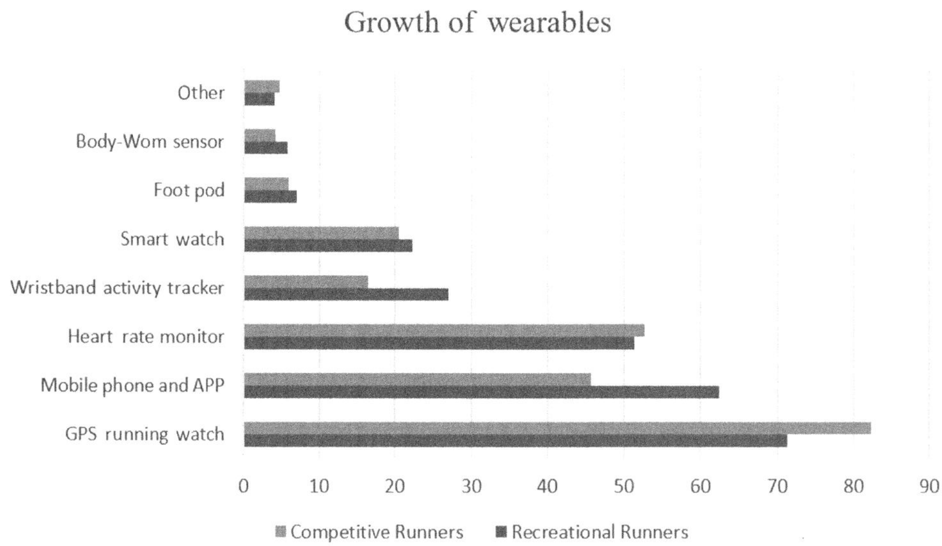

FIGURE 5.3

Growth of wearables.

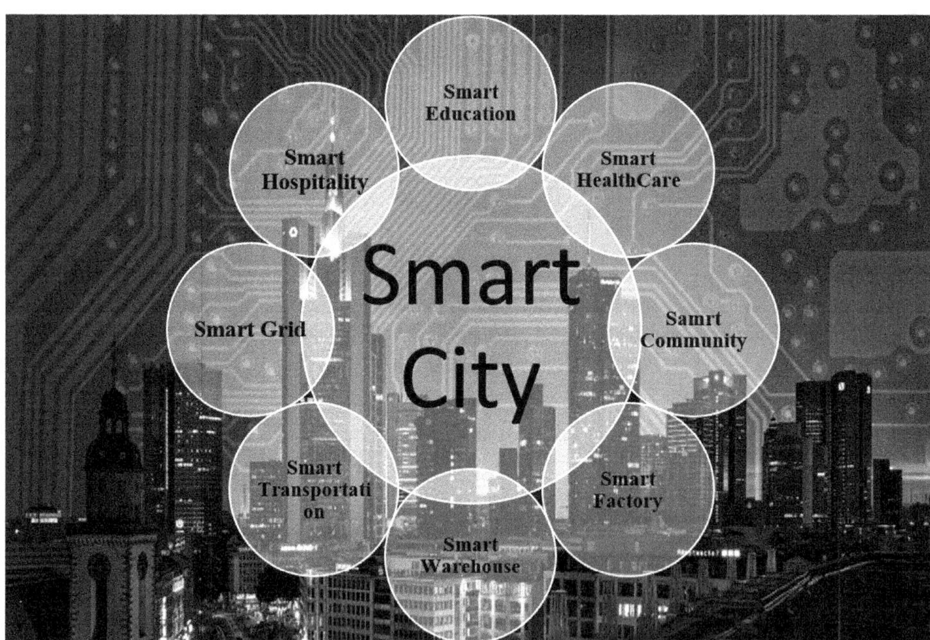

FIGURE 5.4

Smart city.

specific area. Its primary focus is to make an efficient, reliable, and economical supply of electricity. However, if we see the trend, it is not much popular among the masses [82,83].

- *Connected car*: Connected cars can be a revolution for the technology world. Many firms and organizations are working on that idea to bring a revolution in the automobile sector. A connected car is well equipped with Internet services, GPS, and also can send and receive any data. However, many efforts have been on this, but it still needs more focus to provide the customer with better technology at an affordable price [84–86]. Fig. 5.5 represents the scenario of the connected car.
- *Connected health*: The health industry becomes more profound with the increase in the number of competitors. Many advancements have been made in this sector by the health sector in the last decade. From surgical robots to personal fitness sensors at a very affordable price, it developed a lot. However, seeing the trend, it is still unpopular and unreachable to many. With this, we can manage adjustable patient monitoring, enhanced drug management, augmented asset monitoring and tracking, and so on [87–89]. Fig. 5.6 shows the connected health diagnostic deployment.
- *Smart supply chain*: We are already observing this smart change around us while receiving a hassle-free delivery and updates of each step of our delivery process [90–93].

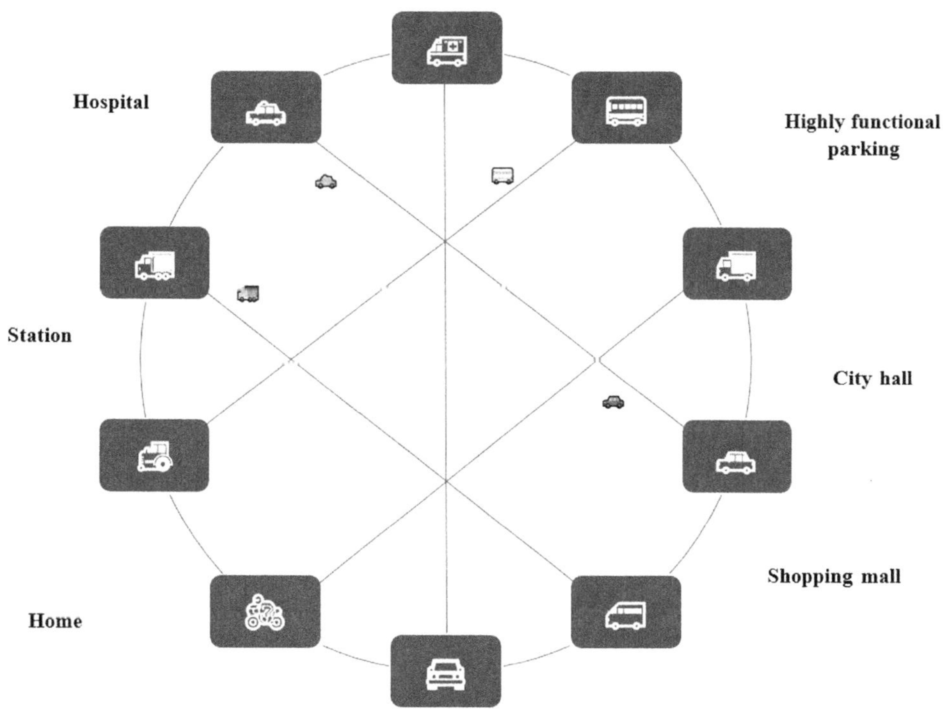

FIGURE 5.5

Connected transport scenario. *MEC*, mobile edge computing; *EPC*, engineering, procurement and construction.

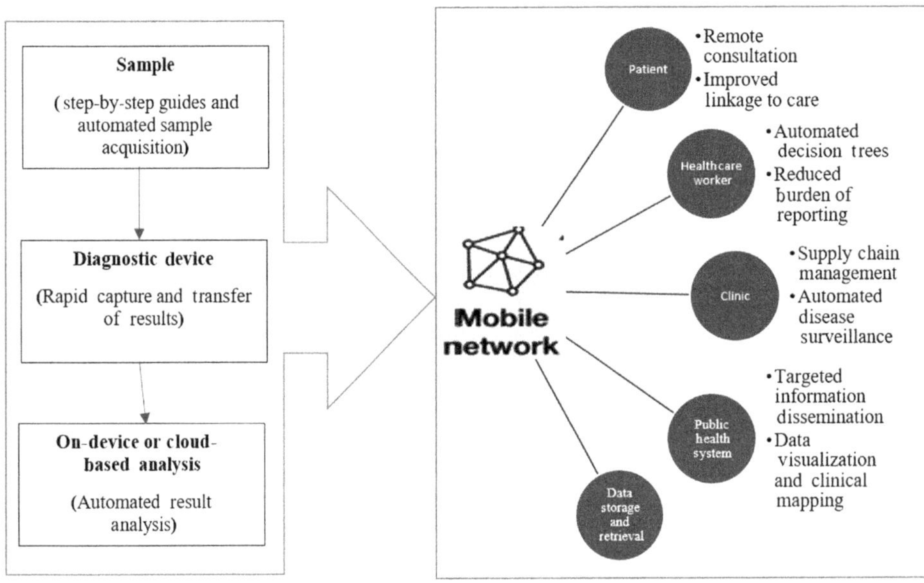

FIGURE 5.6

Connected health diagnostic deployment.

- *Smart farming*: It is an often-overlooked business nowadays. They are making farming easy, reducing the risk of crop loss, intelligent irrigation system, monitoring of moisture contents, and nutrient of the soil. Many Farmers have already accepted that advancement and successfully using smart irrigation systems for their fields, reducing the risk of an overflow of water in areas and also helping to regulate the ON/OFF process of the motor from faraway places using the Internet, Bluetooth, and so on [94–97]. Fig. 5.7 represents the smart farming architecture.

5.7 CONCLUSION AND FUTURE DIRECTIONS

The IoT and the age of AI will bring fundamental changes to existing procedures. Because automation is linked to internal and external research, companies and organizations earn development fees while receiving significant rewards. Importantly, using the IoT and AI for a better future requires better technology. In this chapter, we have discussed the basic introduction of AI and the IoT with the seven-layered Architecture of the IoT system. IoT gadgets are used now a day to reduce human efforts and to perform the complex and tidy tasks, which required a lot of repetition. IoT and AI plays an important role in emerging fields like smart cities which use to have various components, for example, smart homes, wearables, warehouse, factory, community, healthcare, education, hospitality, grid, transportation like connected cares and autopilot cars, and so on, supply chain, smart farming, and so on. All of the aforementioned fields are also the application domains of the IoT

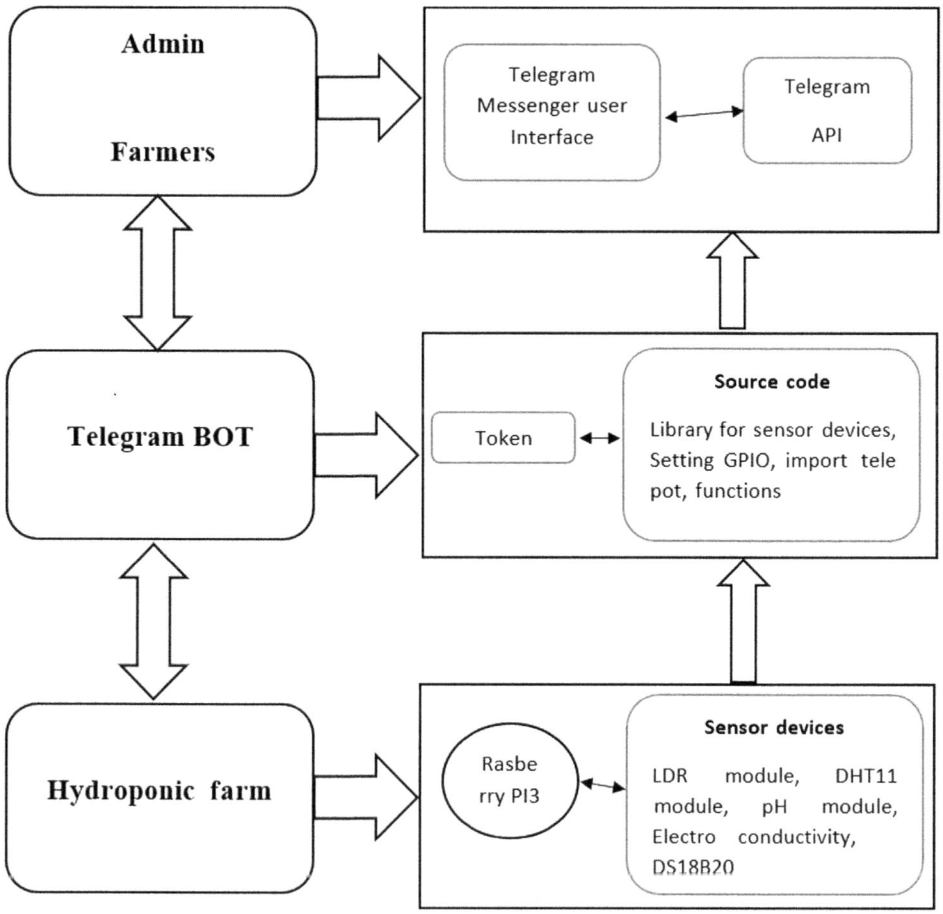

FIGURE 5.7

Smart farming architecture.

and AI, which are being given the most attention in this chapter. IoT is everywhere, so the users use these gadgets for performing some susceptible and secrete tasks using these devices, due to which the security and privacy issue takes place, which is also being discussed in this chapter.

REFERENCES

[1] O. Hamdan, H. Shanableh, I. Zaki, A.R. Al-Ali, T. Shanableh, IoT-based interactive dual-mode smart home automation, in: 2019 IEEE International Conference on Consumer Electronics (ICCE), Las Vegas, NV, 2019, pp. 1–2. doi: 10.1109/ICCE.2019.8661935. <http://ieeexplore.ieee.org/stamp/stamp.jsp?tp = &arnumber = 8661935&isnumber = 8661828>.

[2] G. Salomon, P. Müller, Success factors for the acceptance of smart home technology concepts, in: A. Lochmahr, P. Müller, P. Planing, T. Popović (Eds.), Digitalen Wandel Gestalten, Springer, Gabler, Wiesbaden, 2019. <https://doi.org/10.1007/978-3-658-24651-8_6.3>.

[3] R. Piyare, Internet of Things : ubiquitous home control and monitoring system using android based smart phone, Int. J. Internet Things 2 (1) (2013) 5–11.

[4] M. Amadeo, A. Giordano, C. Mastroianni, A. Molinaro, On the integration of information-centric networking and fog computing for smart home services, in: F. Cicirelli, A. Guerrieri, C. Mastroianni, G. Spezzano, A. Vinci (Eds.), The Internet of Things for Smart Urban Ecosystems. Internet of Things (Technology, Communications, and Computing), Springer, Cham, 2019. <https://doi.org/10.1007/978-3-319-96550-5_4>.

[5] G. Wilson, C. Pereyda, N. Raghunath, G. de la Cruz, S. Goel, S. Nesaei, et al., Robot-enabled support of daily activities in smart home environments, Cognit. Syst. Res. 54 (2019) 258–272.

[6] C.A. Clermont, L. Duffett-Leger, B.A. Hettinga, R. Ferber, Runners' perspectives on 'smart' wearable technology and its use for preventing injury, Int. J. Human-Computer Interact. (2019) 1–10.

[7] S. Paluch, S. Tuzovic, Persuaded self-tracking with wearable technology: carrot or stick? J. Serv. Mark 33 (4) (2019) 436–448. Available from: https://doi.org/10.1108/JSM-03-2018-0091.

[8] B. Reeder, J. Chung, K. Lyden, J. Winters, C.M. Jankowski, Older women's perceptions of wearable and smart home activity sensors, Inform, Health Soc. Care 45 (1) (2020) 96–109. Available from: https://doi.org/10.1080/17538157.2019.1582054.

[9] A. Kubley, D. Chauhan, S.N. Kanakaraj, V. Shanov, C. Xu, R. Chen, et al., Smart textiles and wearable technology innovation with carbon nanotube technology, Nanotube Superfiber Materials, William Andrew Publishing, 2019, pp. 263–311.

[10] L. Anthopoulos, M. Janssen, V. Weerakkody, A unified smart city model (USCM) for smart city conceptualization and benchmarking, Smart Cities Smart Spaces: Concepts, Methodologies, Tools, Appl, IGI Global, 2019, pp. 247–264.

[11] J. Xie, et al., A survey of blockchain technology applied to smart cities: research issues and challenges, IEEE Commun. Surv. & Tutor 21 (3) (2019) 2794–2830. Available from: https://doi.org/10.1109/COMST.2019.2899617. thirdquarter.

[12] Q. Chen, et al., A survey on an emerging area: deep learning for smart city data, IEEE Trans. Emerg. Top. Computational Intell 3 (5) (Oct. 2019) 392–410. Available from: https://doi.org/10.1109/TETCI.2019.2907718.

[13] M. Babar, F. Arif, M. Irfan, Internet of Things–based smart city environments using big data analytics: a survey, Recent Trends and Advances in Wireless and IoT-enabled Networks, Springer, Cham, 2019, pp. 129–138.

[14] M.A. Jan, W. Zhang, M. Usman, Z. Tan, F. Khan, E. Luo, SmartEdge: an end-to-end encryption framework for an edge-enabled smart city application, J. Netw. Computer Appl vol. 137 (2019 Jul 1) 1–10.

[15] C.R. Srinivasan, B. Rajesh, P. Saikalyan, K. Premsagar, E.S. Yadav, A review on the different types of Internet of Things (IoT), J. Adv. Res. Dynamical Control. Syst. 11 (1) (2019) 154–158.

[16] E. Kabalci, Y. Kabalci, Introduction to smart grid architecture, Smart Grids and Their Communication Systems, Springer, Singapore, 2019, pp. 3–45.

[17] M.H. Rehmani, A. Davy, B. Jennings, C. Assi, Software defined networks-based smart grid communication: a comprehensive survey, IEEE Commun. Surv. Tut 21 (3) (2019) 2637–2670. Available from: https://doi.org/10.1109/COMST.2019.2908266. thirdquarter.

[18] S. Yang, Y. Su, Y. Chang, H. Hung, Short-term traffic prediction for edge computing-enhanced autonomous and connected cars, IEEE Trans. Vehicular Technol 68 (4) (April 2019) 3140–3153. Available from: https://doi.org/10.1109/TVT.2019.2899125.

[19] A.C. Pang, E. Au, B. Ai, W. Zhuang, Guest editorial introduction to the special section on fog/edge computing for autonomous and connected cars, IEEE Trans. Vehicular Technol. 68 (4) (2019) 3059–3060.
[20] J. Wang, Y. Du, D. Coleman, M. Peck, S. Myneni, H. Kang, et al., Mobile and connected health technology needs for older adults aging in place: cross-sectional survey study, JMIR Aging 2 (1) (2019) e13864.
[21] S. Steels, T. van Staa, Evaluation protocol of the implementation of a learning healthcare system in clinical practice: the Connected Health Cities programme in the north of England, BMJ Open 9 (6) (2019) e025484.
[22] C.S. Wood, M.R. Thomas, J. Budd, T.P. Mashamba-Thompson, K. Herbst, D. Pillay, et al., Taking connected mobile-health diagnostics of infectious diseases to the field, Nature 566 (7745) (2019) 467.
[23] C. Smileska, N. Koceska, S. Koceski, V. Trajkovik, Development and evaluation of methodology for personal recommendations applicable in connected health, Enhanced Living Environments, Springer, Cham, 2019, pp. 80–95.
[24] S. Gupta, V.A. Drave, S. Bag, Z. Luo, Leveraging smart supply chain and information system agility for supply chain flexibility, Inf. Syst. Front. 21 (3) (2019 Jun 15) 547–564.
[25] A. Calatayud, J. Mangan, M. Christopher, The self-thinking supply chain, Supply Chain Management: An. Int. J. 24 (1) (2019) 22–38.
[26] T. Srisooksai, K. Kaemarungsi, J.I. Takada, K. Saito, Small-fading and wideband propagation characteristics in fruit orchard at 2.4 GHz for wireless network in smart farming application, in: 2019 International Conference on Robotics, Electrical and Signal Processing Techniques (ICREST), IEEE, 2019, pp. 354–358.
[27] M.A. Zamora-Izquierdo, J. Santa, J.A. Martínez, V. Martínez, A.F. Skarmeta, Smart farming IoT platform based on edge and cloud computing, Biosyst. Eng. 177 (2019) 4–17.
[28] A. Serianni, P. Raimondo, N. Palmieri, An energy aware smart station for an UAV fleet in the smart farming application, Autonomous Air and Ground Sensing Systems for Agricultural Optimization and Phenotyping IV, 11008, International Society for Optics and Photonics, 2019, p. 110080H. vol.
[29] A. Luigi, I. Antonio, M. Giacomo, The Internet of Things: a survey, Comp. Netw. 54 (15) (2010) 2787–2805. ISSN 1389-1286.
[30] B. Firner, R.S. Moore, R. Howard, R.P. Martin, Y. Zhang, Poster: smart buildings sensor networks and the Internet of things, Proceedings of the ACM Conference on Embedded Networked Sensor Systems, November 2011, pp. 337–338.
[31] S.B. Calo, M. Touna, D.C. Verma, A. Cullen, Edge computing architecture for applying AI to IoT, IEEE International Conference on Big Data (Big Data), 2017, IEEE, 2017, pp. 3012–3016.
[32] D. Lily, B. Chan, T.G. Wang, A simple explanation of neural network in artificial intelligence, IEEE. Trans. Control. Syst. 247 (2013) 1529–5651.
[33] J. Jagannathan, Current trends: machine learning and AI in IoT, Integrating Internet Things Into Software Engineering Practices, IGI Global, 2019, pp. 181–198.
[34] N. Javaid, A. Sher, H. Nasir, N. Guizani, Intelligence in IoT-based 5G networks: opportunities and challenges, IEEE Commun. Mag. 56 (10) (2018) 94–100.
[35] Y.W. Lin, Y.B. Lin, C.Y. Liu, AItalk: a tutorial to implement AI as IoT devices, IET Netw 8 (3) (2019 Jan 30) 195–202.
[36] R. Nicole, J.H. Lee, The IoT concepts and design, Int. J. Eng. 6 (7) (2015) 16–29.
[37] P.K. Pramanik, S.P. Dutta, P. Choudhury, Beyond automation: the cognitive IoT. Artificial intelligence brings sense to the Internet of Things, Cognitive Computing for Big Data Systems Over IoT, Springer, Cham, 2018, pp. 1–37.
[38] R. Shah, A. Chircu, IOT and AI in healthcare: a systematic literature review, Issues Inf. Syst. 19 (3) (2018).

[39] S. Chen, H. Xu, D. Liu, B. Hu, A vision of IoT: applications challenges and opportunities with china perspective, IEEE Internet Things J. 1 (4) (2014) 349–359.
[40] K. Rabah, Convergence of AI, IoT, big data and blockchain: a review, Lake Inst. J. 1 (1) (2018) 1–18.
[41] Z. Mingjie, H. Jianting, H. Bingsong, L. Wenchao, Building home application system of Internet of Things with home gateway, Tele Commun. Sci. (4)(2010) 44–47.
[42] Z. Wei, Z. Zhe, Access technique of the Internet of Things gateway, J. Nanyang Norm. Univ. (12) (2012) 68–70.
[43] H.-T. Zhang, Y.-K. Zhang, Architecture and core technologies of Internet of Things, J. Changchun Univ. Technol. (Nat. Sci. Ed.) (2)(2012) 176–181.
[44] H. Haikun, D. Jiajia, Discussion on the technology and application of IOT gateway, Telecommun. Sci. (4)(2010) 20–24.
[45] J. Luo, Y. Zhou, X. Deng, Design for gateway in Internet of Things, Telecommun. Science. (2)(2011) 105–110.
[46] J. Hou, C.-D. Wu, Z.-J. Yuan, Y. Zhou, Y.-Z. Zhang, Research of intelligent home security surveillance system based on ZigBee, Mech. Electr. Eng. Mag. (1)(2009) 67–70.
[47] Z.-L. Nan, G.-X. Sun, Design of smart home system "based on ZigBee.", Electron. Des. Eng. (7)(2010) 117–119.
[48] Gan Yong, W. Hua, C. Ya-jun, W. Jun, Design of Zigbee gateway system based on ARM, Commun. Technol. (1)(2009) 199–201.
[49] W. Dun, Y. Zhao, R. Cui, Construction of food safety management system using Internet of things technology, Agriculture Netw. Inf. (7)(2013) 5–8.
[50] W. Tao, A study of the structure of the Internet of Things system and key technology application, Electron. Test. (2)(2014) 142–143.
[51] X. Yan-hong, Investigate the networking model of general technology on Internet of Things, J. Shanxi Univ. Technol. (Nat. Sci. Ed.) (3)(2013) 18–22.
[52] T. Lin, System structure and composition model research on Internet of Things, J. Chifeng Univ. (Nat. Sci. Ed.) (1)(2013) 26–28.
[53] Z. Zhu, J. Wang, The structure of data collection system based on PC, Electr. Meas. & Instrum. (11) (2002) 9–12.
[54] D. Guinard, V. Trifa, Interacting with the SOA-based Internet of things: discovery, query, selection, and on-demand provisioning of Web services, IEEE Trans. Services Computing 235 (3) (2010) 223. Available from: 10.1109/TSC.2010.3.
[55] I. Valerie, G. Nikolaos, H. Sara, Z. Apostolos, V. Panos, A. Marco, et al., Service-oriented middleware for the future Internet: state of the art and research directions, J. Internet Serv. Appl. 45 (1) (2011) 23. Available from: 10.1007/s13174-011-0021-3.
[56] Q. Wei, Z. Jin, G. Li, L.X. Li, Preliminary study of service discovery in Internet of things: feasibility and limitation of SOA, Front. Computer Sci. Technol. 7 (2) (2013) 97–113.
[57] L. Lan, F. Li, B. Wang, et al., An event-driven service-oriented architecture for the Internet of Things, Services Computing Conference, IEEE, 2014, pp. 68–73.
[58] W. Dai, V. Vyatkin, J.H. Christensen, et al., Bridging service-oriented architecture and IEC 61499 for flexibility and interoperability, IEEE Trans. Ind. Inform. 11 (3) (2015) 771–781.
[59] V. Tsiatsis, A. Gluhak, T. Bauge, F. Montagut, J. Bernat, M. Bauer, et al., The SENSEI real world Internet architecture, In: Proceedings of the Future Internet Assembly, Ghent: 256, European Commission, 2010, p. 247.
[60] K.B. Xie, H.M. Chen, C. Li, Refinement of sense-execute model in a software architecture of internet of things [J], J. Softw 25 (8) (2014) 1659–1670.

[61] J. Kiljander, A. D'Elia, F. Morandi, et al., Semantic interoperability architecture for pervasive computing and Internet of Things, IEEE Access 2 (2014) 856–873.

[62] K. Xie, H. Chen, C. Li, PMDA: a physical model driven software architecture for Internet of Things, J. Computer Res. Dev. 50 (6) (2013) 1185–1197.

[63] Y.W. Ma, C.F. Lai, Y.M. Huang, et al., Mobile RFID with IPv6 for hone services, Proceedings of the 13th IEEE Int'l Symposium on Consumer Electronics, IEEE Press, Kyoto, 2009, pp. 169–170. 10.1109/ISCE.2009.5156859.

[64] National Water Resources Monitoring Capacity Building Project Standard, SZY206-2012Ij, Data transmission protocol for monitoring system, NJV1.120121009.

[65] Z.J. Yao, H. Bai, W.D. Yi, Reliability analysis on grid topology in wireless sensor network, J. Computer Res. Deploy. 47 (2) (2010) 55–59.

[66] H.H. Li, J. Fan, L. Chen, Application of grid method in deployment of wireless sensor network, Transducer Microsyst. Technol. 31 (3) (2012) 150–152.

[67] Y.-Q. Mao, S.-B. Shen, Information model and capability analysis of the Internet of Things, J. Softw. 25 (8) (2014) 1685–1695. Available from: 10.13328/j.cnki.Jos.004664.

[68] A. Solanki, A. Nayyar, Green Internet of Things (G-IoT)-ICT technologies, principles, applications, projects, and challenges: green Internet of Things, Handbook of Research on Big Data and the IoT, IGI Global USA, 2018, pp. 379–409. doi:10.4018/978-1-5225-7432-3.ch021.

[69] R. Rameshwar, A. Solanki, A. Nayyar, B. Mahapatra, Green and smart buildings: a key to sustainable global solutions, Green Building Management and Smart Automation, IGI Global USA, 2020, pp. 146–163. doi:10.4018/978-1-5225-9754-4.ch007.

[70] T. Nam, T.A. Pardo, Conceptualizing smart city with dimensions of technology, people, and institutions, in: Proceedings of the 12th Annual International Digital Government Research Conference, New York, June 2011, pp. 282–291.

[71] R. Lea, Smart cities: an overview of the technology trends driving smart cities, IEEE Advancing Technol. Humanity (2017). Mar.

[72] P. Neirotti, A. De Marco, A.C. Cagliano, G. Mangano, F. Scorrano, Current trends in smart city initiatives: some stylised facts, Cities 38 (2014) 25–36.

[73] E. Borgia, The Internet of Things vision: key features, applications and open issues, Comput. Commun. 54 (2014) 1–31.

[74] S. Dirks, C. Gurdgiev, M. Keeling, Smarter cities for smarter growth: how cities can optimize their systems for the talent-based economy, IBM Institute for Business Value, May 2010. [Online]. Available: <https://ssrn.com/abstract = 2001907>.

[75] A. Zanella, N. Bui, A. Castellani, L. Vangelista, M. Zorzi, Internet of Things for smart cities, IEEE Internet Things J. 1 (1) (2014) 22–32.

[76] J.M. Barrionuevo, P. Berrone, J.E. Ricart, Smart cities, sustainable progress, IESE Insight 14 (14) (2012) 50–57.

[77] H. Chourabi, T. Nam, S. Walker, J.R. Gil-Garcia, S. Mellouli, K. Nahon, et al., Understanding smart cities: an integrative framework, 2012 45th Hawaii International Conference on System Science (HICSS), 2012, pp. 2289–2297.

[78] M.D. Lytras, A. Visvizi, Who uses smart city services and what to make of it: toward interdisciplinary smart cities research, 10 (6) (1998).

[79] S.E. Bibri, J. Krogstie, Smart sustainable cities of the future: an extensive interdisciplinary literature review, 31 (2017) 183–212.

[80] G. Robert, Will the real smart city please stand up? Intelligent, progressive entrepreneurial? City 12 (3) (2008) 303–320.

[81] L. Errichiello, R. Micera, Leveraging smart open innovation for achieving cultural sustainability: learning from a new city museum project, 10 (6) (2018) 1–23.

[82] F. Mohammed, A. Idries, N. Mohamed, J. Al-Jaroodi, I. Jawhar. UAVs for smart cities: opportunities and challenges, in: 2014 International Conference on Unmanned Aircraft Systems (ICUAS), 2014, pp. 267–273.
[83] A. Gapchup, A. Wani, A. Wadghule, S. Jadhav, Emerging trends of green IoT for smart world, Int. J. Innov. Res. Comput. Commun. Eng. 5 (2) (2017) 2139–2148.
[84] J. Huang, Y. Meng, X. Gong, Y. Liu, Q. Duan, A novel deployment scheme for green internet of things, IEEE Internet Things J. 1 (2) (2014) 196–205.
[85] S. Rani, R. Talwar, J. Malhotra, S.H. Ahmed, M. Sarkar, H. Song, A novel scheme for an energy efficient internet of things based on wireless sensor networks, Sensors 15 (11) (2015) 28603–28626.
[86] S.-Y. Lien, K.-C. Chen, Y. Lin, Toward ubiquitous massive accesses in 3gpp machine-to-machine communications, IEEE Commun. Mag. 49 (4) (2011).
[87] M. Mozaffari, W. Saad, M. Bennis, M. Debbah, Mobile unmanned aerial vehicles (UAVs) for energy-efficient Internet of Things communications, arXiv preprint arXiv:1703.05401, 2017.
[88] S.H. Alsamhi, O.U. Ma, M.S. Ansari. Predictive estimation of the optimal signal strength from unmanned aerial vehicle over internet of things using ANN, arXiv preprint arXiv:1805.07614, 2018.
[89] V. Sharma, F. Song, I. You, M. Atiquzzaman, Energy efficient device discovery for reliable communication in 5G-based IoT and BSNs using unmanned aerial vehicles, J. Netw. Comput. Appl. 97 (2017) 79–95.
[90] N.H. Motlagh, M. Bagaa, T. Taleb, J.S. Song. Connection steering mechanism between mobile networks for reliable UAV's IoT platform, in: 2017 IEEE International Conference on Communications (ICC), 2017, pp. 1–6.
[91] M. Mozaffari, W. Saad, M. Bennis, M. Debbah. Mobile Internet of Things: can UAVs provide an energy-efficient mobile architecture? in: 2016 IEEE Global Communications Conference (GLOBECOM), 2016, pp. 1–6.
[92] M. Mozaffari, W. Saad, M. Bennis, M. Debbah, Unmanned aerial vehicle with underlaid device-to-device communications: performance and tradeoffs, IEEE Trans. Wirel. Commun. 15 (6) (2016) 3949–3963.
[93] E. Basha, M. Eiskamp, J. Johnson, C. Detweiler, UAV recharging opportunities and policies for sensor networks, Int. J. Distrib. Sens. Netw. 11 (8) (2015) 824260.
[94] M. Alam, D. Moroni, G. Pieri, M. Tampucci, M. Gomes, et al., Real-time smart parking systems integration in distributed ITS for smart cities, J. Adv. Transportation (2018).
[95] S.H. Alsamhi, O.U. Ma, M.S. Ansari, Q. Meng, Greening Internet of Things for smart everythings with a green-environment life: a survey and future prospects, arXiv preprint arXiv:1805.00844, 2018.
[96] I. Guvenc, F. Koohifar, S. Singh, M.L. Sichitiu, D. Matolak, Detection, tracking, and interdiction for amateur drones, IEEE Commun. Mag. 56 (4) (2018) 75–81.
[97] H. Menouar, I. Guvenc, K. Akkaya, A.S. Uluagac, A. Kadri, A. Tuncer, UAV-enabled intelligent transportation systems for the smart city: applications and challenges, IEEE Commun. Mag. 55 (3) (2017) 22–28.

CHAPTER

6 ARTIFICIAL INTELLIGENCE AND MACHINE LEARNING APPLICATIONS IN CLOUD COMPUTING AND INTERNET OF THINGS

Mamata Rath[1], Jyotirmaya Satpathy[2] and George S. Oreku[3]

[1]School of Management (IT), Birla Global University, Bhubaneswar, India [2]Academics Department National Defence Academy, Pune, India [3]Tanzania Industrial Research and Development Organizational, Open University of Tanzania, Tanzania

6.1 INTRODUCTION

A well-developed technology is consistent in current society and most of its credit goes to AI (artificial intelligence). All the modern applications nowadays utilize AI. The competence of reconcile on an appliance to resolve on choices all alone is objective of AI. This chapter shows a concise review on AI and its different developing applications based on Cloud computing and machine learning (ML) alongside constant precedents. A nonexclusive investigation on AI is displayed in this chapter. Intelligence is the mindset and following up in many technologies. This may rely upon the technical knowledge of an individual. AI can likewise be utilized to make forecasts in coming prospect. From AI point of view, all the equipments are intelligently programmed to perform the given task logically and cleverly.

The fast development of the Internet and its associated equipments makes the things connected with different technology in the complete globe. Nowadays, the globe is running on the Internet of Things (IoT), with the expanded communication capacity and best technique for correspondence and transmission lines a large number of things are associated with the Internet. The issues and disturbances in brilliant sensor innovation have reduced a lot due to intelligent approaches used in them. Few challenges in those systems draws attention of lot of the clients and when we observe we can find that the vast majority of the equipments are associated with the Internet. Internet-associated sensors and equipments produce exponential information. Purposely or unwittingly IoT is creating heaps of information. This information is critical in basic leadership framework, yet the issue is the way to isolate this information for the future investigation purposes. The IoT offers building groups an inventive method to gather information and watch the status of their items, administrations, and hardware in the field. Machine supervision systems are utilized to gain from these information to make the gadget or thing savvy. For instance, utilizing the machine

Artificial Intelligence to Solve Pervasive Internet of Things Issues. DOI: https://doi.org/10.1016/B978-0-12-818576-6.00006-X

supervision to distinguish the irregularities in wearable and taking vital activities like calling specialist and rescue vehicle naturally when it essential.

With the various IoT equipments, the cloud-driven information handling neglects to meet the necessity of all IoT applications. The restricted calculation and correspondence limit of the cloud require the edge figuring, that is, beginning the IoT information, handling at the edge, and changing the associated equipments to insightful equipments. Machine supervision-based methods are the key methods for data surmising, ought to stretch out to the cloud-to-things continuum as well. The responsibility of ML in IoT and cloud starting from the cloud security to IoT equipment's safety is a major contribution. Various uses of ML for application information handling and the board assignments are examined in this research. The sophisticated usages of ML in IoT are presented by their application space, input information type, developed ML systems, and in systems where they have a place in the cloud-to-things continuum. The difficulties and research patterns toward effective ML on the IoT edge are talked about.

A tremendous technical and advanced field of AI requires multidisciplinary aptitude where the basic objective is to robotize all the human exercises that directly require human knowledge. The serious issue is to build up a strategy, which works precisely the way how a human cerebrum functions [1]. The engineering of manmade reasoning must stress on assessment and update the idea of configuration process. Information science is likewise inclining now and logically arrangements to take care of complex issues. Information is isolated into littler parts and its patterns, practices are comprehended. The fundamental issue in information science is to deal with substantial amounts of information. In spite of the fact that there is huge increment as far as research open doors few difficulties like absence of process control, individuals control still remains a major test.

Computerized reasoning or otherwise known as AI, a prominent term in the innovative world is identified with AI and profound adapting definitely. AI utilizes calculations to find designs and produce bits of knowledge from the information they are chipping away at. Then again, profound learning is a subset of AI, one that conveys AI closer to the objective of empowering machines to think and function as people as could be expected under the circumstances. There are many challenges in the innovation of AI, which are illustrated in Fig. 6.1 like building trust, interference with human being and AI, investment, software malfunction, and higher expectations of human community. Fig. 6.2 shows basic application domains of AI

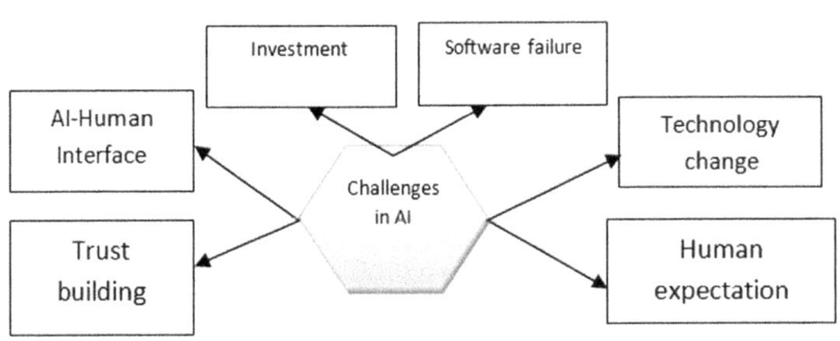

FIGURE 6.1

Challenges in artificial intelligence (AI).

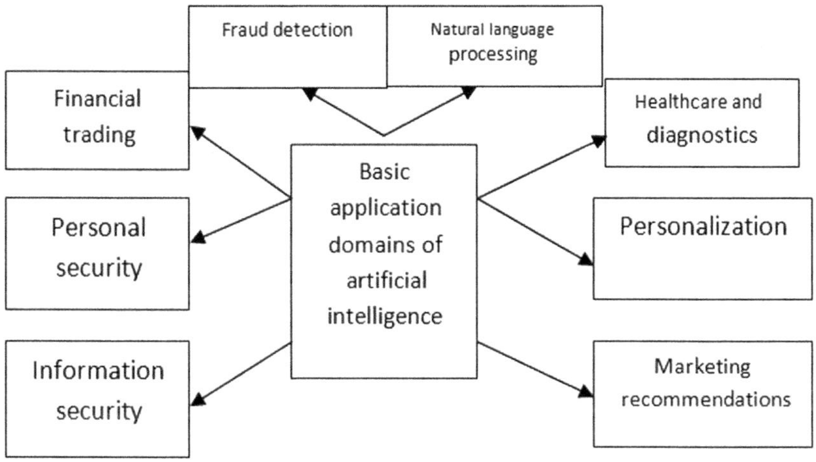

FIGURE 6.2

Basic application domains of artificial intelligence.

6.1.1 DATA SECURITY AND PERSONAL SECURITY

As per institutional intelligence organization Deep Instinct, each bit of new malware will in general have nearly a similar code as past variants, just somewhere in the range of 2%–10% of the records change from cycle to emphasis. Their learning model has no issue with the 2%–10% varieties, and can foresee which documents are malware with incredible precision. In different circumstances, AI calculations can search for examples in how information in the cloud is gotten to, and report abnormalities that could foresee security breaks.

AI is demonstrating that it tends to be a resource for help wipe out false alerts and spot things human screeners may miss in security screenings at air terminals, arenas, shows, and different scenes. That can accelerate the procedure altogether and guarantee more secure occasions.

6.1.2 PERSONALIZATION AND FRAUD DETECTION

Maybe you have had the involvement in which you visit an online store and take a gander at an item yet do not get it and afterward observe computerized promotions over the web for that precise item for quite a long time a short time later. That sort of showcasing personalization is only a hint of something larger. Organizations can customize which messages a client gets, which direct mailings or coupons, which offers they see, which items appear as "suggested," and so on, all intended to lead the customer all the more dependably toward a deal.

AI is improving and better at spotting potential instances of extortion crosswise over a wide range of fields. PayPal, for instance, is utilizing AI to battle illegal tax avoidance. The organization has devices that think about a great many exchanges and can correctly recognize real and false exchanges among purchasers and merchants.

6.1.3 NATURAL LANGUAGE PROCESSING AND ONLINE SEARCH

Maybe the most well-known utilization of AI, Google, and its rivals are always improving what the web crawler gets it. Each time you execute an inquiry on Google, the program observes how you react to the outcomes. In the event that you click the top outcome and remain on that page, we can accept you got the data you were searching for and the pursuit was a triumph. On the off chance that, then again, you snap to the second page of results, or type in another hunt string without clicking any of the outcomes, we can gather that the web search tool did not present the outcomes you needed the program can gain from that oversight to convey a superior outcome later on. AI calculations with natural language can remain in for client administration operators and all the more rapidly course clients to the data they need. It is being utilized to decipher cloud legalese in contracts into plain language and help lawyers sort through substantial volumes of data to plan for a case.

6.2 APPLICATION OF MACHINE LEARNING IN DIFFERENT SECTORS OF SOCIETY

There are awesome research work going on in smart farming using developed technology like IoT and ML. Varghese and Sharma [2] built up a reasonable framework, which when applied will give an understanding into the constant state of the crop. The framework use IoT and machine figuring out how to create a reasonable shrewd cultivating module. This framework utilizes cutting-edge techniques so as to improve the precision of the outcomes and robotize the observing of crops accordingly requiring insignificant human intercession. IoT is utilized to associate the ground module, which incorporates the sensors to the cloud framework. In the cloud, ML-based constant examination is performed to foresee the future state of the crops dependent on its past information.

6.2.1 WATER QUALITY IMPROVEMENT USING MACHINE LEARNING TECHNIQUE

As per World Economic Forum analysis, drinking water quality problem has been ranked as an major issue. By concentrating on the above issue, Koditala and Pandey [3] proposes a better water quality checking framework utilizing developing advances, for example, IoT, ML, and Cloud computing, which can supplant conventional method for quality observing. This aids in sparing individuals of rustic zones from different hazardous infections, for example, fluorosis, bone disfigurements, and so on. Proposed show [3] produced for water quality checking has an ability to control temperature of water and alters it in order to suit condition temperature. Table 6.1. Details of studied research work based on IoT, ML, and Cloud computing.

6.2.2 REVERSE ENGINEERING OF DATA THROUGH NEURAL NETWORK

In current advanced technology, machine-based education in Cloud platform and administrations are accomplished well known. A customer sends input information to a cloud, which performs ML calculations dependent on neural systems (NN) utilizing its groundbreaking server, and returns the outcome back to the customer. In this situation, the server can without much of a stretch gather the clients' delicate information, raising the protection issues, and making clients hesitant to utilize

Table 6.1 Details of Studied Research Work Based on IoT, Machine Learning, and Cloud Computing.

S. No.	Literature	Year	Approach/Concept	Technology Used	Challenges/Issues Addressed
1	Varghese and Sharma	[2]	Smart farming for farmers using IoT and MACHINE LEARNING	IoT MACHINE LEARNING	Saving of crops from being wasted using automated system and reduced human interference
2	Bosse	[4]	Mobile multiagent system for IoT and cloud using java script	CLOUD COMPUTING MACHINE LEARNING Java script	Security of Cloud-based service MaaS (MACHINE LEARNING as a service)
3	Ara et al.	2017	IoT and MACHINE LEARNING are used to look up intelligent diabetes supervision system	IoT MACHINE LEARNING	Delay in a Cloud computing health care service by proposing sophisticated digital device in the system
4	Nóbrega et al.	[5]	Animal supervise mechanism using IoT	IoT	Selection of animal grazing area appropriately by studying and analyzing their posture and movement
5	Singh et al.	[5]	Functional MACHINE LEARNING	MACHINE LEARNING	Classification of milky-way Galaxy RR Lyrae and variable starts
6	Moon et al.	[6]	Information processing in Cloud environment	Cloud, IoT	Privacy issue of personal data to use cloud service
7	Tang et al.	[7]	Deep education on IoT devices and training	Deep learning, IoT	Programming IoT devices intelligently using MACHINE LEARNING approach
8	Alrashdi et al.	[8]	AD-IoT with incongruity detection of IoT cyber assail in smart metropolis using mechanism of LEARNING	MACHINE LEARNING IoT	Detection of Infected IoT devices in smart traffic and smart applications
9	Elgamal et al.	[9]	IoT and its use in Cloud resources	Cloud computing IoT	Type of operation execution on edge devices on cloud
10	Rath	[10]	QoS maintenance with supplementary security in Cloud-based work out	Cloud Computing	Security protocol for value-added services in cloud computing
11	Oyekanlu	[11,12]	Osmotic combined computing for MACHINE LEARNING and cyber safety submission in manufacturing and IoT	MACHINE LEARNING Security	Gaussian Mixer model for solution analysis in IIoT
12	Malpe and Tugaonkar	[13]	Machine education trends in health check and science	MACHINE LEARNING Healthcare	MACHINE LEARNING algorithm used and compared for kidney, pancreas, and cancer-related problem

(Continued)

Table 6.1 Details of Studied Research Work Based on IoT, Machine Learning, and Cloud Computing. *Continued*

S. No.	Literature	Year	Approach/Concept	Technology Used	Challenges/Issues Addressed
13	Serra et al.	[14]	Activity based on IoT data analytics with MACHINE intelligence and SDN	MACHINE LEARNING SDN	Use of IoT data analytics in a collaborative platform
14	Varma et al.	[15]	Android portable security by MACHINE-based algorithms	MACHINE LEARNING Android Security	Detection of android malware based on machine learning approach
15	Vuppalapati et al.	[16]	Cognitive protected shield: A MACHINE LEARNING-based guard for store constrained IoT device	MACHINE LEARNING IoT	Security challenges of IoT devices used in electrical grid with Cloud computing memory management
16	Morshed et al.	[17]	Deep Osmosis—Holistic dispersed deep learning in osmotic calculation	Deep learning	Reliable transfer of medical data during medical diagnosis and check-up
17	Gurulakshmi	2018	Investigation of IoT bots next to DDoS attack with MACHINE LEARNING advance	Security MACHINE LEARNING IoT	Mirai attack due to flow of large amount of data shared between client and server using SVM
18	Pacheco et al.	[18]	Smart division of rooms based on deep education and IoT	Deep learning IoT	Deep learning model of person reorganization in smart classroom

the administrations. Research work by Jeong et al. [19] propose another way to deal with ML benefits that can lessen the security concern, along these lines making the administrations more secure. The fundamental thought is that as opposed to sending the basic information, the customer sends the in part handled component information got from the beginning period of the NN, and the member of staff serving at table keeps on executing the remainder of the NN. Because it is not easy to figure out the first in turn out of the component information, particularly when the element information is gotten from the later chapter of NN, its protection matter can be diminished. Then again, incomplete preparing at the customer can influence the execution, as more calculations are performed at the customer.

6.2.3 CLOUD-BASED SECURED JAM SYSTEM

There are colossal impact of IoT in current perceptive applications. It gets genuine in the present everyday life just as it is ending up some portion of unavoidable and pervasive processing systems offering conveyed and straightforward administrations. A bound together and regular information preparing and specialized strategy is required to blend the device, sensor scheme, and Cloud-oriented situations consistent, that can be satisfied by the portable operator-based processing worldview, examined in this work. Presently, convenience, asset requirements, safety, and versatility of manager dispensation system are basic subject for the arrangement of variable operator objects in solid assorted systems together with the www, tended to in such exertion. To improve the advancement and arrangement of MAS it is attractive to execute operators legitimately in JavaScript, which is a notable and open boundless utilized computing and technology are accessible on every host stage together with WEB programs. The tale-planned systems by Bosse [4] is proficient to implement operators in a domain with full executable indemnity and Machine knowledge as a management. Specialists can relocate among various technical hubs consistent saving such information and direct position by utilizing an important content change in an every encompassing group. A disseminated group arrangement layer gives this hub network and safekeeping in the web, finished by a domain Name scheme offering a hierarchical diagram arrangement. Operator approval plus stage defense is guaranteed through ability-based admission and distinctive specialist benefit levels.

6.2.4 FEATURE EXTRACTION OF VIDEO STREAMING

In current modern technical society, the fast pervasiveness of cell phones prompts enhanced utilizations of astute IoT frameworks (counting associated gadgets and cloud registering) for administrations. Better execution and proficiency can be accomplished from well arranging and using processing assets. Highlight extraction is the principal task alongside live video streaming, and the ML systems are included to decide the fall occasion. Amid incorporating different IoT parts, this framework has four creating stages, and is steadily improved for better execution. This framework goes for giving background as a source of perspective to other homecare or therapeutic administrations later on.

6.2.5 INTERACTIVE DATA FEED TO CLOUD

Today expansive measure of information are put away in medicinal services segment, for example, scientific and pharmaceutical and scholastic investigation documents. Many of organized and shapeless information be constantly gushing from important sensors, for example, action recognizers, nonstop sugar checking gadgets, and solutions. Current day AI technology based healthcare applications are serving various human beings to live added advantageous lives by taking simple way in to guidance plus data starting from wellbeing experts. The emergence of advanced gadgets and complex investigation has pulled in conventional organizations into the computerized upset.

6.2.6 INTERNET OF THINGS NETWORK FOR ANIMAL CARE

The proper allocation of animals in better farms requires extra help to the animal cultivation exercises. Such help must incorporate the checking and the molding of creature's area and conduct, uncommonly their bolstering stance. With such a framework, it is conceivable to enable sheep to brush in developed territories (e.g., vineyards and plantations) without jeopardizing them. There is a proposition of a creature conduct checking stage [5], in light of IoT advances. It incorporates an IoT neighborhood system to assemble information from creatures and a cloud stage, with handling and capacity abilities, to self-rulingly shepherd ovine inside vineyard zones. The cloud stage additionally joins AI highlights, permitting the extraction of important data from the information accumulated by the IoT arrange. Along these lines, other than the stage portrayal, a few outcomes are displayed in regards to the AI stage. In particular, this stage was assessed for identifying and characterizing conditions regarding creature's stance, with primer promising outcomes. Since a few calculations were tried, such research incorporates a correlation of those calculations.

In current technological scenario, appliance knowledge is being utilized broadly in different fields. From wellbeing, physical frameworks, and cosmology, the extent of mechanizing process and examining information has no restrictions, included by the complete quantity of calculation ability. Singh et al. [20] in their examination article utilize different appliance education based methods to execute regulated arrangement of a standout amongst the most important and novel impressions of the famous satellite. They illustrate about the execution of every calculation plus employ different blunders to get better execution. Moreover, we will imagine and examine the information to mine important data.

6.2.7 SUPERVISED CLASSIFICATION USING MACHINE LEARNING

In modern summit of time, huge information investigation from IoT has been getting more consideration. A few cloud stages have furnished ML administration with a pre-prepared model to comprehend IoT information. Notwithstanding, it is important to move individual information so as to utilize the cloud administration, and system issues may keep the client from getting investigation results at a fitting time. To beat these issues, information and investigation task are moving to the edge stage. Be that as it may, most edge gadgets do not have enough ability to process and prepare a lot of information. Moon et al. [6] outline new blueprint structure that can investigate IoT information by conveying examination job. The proposed structure is intended to expand the assets of the cloud to produce the model and to utilize the model at the edge to empower prompt and momentary actuator operation. Also exhibited a contextual analysis to check this system.

6.2.8 DEEP LEARNING EMPOWERED INTERNET OF THINGS

Deep learning can empower IoT gadgets to translate unstructured media information and cleverly respond to both client and natural occasions yet has requesting execution and power prerequisites. The researcher ([7]) investigates two different ways to effectively coordinate deep learning with low-control IoT items.

6.2.9 MACHINE LEARNING ALGORITHM FOR INDUSTRIAL INTERNET OF THINGS

To execute engine knowledge calculations and other valuable calculations in modern manufacturing-based Internet of everything, novel registering methodologies are expected to forestall expenses related with introducing best in class edge logical gadgets. An appropriate methodology may incorporate community oriented edge processing utilizing accessible, asset obliged IoT edge logical equipment. Oyekanlu [11,12] plan communitarian figuring strategy is utilized to build a well-known and extremely valuable waveform for IoT examination, the Gaussian mixture model (GMM). GMM parameters are found out in the cloud, yet the GMMs are developed at the Industrial Internet of Things (IIoT) edge layer. GMMs are developed utilizing C28x, a universal, minimal effort, inserted computerized flag processor (DSP) that is generally accessible in numerous prior IIoT foundations and in many edge explanatory gadgets. A few GMMs including 2-GMM and

3-GMMs are built utilizing the C28x DSP and Embedded C to demonstrate that GMM plans could be accomplished in type of an osmotic microservice from the IIoT edge to the IIoT fog layer. Planned GMMs are assessed utilizing their differential and zero intersections and are found to fulfill significant waveform structure criteria. At the fog layer, built GMMs are then connected for curiosity discovery, an IIoT digital security and flaw observing application and are observed to have the capacity to distinguish peculiarities in IIoT machine information utilizing Hampel identifier, 3-Sigma rule, and the Box plot rule. The osmotic communitarian figuring technique pushed in this paper will be essential in guaranteeing the likelihood of moving numerous mind-boggling applications, for example, curiosity location and other ML-based digital security applications to edges of expansive scale IoT systems utilizing ease generally accessible DSPs.

6.2.10 SCALABLE DYNAMIC PROGRAMMING

The promising technology of Web of Things applications produces huge measures of continuous spilling information. IoT information proprietors endeavor to make expectations/inductions from these huge floods of information frequently during implementing engine education, and picture preparing activities. A normal organization of these submissions incorporates periphery gadgets to give handling/stockpiling activities closer to the area where the gushing information is caught. A significant test for IoT applications is planned, which tasks to carry out on a periphery equipment and some activities ought to get completed on the Internet. A versatile unique programming calculation called DROPLET has been planned, to parcel tasks in IoT projects crosswise over distributed edge and Internet-based assets, whereas limiting fruition occasion of the start toward finish activities. DROPLET assesses utilizing three certifiable applications. The outcome demonstrates that DROPLET finds an apportioning of activities having generally speaking fulfilment time inside 4% of the ideal for these applications. It additionally scales to a great many activities and beats nearest possible value in the writing, by being multiple rounds quicker in running instance while discovering apportioning of tasks by means of all out fulfilment occasion that is 20% improved for the substantial request that they approached. DROPLET has been examined toward demonstrate that it balances with complete figure of activities in correct time. ML is administering the world because of its exactness and auspicious expectations for the given arrangement of issues. ML is very utilized for wellbeing observing to diminish the death rate and upgrade the future. Organs, for example, kidneys and pancreas are very influenced in the kept running of life. Malignant growths like bosom disease have appeared in the check since a decade ago. This prompts develop new procedures in the field of restorative sciences, which can give precise and convenient forecasts to decrease the death rate. Malpe and Tugaonkar [13] present near investigation of the ebb and flow look into utilizing different ML calculations and huge information systems to deal with tremendous volume of information.

6.3 ARTIFICIAL INTELLIGENCE WITH MULTIPLE APPLICATION FIELDS

Artificial intelligence system (AIS) fulfils the requirements of Industry 4.0 with the new preparation to smoothen capability of distributed computing. There are two types of manufacturing tasks can be achieved through the Cloud. One is atomic services, which consist of the single simple task. Whereas, the composite manufacturing service consists of the collaboration among the different manufacturing companies, by utilizing the combination of the basic manufacturing services.

6.3.1 ANALYTICAL DEEP KNOWLEDGE AT IIOT NETWORK SYSTEM OF IIOT SYSTEMS

Difficulties related with creating examination arrangements at the perimeter of extensive level Engineering Interne based devices organizes near where information be produced as a rule includes creating investigation arrangements from ground up.

In any case, this methodology builds IoT improvement expenses and framework complexities, defer time to showcase, and eventually brings down upper hands related with conveying cutting edge IoT plans. To defeat these difficulties, existing, broadly accessible, equipment can be used to effectively partake in appropriated edge figuring for IIoT frameworks. Oyekanlu [11,12] present osmotic processing advance is utilized to show how dispersed is this technique for registering and existing ease equipment might be used to fathom complex, figure serious reasonable artificial aptitude and thoughtful taking in issue as of the perimeter, through the fog, to the organization of IIoT frameworks. Fig. 6.3 portrays demonstrates various service-oriented applications in Cloud.

6.3.2 DYNAMICALLY MANAGING RESOURCE USING ARTIFICIAL INTELLIGENCE AND INTERNET OF THINGS

Developing concept of Augmented reality, IoT edge equipments like smart glasses, show guarantee being used as assistive gadgets that include human(s) insider savvy. Attributable to the frequently security basic nature of the assistive application, for example, directing a visually impaired individual, both the execution and reliability of the start to finish framework is basic. Among the numerous specialized obstacles in acknowledging such applications, constraints in execution affirmations from the cloud-facilitated backend servers that help these applications, and the more extended and regularly erratic start to finish arrange delay connecting the last client and the server can be unfavorable to the reaction instance necessities of these submissions, those can bargain the betterment of the end clients. Shekhar and Gokhale [21] present an inventive thought that tends to these

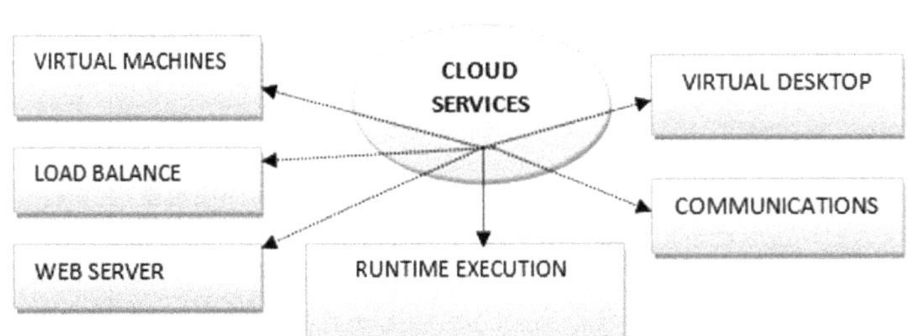

FIGURE 6.3

Different Cloud services by Cloud computing.

difficulties by powerfully overseeing assets over the cloud-edge range. It we depicts the Dynamic Data Driven Cloud and Edge Systems. Chen et al. [22] portray distributed computing in provincial healthcare, service apply autonomy, distributed storage information administrations, astute transportation, smart city, keen power utilization, versatile Internet, full of feeling registering, and IOT industry.

Table 6.2 illustrates different research & innovation in the field of AI. Cheng [23] presents a plan for intelligent animal care and management system whereas Shekhar [28] presents dynamic data driven cloud system. Similarly Kiss et al. [24] describes IoT applications on 5G Edge. Mendki [25] propose Docker container-based analytics at IoT edge video and analytics usecase. L. Carnevale et al. [29] perform a survey on cloud, edge and IoT.

6.3.3 DATA ENCRYPTION IN CLOUD-BASED INTERNET OF THINGS

The assessment and output demonstrated by Shen et al. [30] about a cloud-based policy called P3 with secured framework contrasted incredibly got precision with restrained difficulties. Table 6.3. Illustrates research in the field of AI with Cloud computing services.

Table 6.2 Research and Innovation in the Field of Artificial Intelligence with Internet of Things.

S. No.	Literature	Year	Approach Used
1	Cheng	[23]	Intelligent Animal Care Management and architecture using IoT and AI
2	Kiss et al.	[24]	IoT applications on 5G Edge
3	Mendki	[25]	Docker storage place-based analytics at IoT edge video and analytics process
4	Oyekanlu	[11,12]	Osmatic computing approach to use deep learning at IIoT (Industrial IoT)
5	Krishna et al.	[1]	AI methods for data science and data analytics
6	Al-Rakhami et al.	[26]	Cost competent edge aptitude support using storage containers
7	Jeong et al.	[19]	Cloud foundation with machine knowledge for internet devices with seclusion and safety
8	Lee et al.	[27]	Cloud power-based planning for context conscious IoT support
9	Son et al.	2018	Cloud of Things based on linked data
10	Wang et al.	2017	IoT Surveillance system for fall detection
11	Koditala and Pandey	[3]	Water superiority monitoring structure by means of IoT and Machine-based knowledge
12	Wu et al.	2018	Local area service based on IoT and Cloud intellect switch
13	Peralta et al.	2017	Fog based methods and competent IoT scheme for new manufacturing system
14	Muhammed et al.	2018	Personalized ubiquitous cloud and edge supported healthcare system for smart cities
15	Yamakami et al.	2018	Edge-based artificial engine with edge cloud coordination

Table 6.3 Investigate in the Area of Artificial Aptitude with Cloud Computing Services.

S. No.	Literature	Year	Approach Used
1	Shekhar	[28]	Dynamic data-driven cloud system
2	Carnevale et al.	[29]	Cloud, edge and IoT—survey and smart architecture for osmatic computing
3	Chen et al.	[22]	Use of cloud computing in healthcare application
4	Shekhar and Gokhale	[21]	IoT application through dynamic cloud-edge resource management
5	Wan et al.	[31]	AI for cloud based smart factory
6	Shen et al.	[30]	Secured expression investigation for bright dispensation of secured data
7	Yang et al.	[32]	Intelligent grid based cloud computing and risk analysis
8	Skouby and Lynggaard	[33]	Elegant house and stylish city approach executed by 5G, artificial intelligence
9	Power and Weinman	[34]	Revenue growth as the primary benefit of Cloud

6.3.4 CLOUD OF THINGS AND INTERNET OF THINGS FOR SMART-CITY SOLUTION

Our smart-city strategies focus on automation of maximum applications in smart ways. The agents used in AI help to modernize the systems in a way that they are easily operated and apprehend by normal citizens of city. In cloud of things machines, the cloud services are considered extremely important and their services exclusively focus on customer service. So, AI and ML approaches help in enhanced perceptive of those services. In public cloud, the platform is made protected and controlled by service providers.

6.3.5 CLOUD COMPUTING HELPS IN REVENUE GROWTH

It is over and over again observed as a strategic method to lessen costs, when its most significant advantage is as a vital method to develop incomes. Such income development can come to fruition in an assortment of ways, for example, through quicker advancement of new items, procedures, and client communications; distinguishing more clients and shutting more buys; and improving client connections through more focused on offers and better administration and encounters. Organizations that obviously comprehend the general extent of cost investment funds and income development and situate toward the previous performance will better exploit the cloud and associated advancements, for example, vast information, simulated aptitude, the smart technology, business techniques, and in this manner reinforce their upper hand and client esteem [34].

IoT design comprises of various layers of technologies supporting Internet and IoT-based devices. This design forms a basis to represent how different technologies identify with one another and to convey the adaptability, seclusion, and setup of IoT organizations in various situations. Fig. 6.4 represents the IoT-based applications managed at the top most application layer. Such applications are implemented in smart cities like smart energy, smart transport, smart

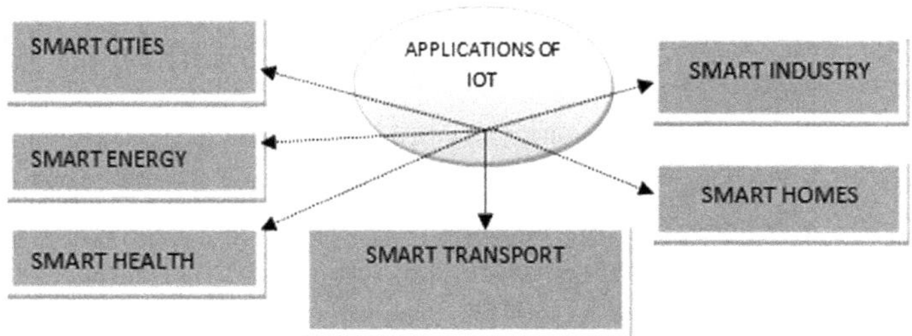

FIGURE 6.4

IoT applications in application layer. *IoT*, Internet of Things.

healthcare-based, industrial applications, and appliances used in smart homes. Table 6.4 demonstrates about various ML/AI-based applications in IoT environment with their related domains.

6.3.6 APPLICATIONS OF INTERNET OF THINGS IN DIFFERENT SECTORS

The facilities obtainable by the magnificent Internet take it believable to put up different submission dependent on it, of which just a couple of uses are as of now expressed. In the supplementary parts, a piece of the imperative precedent uses of Internet are speedily talked on.

Nowadays more and more transport mediums are equipped with advanced sensor systems and real-time-based applications with high computing ability. In vehicular industry, there is high involvement of IoT devices and safety measures, which makes these equipments more efficient for use. RFID (radiofrequency identification technology) are used to scan encrypted information in labels on the vehicles for security checking in transport management systems. Due to these implementations, there is better control and coordinations among vehicles in VANET. During use of V2V (vehicle-to-vehicle) communication and communication with RSU (road side units), there is absolutely no major issue found.

As indicated in Fig. 6.5, a basic service offered by IoT is proper spectrum distribution and access by telecom companies to manage data services. Mobility being a great challenge for ubiquitous devices, spectrum allocation with consistent connectivity is basically performed by service providers. Fig. 6.5 further shows spectrum distribution in internet as a basic telecomm service. In this architecture spectrum distribution to customers through ISP and during the satellite communication are evenly distributed using ML methods. The preferred standpoint picked up is in counteractive action and simple checking of maladies, specially appointed analysis and giving brief therapeutic consideration in instances of mishaps.

Sophisticated life style: IoT-based projects and administrations will importantly affect free living by offering help for a maturing populace by distinguishing the exercises of day by day living utilizing wearable and encompassing sensors, observing social collaborations utilizing wearable and surrounding sensors, checking unending ailment utilizing wearable essential signature based sensing nodes and in body sensors. In accordance with the development of model location and AI calculations, the important components in a patient's situation would probably keep track on the care for the patient.

Table 6.4 Machine Learning/Artificial Intelligence-Based Contributed Applications in Internet of Things (IoT) Environment with Their Related Domains.

Authors	Year	Related Domain of IoT
Ahamed and Farid	[35]	Data acquisition, decision support systems, diseases, electronic health records, healthcare, Internet of Things, supervision (artificial intelligence), patient treatment
Kumar and Lim	[36]	Firewalls, Internet of Things, invasive software, supervision (artificial intelligence)
Reddy et al.	[37]	Data handling, decision making, intelligent sensors, Internet, Internet of Things, supervision (artificial intelligence)
Aseeri et al.	[38]	Cryptography, Internet of Things, supervision (artificial intelligence), neural nets
Zhu, Song, Jiang, Song	[39]	Cognitive radio, Internet of Things, supervision (artificial intelligence), scheduling
Gurulakshmi, Nesarani	[40]	Computer network security, Internet of Things, supervision (artificial intelligence), network servers, support vector machines
Park and Saad	[41]	Internet of Things, supervision (artificial intelligence), resource allocation
Kang and Kim	[42]	Data compression, inference mechanisms, Internet, Internet of Things, supervision (artificial intelligence), quality of service
Anthi et al.	[43]	Computer network security, data privacy, Internet of Things, invasive software, supervision (artificial intelligence)
Suresh et al.	[44]	Internet of Things, supervision (artificial intelligence), low-power electronics, wide area networks
Anderson	[45]	Supervision (artificial intelligence), natural gas technology, neural nets, oil technology, petroleum, petroleum industry support vector machines
Hribar and DaSilva	[46]	Internet of Things, telecommunication power management
Ara and Ara	[47]	Competitive intelligence, data analysis, diseases, healthcare, Internet of Things, supervision (artificial intelligence), medical information systems, patient monitoring
Siryani et al.	[48]	Bayes methods, belief networks, decision support systems, decision trees, Internet of Things, supervision (artificial intelligence), pattern classification, power engineering computing, power meters, smart meters, statistical analysis
Yu et al.	[49]	Data analysis, information retrieval, Internet of Things, meta data, semantic Web
Chen et al.	[50]	CMOS integrated circuits, copy protection, current mirrors, Internet of Things, supervision (artificial intelligence), logic design, matrix algebra, multiplying circuits
Samie et al.	[51]	Cloud computing, Internet of Things, supervision (artificial intelligence)
Sharaf-Dabbagh and Saad	[52]	Cyber-physical systems, data privacy, Internet of Things, supervision (artificial intelligence), message authentication
Onal et al.	[53]	Big Data, Internet of Things, supervision (artificial intelligence), protocols
Tang et al.	[7]	Internet of Things, supervision (artificial intelligence)
Polyakov et al.	[54]	Cloud computing, human computer interaction, Internet of Things, supervision (artificial intelligence), social networking
Roukounaki et al.	[55]	Computer network security, Internet of Things, supervision (artificial intelligence)
Ganz et al.	[56]	Data mining, Internet of Things, supervision (artificial intelligence), semantic web
CaÃ ± edo and Skjellum	[57]	computer network security, Internet of Things, internetworking, supervision (artificial intelligence), neural nets, radio networks
Mamdouh et al.	[58]	Internet of Things, supervision (artificial intelligence), security of data, wireless sensor networks

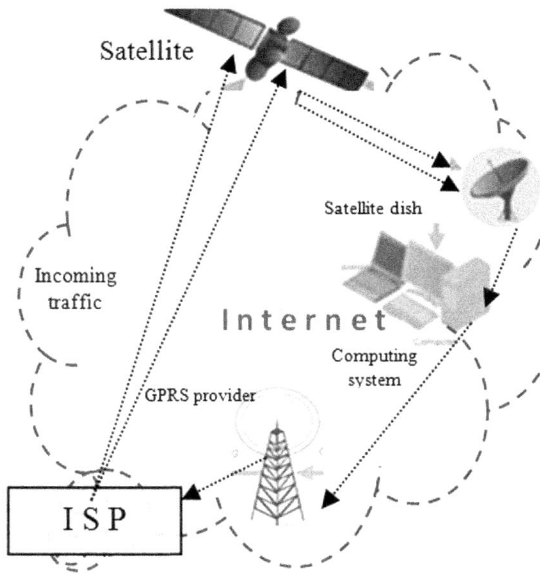

FIGURE 6.5

Spectrum distribution as a telecomm service.

6.4 APPLICATION OF INTERNET OF THINGS IN MEDICAL FIELD

For medical items, safety and protection is of generally great importance. IoT worldview, connecting savvy names to medicines, following them from beginning to end the store network and checking their grade with sensing nodes have plentiful potential benefits. For instance, things which need explicit capacity conditions, for example upkeep of a linked chain, can be incessantly observed and disposed of if situations used to destroy amid travel support. Medication following and e-families take into account the discovery of fake items and keep the production network free of fraudsters. The shrewd marks on the medications can likewise straightforwardly advantage patients, for example by empowering putting away of the bundle embed, educating customers of measurements and cessation dates, and confirming the realness of the prescription.

Retail, business people, and inventory network management: IoT can give a few focal points in trade and production network organization activities. Consider for instance, taking RFID-prepared things in addition to shrewd racks that track the present things progressively, a vendor can advance numerous projects. For instance, he may be able to make programmed inspection of products acceptance, continuous observing of storage, following shortage or the location of robbery. Moreover, IoT can assist building the information from the trade location accessible for upgrading the coordination of the entire store network. In the event that makers make out the store and deals information as of vendor, they can deliver and dispatch the correct amounts of products, thus maintaining a strategic distance from the circumstance of overcreation or underproduction. By connecting things with data innovation, either through installed shrewd gadgets or using one of a kind identifiers and information bearers that can cooperate with a wise supporting network framework and data frameworks, generation procedures can be enhanced and the whole lifecycle of items, from creation to transfer can be checked.

Operation processing plant: In many furnace of the lubricate and gaseous factories, malleable models are frequently utilized that think about conceivable outcomes for fitting and play new ID techniques joined with detecting/impelling coordinated with the IoT foundation and incorporate the remote observing of oil work force in basic coastal and seaward activities, compartment following, following of drill string parts pipes, checking and overseeing of fixed gear and so forth.

Environment observing: Utilization of remote recognizable gadgets and former engineering methods are used in green applications and natural preservation is a standout among the most encouraging business sector fragments later on. There will be an expanded use of remote recognizable gadgets in earth benevolent projects worldwide.

There are many more IoT frameworks developed for travel, transport, and transfer. These platforms improves the mechanism of screening of travelers, checking, and boarding business bearers along with the products moved by the universal freight framework that help the security approaches of the administrations and the transportation business, to satisfy the expanding need for security in the globe. Observing roads turned parking lots through phones of the clients and organization of insightful transport frameworks (ITF), we can conclude that there are fast development in smart cities with the automation of smart transportation, smart parking, smart travel with safety and security embedded mechanism.

Agriculture and reproducing: The guidelines for detect ability of horticultural creatures and their developments need the utilization of innovations like IoT, creating conceivable the continuous recognition of creatures, for instance amid flare-ups of infectious ailment. Moreover, in numerous cases, nations give sponsorships relying upon the quantity of creatures in a group and different necessities for the welfare of domestic animals. As the assurance of the quantity is troublesome, so more chance of false mechanism can be apprehended. Great recognizable smart frameworks can help limit this misrepresentation. In this way, with the use of recognizable proof frameworks, creature ailments can be controlled, reviewed, and counteracted.

6.5 CONCLUSION

It is over and over again observed as a strategic analysis that the technology is growing day by day with newer innovations, smart-city applications making the life of people more comfortable and higher services are offered to maintain a superior lifestyle. Still it cannot be denied that parallel to the development there are also some technical challenges faced by applications using technologies like AI, IoT, and ML. Organizations also use ML for general extent of cost investment funds and income development with better adventure of the Cloud and related advancements. With Cloud-based system, the valuable data and information are stored securely and safely, hence making the business transactions more reliable. Currently, maximum basic operations and facilities are handled prominently by machine-driven applications with strong support of AI, ML, and IoT. The aforementioned review on the said topic presents a justified report on various modern applications as well as contributions by research experts on these majestic research areas. This basic information with analytical study will provide a guideline to perform better research study on this excellent area of IoT, ML, and AI.

REFERENCES

[1] C.V. Krishna, H.R. Rohit, Mohana, A review of artificial intelligence methods for data science and data analytics: applications and research challenges, in: 2018 Second International Conference on I-SMAC (IoT in Social, Mobile, Analytics and Cloud) (I-SMAC)I-SMAC (IoT in Social, Mobile, Analytics and Cloud) (I-SMAC), Palladam, 2018, pp. 591–594. Available from: https://doi.org/10.1109/I-SMAC.2018.8653670.

[2] R. Varghese, S. Sharma, Affordable smart farming using IoT and machine learning, in: 2018 Second International Conference on Intelligent Computing and Control Systems (ICICCS), Madurai, 2018, pp. 645–650. Available from: https://doi.org/10.1109/ICCONS.2018.8663044.

[3] N.K. Koditala, P.S. Pandey, Water quality monitoring system using IoT and machine learning, in: 2018 International Conference on Research in Intelligent and Computing in Engineering (RICE), San Salvador, 2018, pp. 1–5. Available from: https://doi.org/10.1109/RICE.2018.8509050.

[4] S. Bosse, Mobile multi-agent systems for the Internet-of-Things and Clouds using the JavaScript agent machine platform and machine learning as a service, in: 2016 IEEE Fourth International Conference on Future Internet of Things and Cloud (FiCloud), Vienna, 2016, pp. 244–253. Available from: https://doi.org/10.1109/FiCloud.2016.43.

[5] L. Nóbrega, A. Tavares, A. Cardoso, P. Gonçalves, Animal monitoring based on IoT technologies, in: 2018 IoT Vertical and Topical Summit on Agriculture - Tuscany (IOT Tuscany), Tuscany, 2018, pp. 1–5. Available from: https://doi.org/10.1109/IOT-TUSCANY.2018.8373045.

[6] J. Moon, S. Cho, S. Kum, S. Lee, Cloud-edge collaboration framework for IoT data analytics, in: 2018 International Conference on Information and Communication Technology Convergence (ICTC), Jeju, 2018, pp. 1414–1416. Available from: https://doi.org/10.1109/ICTC.2018.8539664.

[7] J. Tang, D. Sun, S. Liu, J. Gaudiot, Enabling deep learning on IoT devices, Computer 50 (10) (2017) 92–96. Available from: https://doi.org/10.1109/MC.2017.3641648.

[8] I. Alrashdi, A. Alqazzaz, E. Aloufi, R. Alharthi, M. Zohdy, H. Ming, AD-IoT: anomaly detection of IoT cyberattacks in smart city using machine learning, in: 2019 IEEE 9th Annual Computing and Communication Workshop and Conference (CCWC), Las Vegas, NV, 2019, pp. 0305–0310. Available from: https://doi.org/10.1109/CCWC.2019.8666450 .

[9] T. Elgamal, A. Sandur, P. Nguyen, K. Nahrstedt, G. Agha, DROPLET: distributed operator placement for IoT applications spanning edge and cloud resources, in: 2018 IEEE 11th International Conference on Cloud Computing (CLOUD), San Francisco, CA, 2018, pp. 1–8. Available from: https://doi.org/10.1109/CLOUD.2018.00008.

[10] M. Rath, Resource provision and QoS support with added security for client side applications in cloud computing, M. Int. J. Inf. Tecnol. 11 (2019) 357. Available from: https://doi.org/10.1007/s41870-017-0059-y.

[11] E. Oyekanlu, Osmotic collaborative computing for machine learning and cybersecurity applications in industrial IoT networks and cyber physical systems with gaussian mixture models, in: 2018 IEEE Fourth International Conference on Collaboration and Internet Computing (CIC), Philadelphia, PA, 2018, pp. 326–335. Available from: https://doi.org/10.1109/CIC.2018.00051.

[12] E. Oyekanlu, Distributed osmotic computing approach to implementation of explainable predictive deep learning at industrial IoT network edges with real-time adaptive wavelet graphs, in: 2018 IEEE First International Conference on Artificial Intelligence and Knowledge Engineering (AIKE), Laguna Hills, CA, 2018, pp. 179–188. Available from: https://doi.org/10.1109/AIKE.2018.00042.

[13] V. Malpe, P. Tugaonkar, Machine learning trends in medical sciences, in: 2018 Second International Conference on I-SMAC (IoT in Social, Mobile, Analytics and Cloud) (I-SMAC)I-SMAC (IoT in Social, Mobile, Analytics and Cloud) (I-SMAC), Palladam, 2018, pp. 495–499. Available from: https://doi.org/10.1109/I-SMAC.2018.8653756.

[14] J. Serra, L. Sanabria-Russo, D. Pubill, C. Verikoukis, Scalable and flexible IoT data analytics: when machine learning meets SDN and virtualization, in: 2018 IEEE 23rd International Workshop on Computer Aided Modeling and Design of Communication Links and Networks (CAMAD), Barcelona, 2018, pp. 1–6. Available from: https://doi.org/10.1109/CAMAD.2018.8514997.

[15] P.R.K. Varma, K.P. Raj, K.V.S. Raju, Android mobile security by detecting and classification of malware based on permissions using machine learning algorithms, in: 2017 International Conference on I-SMAC (IoT in Social, Mobile, Analytics and Cloud) (I-SMAC), Palladam, 2017, pp. 294–299. Available from: https://doi.org/10.1109/I-SMAC.2017.8058358.

[16] J.S. Vuppalapati, et al., Cognitive secure shield - a machine learning enabled threat shield for resource constrained IoT devices, in: 2018 17th IEEE International Conference on Machine Learning and Applications (ICMLA), Orlando, FL, 2018, pp. 1073–1080. Available from: https://doi.org/10.1109/ICMLA.2018.00175.

[17] A. Morshed, P.P. Jayaraman, T. Sellis, D. Georgakopoulos, M. Villari, R. Ranjan, Deep osmosis: holistic distributed deep learning in osmotic computing, IEEE Cloud Comput. 4 (6) (2017) 22–32. Available from: https://doi.org/10.1109/MCC.2018.1081070.

[18] A. Pacheco, P. Cano, E. Flores, E. Trujillo, P. Marquez, A smart classroom based on deep learning and osmotic IoT computing, in: 2018 Congreso Internacional de Innovación y Tendencias en Ingeniería (CONIITI), Bogota, 2018, pp. 1–5. Available from: https://doi.org/10.1109/CONIITI.2018.8587095.

[19] H. Jeong, H. Lee, S. Moon, Work-in-progress: cloud-based machine learning for IoT devices with better privacy, in: 2017 International Conference on Embedded Software (EMSOFT), Seoul, 2017, pp. 1–2. Available from: https://doi.org/10.1145/3125503.3125626.

[20] J. Singh, G. Joshi, H. Tiwari, I. Tiwari, Applied Machine Learning for and Analysis of RR Lyrae Variable Stars, in: 2018 Second International Conference on I-SMAC (IoT in Social, Mobile, Analytics and Cloud) (I-SMAC)I-SMAC (IoT in Social, Mobile, Analytics and Cloud) (I-SMAC), Palladam, 2018, pp. 247–252. Available from: https://doi.org/10.1109/I-SMAC.2018.8653790.

[21] S. Shekhar, A. Gokhale, Poster abstract: enabling IoT applications via dynamic cloud-edge resource management, in: 2017 IEEE/ACM Second International Conference on Internet-of-Things Design and Implementation (IoTDI), Pittsburgh, PA, 2017, pp. 331–332.

[22] S. Chen, et al., Keynote speech: the usage of cloud computing in China regional healthcare, in: 2011 IEEE International Conference on Cloud Computing and Intelligence Systems, Beijing, 2011, pp. i–v. Available from: https://doi.org/10.1109/CCIS.2011.6045151.

[23] Y. Cheng, A development architecture for the intelligent animal care and management system based on the Internet of Things and artificial intelligence, in: 2019 International Conference on Artificial Intelligence in Information and Communication (ICAIIC), Okinawa, 2019, pp. 078–081. Available from: https://doi.org/10.1109/ICAIIC.2019.8669015.

[24] P. Kiss, A. Reale, C.J. Ferrari, Z. Istenes, Deployment of IoT applications on 5G edge, in: 2018 IEEE International Conference on Future IoT Technologies (Future IoT), Eger, 2018, pp. 1–9. Available from: https://doi.org/10.1109/FIOT.2018.8325595.

[25] P. Mendki, Docker container based analytics at IoT edge video analytics use case, in: 2018 Third International Conference On Internet of Things: Smart Innovation and Usages (IoT-SIU), Bhimtal, 2018, pp. 1–4. Available from: https://doi.org/10.1109/IoT-SIU.2018.8519852.

[26] M. Al-Rakhami, M. Alsahli, M.M. Hassan, A. Alamri, A. Guerrieri, G. Fortino, Cost efficient edge intelligence framework using Docker containers, in: 2018 IEEE 16th International Conference on Dependable, Autonomic and Secure Computing, 16th International Conference on Pervasive Intelligence and Computing, Fourth International Conference on Big Data Intelligence and Computing and Cyber Science and Technology Congress (DASC/PiCom/DataCom/CyberSciTech), Athens, 2018, pp. 800–807. Available from: https://doi.org/10.1109/DASC/PiCom/DataCom/CyberSciTec.2018.00138.

[27] T. Lee, B.M. Lee, W. Noh, Hierarchical Cloud computing architecture for context-aware IoT services, IEEE Trans. Consum. Electron. 64 (2) (2018) 222–230. Available from: https://doi.org/10.1109/TCE.2018.2844724.

[28] S. Shekhar, Dynamic data driven Cloud systems for Cloud-hosted CPS, in: 2016 IEEE International Conference on Cloud Engineering Workshop (IC2EW), Berlin, 2016, pp. 195–197. Available from: https://doi.org/10.1109/IC2EW.2016.38.

[29] L. Carnevale, A. Celesti, A. Galletta, S. Dustdar, M. Villari, From the Cloud to Edge and IoT: a smart orchestration architecture for enabling osmotic computing, in: 2018 32nd International Conference on Advanced Information Networking and Applications Workshops (WAINA), Krakow, 2018, pp. 419–424. Available from: https://doi.org/10.1109/WAINA.2018.00122.

[30] M. Shen, B. Ma, L. Zhu, X. Du, K. Xu, Secure phrase search for intelligent processing of encrypted data in Cloud-based IoT, IEEE Internet Things J. (2018). Available from: https://doi.org/10.1109/JIOT.2018.2871607.

[31] J. Wan, J. Yang, Z. Wang, Q. Hua, Artificial intelligence for Cloud-assisted smart factory, IEEE Access. 6 (2018) 55419–55430. Available from: https://doi.org/10.1109/ACCESS.2018.2871724.

[32] Y. Yang, L. Wu, S. Yang, The structure of intelligent grid based on Cloud computing and risk analysis, in: 2012 Fourth International Conference on Intelligent Human-Machine Systems and Cybernetics, Nanchang, Jiangxi, 2012, pp. 123–126. Available from: https://doi.org/10.1109/IHMSC.2012.126.

[33] K.E. Skouby, P. Lynggaard, Smart home and smart city solutions enabled by 5G, IoT, AAI and CoT services, in: 2014 International Conference on Contemporary Computing and Informatics (IC3I), Mysore, 2014, pp. 874–878. Available from: https://doi.org/10.1109/IC3I.2014.7019822.

[34] B. Power, J. Weinman, Revenue growth is the primary benefit of the Cloud, IEEE Cloud Comput. 5 (4) (2018) 89–94. Available from: https://doi.org/10.1109/MCC.2018.043221018.

[35] F. Ahamed, F. Farid, Applying Internet of Things and machine-supervision for personalized healthcare: issues and challenges, in: 2018 International Conference on Machine Supervision and Data Engineering (iCMLDE), 2018, pp. 19–21, isbn-978-1-7281-0404-1978-1-7281-0405, Available from: https://doi.org/10.1109/iCMLDE.2018.00014.

[36] A. Kumar, T.J. Lim, EDIMA: early detection of IoT malware network activity using machine supervision techniques, in: 2019 IEEE Fifth World Forum on Internet of Things (WF-IoT), 2019, pp. 289–294, isbn-978-1-5386-4980-0978-1-5386-4981, Available from: https://doi.org/10.1109/WF-IoT.2019.8767194.

[37] R.R. Reddy, C. Mamatha, R.G. Reddy, A review on machine supervision trends, application and challenges in Internet of Things, in: 2018 International Conference on Advances in Computing, Communications and Informatics (ICACCI), 2018, pp. 2389–2397, isbn-978-1-5386-5314-2978-1-5386-5315, Available from: https://doi.org/10.1109/ICACCI.2018.8554800.

[38] A.O. Aseeri, Y. Zhuang, M.S. Alkatheiri, A machine supervision-based security vulnerability study on XOR PUFs for resource-constraint Internet of Things, in: 2018 IEEE International Congress on Internet of Things (ICIOT), 2018, pp. 49–56, isbn-978-1-5386–7244-0978-1-5386–7245, Available from: https://doi.org/10.1109/ICIOT.2018.00014.

[39] J. Zhu, Y. Song, D. Jiang, H. Song, A new deep-Q-supervision-based transmission scheduling mechanism for the cognitive Internet of Things, IEEE Internet Things J. 5 (4) (2018) 2375–2385. Available from: https://doi.org/10.1109/JIOT.2017.2759728.

[40] K. Gurulakshmi, A. Nesarani, Analysis of IoT bots against DDOS attack using machine learning algorithm, in: 2018 Second International Conference on Trends in Electronics and Informatics (ICOEI), Tirunelveli, 2018, pp. 1052–1057, isbn-978-1-5386-3570-4978-1-5386-3569-8978-1-5386-3571, Available from: https://doi.org/10.1109/ICOEI.2018.8553896.

[41] T. Park, W. Saad, Distributed supervision for low latency machine type communication in a massive Internet of Things, IEEE Internet Things J. 6 (3) (2019) 5562–5576. Available from: https://doi.org/10.1109/JIOT.2019.2903832. 2327–4662, 2372–2541 isbn.

[42] W. Kang, D. Kim, Poster abstract: DeepRT: a predictable deep supervision inference framework for IoT devices, in: 2018 IEEE/ACM Third International Conference on Internet-of-Things Design and Implementation (IoTDI), 2018, pp. 279–280, isbn-978-1-5386–6312–7978-1–5386–6313, Available from: https://doi.org/10.1109/IoTDI.2018.00039.

[43] E. Anthi, L. Williams, P. Burnap, Pulse: an adaptive intrusion detection for the Internet of Things, in: Living in the Internet of Things: Cybersecurity of the IoT, 2018, pp. 1–4, isbn-978-1-78561-843, Available from: https://doi.org/10.1049/cp.2018.0035.

[44] V.M. Suresh, R. Sidhu, P. Karkare, A. Patil, Z. Lei, A. Basu, Powering the IoT through embedded machine supervision and LoRa, in: 2018 IEEE Fourth World Forum on Internet of Things (WF-IoT), 2018, pp. 349–354, isbn-978-1-4673–9944-9978-1-4673–9945, Available from: https://doi.org/10.1109/WF-IoT.2018.8355177.

[45] R.N. Anderson, Petroleum analytics supervision machine for optimizing the Internet of Things of today's digital oil field-to-refinery petroleum system, in: 2017 IEEE International Conference on Big Data (Big Data), 2017, pp. 4542–4545, isbn-978-1-5386-2715-0978-1-5386-2714-3978-1-5386-2716, Available from: https://doi.org/10.1109/BigData.2017.8258496.

[46] J. Hribar, L. DaSilva, Utilising correlated information to improve the sustainability of Internet of Things devices, in: 2019 IEEE Fifth World Forum on Internet of Things (WF-IoT), 2019, pp. 805–808, isbn-978-1-5386-4980-0978-1-5386-4981, Available from: https://doi.org/10.1109/WF-IoT.2019.8767256.

[47] A. Ara, A. Ara, Case study: integrating IoT, streaming analytics and machine learning to improve intelligent diabetes management system, in: 2017 International Conference on Energy, Communication, Data Analytics and Soft Computing (ICECDS), Chennai, 2017, pp. 3179–3182. isbn-978-1-5386-1887-5978-1-5386-1888, Available from: https://doi.org/10.1109/ICECDS.2017.8390043.

[48] J. Siryani, B. Tanju, T.J. Eveleigh, A machine supervision decision-support system improves the Internet of Things™ smart meter operations, IEEE Internet Things J. 4 (4) (2017) 1056–1066. Available from: https://doi.org/10.1109/JIOT.2017.2722358. 2327–4662, 2372–2541 isbn.

[49] J. Yu, S. Kwon, H. Kang, S. Kim, J. Bae, C. Pyo, A framework on semantic thing retrieval method in IoT and IoE environment, in: 2018 International Conference on Platform Technology and Service (PlatCon), 2018, pp. 1–6, isbn-978-1-5386-4710-3978-1-5386-4709-7978-1-5386-4711, Available from: https://doi.org/10.1109/PlatCon.2018.8472745.

[50] Y. Chen, Z. Wang, A. Patil, A. Basu, A current mirror cross bar based 2.86-TOPS/W machine learner and PUF with <2.5% BER in 65 nm CMOS for IoT application, in: 2019 IEEE International Symposium on Circuits and Systems (ISCAS), 2019, pp. 1–4, 2158–1525, isbn-978-1-7281-0397, Available from: https://doi.org/10.1109/ISCAS.2019.8702441.

[51] F. Samie, L. Bauer, J. Henkel, From Cloud down to things: an overview of machine supervision in Internet of Things, IEEE Internet Things J. 6 (3) (2019) 4921–4934. Available from: https://doi.org/10.1109/JIOT.2019.2893866. 2327–4662, 2372–2541 isbn.

[52] Y. Sharaf-Dabbagh, W. Saad, Demo abstract: cyber-physical fingerprinting for Internet of Things authentication, in: 2017 IEEE/ACM Second International Conference on Internet-of-Things Design and Implementation (IoTDI), 2017, pp. 301–302, isbn-978-1-4503–4966–6978-1–4673–9146.

[53] A.C. Onal, O. Berat Sezer, M. Ozbayoglu, E. Dogdu, MIS-IoT: modular intelligent server based Internet of Things framework with Big Data and machine supervision, in: 2018 IEEE International Conference on Big Data (Big Data), 2018, pp. 2270–2279, isbn-978-1-5386-5035-6978-1-5386-5034-9978-1-5386-5036, Available from: https://doi.org/10.1109/BigData.2018.8622247.

[54] E.V. Polyakov, M.S. Mazhanov, A.Y. Rolich, L.S. Voskov, M.V. Kachalova, S.V. Polyakov, Investigation and development of the intelligent voice assistant for the Internet of Things using machine supervision, in: 2018 Moscow Workshop on Electronic and Networking Technologies (MWENT), 2018,

pp. 1–5, isbn-978-1-5386-3498-1978-1-5386-3499, Available from: https://doi.org/10.1109/MWENT.2018.8337236.

[55] A. Roukounaki, S. Efremidis, J. Soldatos, J. Neises, T. Walloschke, N. Kefalakis, Scalable and configurable end-to-end collection and analysis of IoT security data: towards end-to-end security in IoT systems, in: 2019 Global IoT Summit (GIoTS), 2019, pp. 1–6, isbn-978-1-7281-2171-0978-1-7281-2170-3978-1-7281-2172, Available from: https://doi.org/10.1109/GIOTS.2019.8766407.

[56] F. Ganz, D. Puschmann, P. Barnaghi, F. Carrez, A practical evaluation of information processing and abstraction techniques for the Internet of Things, IEEE Internet Things J. 2 (4) (2015) 340–354. Available from: https://doi.org/10.1109/JIOT.2015.2411227. 2327–4662, 2372–2541 isbn.

[57] J. CaÃ ± edo, A. Skjellum, Using machine supervision to secure IoT systems, in: 2016 14th Annual Conference on Privacy, Security and Trust (PST), 2016, pp. 219–222, isbn-978-1-5090-4379-8978-1-5090-4380, Available from: https://doi.org/10.1109/PST.2016.7906930.

[58] M. Mamdouh, M.A.I. Elrukhsi, A. Khattab, Securing the Internet of Things and wireless sensor networks via machine supervision: a survey, in: 2018 International Conference on Computer and Applications (ICCA), 2018, pp. 215–218, isbn-978-1-5386-4371-6978-1-5386-4370-9978-1-5386-4372, Available from: https://doi.org/10.1109/COMAPP.2018.8460440.

FURTHER READING

Hsu et al., 2017 C.C. Hsu, M.Y. Wang, H.C.H. Shen, R.H. Chiang, C.H.P. Wen, FallCare + : an IoT surveillance system for fall detection, in: 2017 International Conference on Applied System Innovation (ICASI), Sapporo, 2017, pp. 921–922. Available from: https://doi.org/10.1109/ICASI.2017.7988590.

Rath, 2018 M. Rath, A methodical analysis of application of emerging ubiquitous computing technology with fog computing and IoT in diversified fields and challenges of Cloud computing, Int. J. Inf. Commun. Technol. Hum. Dev. 10 (2) (2018). Available from: https://doi.org/10.4018/978-1-5225-4100-4.ch002. ISSN: 1935–5661; EISSN: 1935-567X, IGI Global Publishing, Hershey, USA.

CHAPTER

KNOWLEDGE REPRESENTATION FOR CAUSAL CALCULI ON INTERNET OF THINGS

7

Phillip G. Bradford[1], **Himadri N. Saha** [2] **and Marcus Tanque**[3]

[1]*University of Connecticut Stamford, Stamford CT, United States* [2]*Department of Computer Science, Surendranath Evening College, Calcutta University, Kolkata, India* [3]*Independent Researcher, United States*

7.1 INTRODUCTION

The chapter focuses on introducing knowledge representation (KR) for Causal Calculi (CCs) on the Internet of Things (IoT). Together KR and CC support functions of IoT intelligent devices, sensors, and related systems. KR and CC may be deployed as part of the IoT ecosystem to determine likely causes using distributed and decentralized nodes. KR and CC may determine causes about incidents in IoT distributed systems. These decentralized systems can gather a great deal of disparate data. Causality helps understand anomalies found in and about these systems.

In some cases, statistical deviations or correlations may be explained by causal relationships. For instance, a network problem can be troubleshot by changing network parameters and observing what affects their results. Imposing certain values to a network parameter, we may notice specific behavioral changes in the system. Of course, correlation is not causation. Performing experiments on network parameters might suggest causality from corresponding network behavior. A good deal of causality is based on experimentation. IoT systems can be experimented upon and they can also be structured for experimentation.

7.2 BACKGROUND

This chapter addresses the roles KR and CC may play in the IoT ecosystem. The study discusses methods and theories involving Pearl's Do-Calculus, Shafer's probability trees, and the Halpern–Pearl model. These methods require some form of basic KR.

Causal methods may enhance IoT functionality and IoT operation. Computing causality seems expensive for a small IoT device. The causal systems being reviewed in this chapter require logical computation enhanced with probability. These logical and probabilistic systems are essentially run small experiments to gain insight into causality. Such work is expensive for individual devices. Nonetheless, causality can enhance an IoT system. An IoT devices and systems may be structured to maximize and determine causality of things they monitor.

Artificial Intelligence to Solve Pervasive Internet of Things Issues. DOI: https://doi.org/10.1016/B978-0-12-818576-6.00007-1

Independently Pearl's system and Shafer's system can be implemented in Python or ProbLog. Both of these systems can run on modest IoT device and systems; hence, the computation of causality is often expensive. In distributed IoT networks, there may be an opportunity for causality computation.

7.3 KNOWLEDGE REPRESENTATION

KR represents systems, situations or abstractions needed to compute things. Over the years, KR played an essential role for AI and ML [1,2].

In this chapter, we focus on discussing these applications than philosophical views, which might yield tenable scientific results [3,4]. For instance, inn knowledge management systems, KR may rely on knowledge accumulation. In AI/ML environment, knowledge management systems are powered by distributed computers, such as IoT devices and systems. These machines may contain standard targets needed to support KR when integrated with AL/ML applications [4].

7.4 DYNAMIC KNOWLEDGE REPRESENTATION

Dynamic knowledge representation (DKR) is an abstract or mathematical functions. In AL/ML, DRK techniques involve computer simulation and other mathematical agent-based modeling theories. When computer models are applied to AI/ML, these models often use DKR. Knowledge in DKR may be defined as linear or nonlinear functions. In computational modeling, DKR is applied to information model or knowledge [5].

DKR can be used in connectionist neural networks, to produce neuronal structure results. These conclusions can be in a form of real vectors or valued synaptic weights. It is hard to integrate neural networks with the CC methods discussed in this chapter. It is challenging to use these neural coding systems to explicitly understand causal models, hence, it is plausible neural networks may be applied to discover casual relationship.

7.5 INTERSECTION OF KNOWLEDGE REPRESENTATION, CAUSAL CALCULUS, AND INTERNET OF THINGS

KR and CC have contributed to the advance of AI and ML [6]. If properly integrated, KR can provide IoT devices and sensors the ability to work with AI and ML [6]. Despite many uncertainties on how KR can be applied to distributed IoT networks. For many years, several IT solutions have been applied to help solve technical issues affecting devices and systems. KR has contributed to the implementation of semantic web techniques in recent years.

Predicate Calculus, or first-order logic (FOL), is a basic symbolic logic where each structure comprises functions or variables. These functions or variables represent physical or abstract things. These functions have broader expressivity given their variable parameters. Further, these variables and functions are joined by logical AND as well as OR binary operations. Predicate Calculus are equally viewed as a NOT function. Predicate Calculus is a practical tool for reasoning and analysis. Besides, it has very broad and deep applicability [6].

CC involves methods for connecting causes and effects. These methods rely on predicate Calculus and probability theory. Analyzing these causes and effects is a preferred way to view CC methods, such as causes and effects for logical experimentation. These experiments are augmented with probabilities or likelihoods. This concept similarly can be applied to randomized controlled experiments. We are interested in how these controlled experiments may be applied to IoT devices and systems. In the next decade, more devices will be connected to the distributed IoT networks [7]. The use of CC in this context, will likely become more important. In distributed IoT environment, devices and sensors are interconnected, thus the use of real-time causality may be necessary [7]. As millions of devices and sensors are being added to distributed IoT networks, both causality and IoT systems will scale significantly [7]. The ability for these devices and sensors to share data in a decentralized environment is key to continuous adoption of IoT solutions [8,9]. In this context, IoT systems will be implemented to determine causal data. Nonetheless, the general desire to understand cause and effect is growing.

There is a continuous interaction between IoT devices and sensors. In IoT, physical networks are designed to perform calculations while embedded in edge computing environment [7]. In distributed networks, IoT devices and systems collect and process data in real-time and drive automated processes.

KR and CC have played a critical role in supporting large-scale application deployment for IoT solutions. As the global technology landscape evolves, AI, ML, and IoT capabilities have experienced continuous integration of processes. This concerted effort gives AI/ML and IoT a unified operational capability [6].

7.6 FIRST-ORDER LOGIC AND PREDICATE CALCULUS FOR KR, CC, AND IOT

In KR, FOL represents basic logical relationships. FOL is a foundational formal system that anchors a great deal of mathematics and computer science. Hence, FOL systems can be applied in a broad range of areas from engineering through philosophy.

In computer science, for example, FOL has many applications. It can be used in the semantics of programming languages and AI theorem-proving while relying on variables and connectors, which can be interpreted as nouns or verbs. These may be interpreted as nouns or verbs. If x represents a car, then $\bar{x}$ represents anything but a car. These variables are also connected using the logical connectives $\wedge$ as well as $\vee$. The connective $\wedge$ is a logical ***AND*** of its two arguments. The connective $\vee$ is the logical ***OR*** of its two arguments.

For example, suppose x is battery power, whereas y is the capacity to send out a beacon signal. We might represent the statement:

> If a device has battery power, then the device can send a beacon.

As $x \rightarrow y$. The implication $x \rightarrow y$ is logically equivalent to $\bar{x} \vee y$. That is x is true indicates there is battery power and y is true indicating there is capacity to send out a beacon. Supposing is x is true, the equation $\bar{x} \vee y$ holds when the sensor can send a beacon. Similarly, if x is false, then such device may or may not send out a beacon. That is, the implication $x \to y$ holds, it is just that there is no battery power so y is irrelevant.

In AI/ML and IoT, predicate calculus comprises quantify statements about device's and sensor's logical interaction within their ecosystem. Existential quantification of FOL expressions may qualify several variables with the there-exists symbol: $\exists$. According to $\exists x{:}\bar{x} \vee y$ there is a sensor with battery power, which is capable of broadcasting a beacon signal. Universal quantification of FOL expressions may qualify variables with the four-all symbol: $\forall$. For instance, the $\forall x{:}\bar{x} \vee y$ equation indicates that all sensors with battery power can be broadcast with beacon signals.

In IoT, higher-order logic (HOL) plays a key role in a number of ways. HOL is the basis of programming languages and theorem proving systems. It makes inferences about the provability of other statements. This impacts proving statements on IoT devices/systems. HOL may be used to make statements or deductions about IoT devices and systems. A second-order logic, with a lot of computational operations may support the functioning of devices and sensors.

7.7 STRUCTURAL CAUSAL MODEL

A construction or structural causal model (SCM) illustrates causal processes. These SCM focuses on extended research designs by providing clear rules [10]. These guidelines stem from independent variables that should be counted in or measured. Such instructions can be applied in AI, KR, and IoT to reduce the sound or device and sensor performance. When used based on well-defined rules, SCM determines whether independent variables can be included or controlled for, when integrated with devices and systems. This concept, however, can be sustained by the process of a randomized controlled trial [10]. In addition to these developments, the need for interventional studies when dealing with controlled trial randomization can be untimely [6,10]. It may not offer the necessary hypotheses to support CM processing and testing environments. When sharing with external validity, SCM can play an important part. It can be applied to determine outcomes and ensure whether there is parallelism to other unaffected populations. SCM has transcended to other technologically advanced domains, such as machine learning (ML), deep learning, signal processing, IoT, and AI [10].

Pearl describes SCM an ordered triple (***U***, ***V***, ***E***). Here ***U*** contains exogenous variables. Each of these sequential concepts can be defined by external elements—through ***V*** focuses on describing endogenous variables, which coexist within the model that describes its values. Thus ***E*** is determined by structural equations used to interpret the value of endogenous variables. SCM comprises other areas such as the "ladder of causation." Pearl studied metamodel: a method that involves a triad abstraction, commonly known as "ladder of causation." The ladder of causation is a three-level abstraction or scientific method that can be applied to all three domains, notably AI, KR, and IoT. The ubiquitous subject process has data patterns that can be represented as parallels. In ML, AI, and deep learning, and the "ladder of causation" can determine the prediction and the desired outcomes. A small deliberative action can activate a small deliberative action. The process describes "causal relationship(s)." Besides, Pearl explained the relationship between "causality and correlation." In his classical analyses, he concludes that there is a distinction between these two variables, namely "causality and correlation." In AI, the mathematical equation can describe these methods:

$$P[Y|X] > P[Y].$$

Registered in a representative setting (X and Y) can be set to accommodate the desired condition [11,12].

A workflow incorporating steps where each subprocess goes through to output the event desired outcome is shown to a lower place. For example, if the parameters of the process are considered as a Gaussian distribution, it is possible to find out the probability of the system being abnormal using standard tests. The benefit of viewing the parameters as a Gaussian function is to produce an algorithm for catching purposes.

The steps, in brief, are as follows:

- Read parameters from the systems that are likely to be indicative of unusual cases. For the design, we can use principal component analysis.
- We fit the parameters, such as the variance and the mean of the data considered.
- Given a new example X, we compute the $P[X]$ where $P[X]$ resembles the product of the Gaussian probabilistic values of all considered examples.
- We conclude that $P[X]$ is anomalous if it is less than some threshold value.

A multivariate Gaussian distribution can be used in systems where multiple features can be considered when detecting anomalies. After determining the Gaussian probabilities of each function, we considered all the features, before assuming that they are on a Gaussian plane. Similarly, Causality and Granger causality can be used to detect anomalies in time-series data in the IoT system.

Deep learning tools such as temporal analysis give an efficient means of anomaly detection. An IoT device/system can be anomalous about different attacks. The sources of attacks may include the following:

- *System vulnerability*—An IoT device and sensor, once connected to the Internet, can be vulnerable to cyber-attacks. These attacks may result in the whole system collapsing.
- *Data leakage*——Anomaly can occur if the data is not protected. Sensitive information from databases, web browsers, servers can lead to an external entity.

An anomaly in the IoT system can be tracked at the subject root if found the type. Generally, a defect can be contextual, point-wise, or collective.

7.8 PEARL'S DO-CALCULUS

Pearl is the originator of Bayesian-based CC. His work in this area was cited in his 2011 Turning Award. Pearl's CC is often called the Do-Calculus and it is based on properties of the Bayesian networks. Bayesian networks are directed acyclic graphs. Representing DAGs on IoT systems is natural using current techniques. Many IoT networks are DAGs themselves. Some IoT networks are cyclic. In these cases, we can find spanning trees of these cyclic graphs. These spanning trees should be carefully selected so they are reasonable Bayesian networks.

7.9 PEARL'S DO-OPERATOR

Pearl's CC is based on Bayesian networks. Pearl's work uses controlled situations. The basic idea of the Do-Operator is to break up an experiment by performing an exogenous test. An exogenous test is an experiment from outside of a system. An instance of this method is diagnosing an automobile's engine failure. For example, the cause of a car engine failure may be discovered using Bayesian analysis. Suppose, the leading case of such a failure is a battery failure. A newly tested battery can be swapped to see if it is the reason for the engine failure. Using exogenous test and other mathematical tools, Pearl discussed a method involving data and assumptions.

The conditional statement $P[H=h|E=e]$ indicates the probability that $H=h$ given that the event $E=e$ has occurred. The random variable E may be limited to two possible values $\{e_1, e_2\}$. Pearl's Do-Operator allows us to select one of these values to access the probability of $H=h$. That is, we have done an experiment exogenously setting E to e_2. This is represented as $P[H=h|do(E=e_2)]$. This may be represented on an IoT device by indicating what conditions were set prior to the event where $H=h$. In fact, the basic representation could be expressed as $P[H=h]$ with the text "$do(E=e)$" indicating this is a interventional setting of E to e.

7.10 PEARL'S BAYESIAN NETWORKS

Consider two random variables H and E that denote the hypothesis and evidence, respectively. When variables H and E, are not independent, Bayes' theorem states:

$$P\left[\bigwedge_{i=1}^{s} H_i=h_i \middle| \bigwedge_{i=1}^{t} E_i=e_i\right] = \frac{P\left[\wedge_{i=1}^{t} E_i=e_i | \wedge_{i=1}^{s} H_i=h_i\right] P\left[\wedge_{i=1}^{s} H_i=h_i\right]}{P\left[\wedge_{i=1}^{t} E_i=e_i\right]}.$$

Suppose, the hypotheses are not independent of the evidence. However, the evidence may support the hypotheses.

Actually, when variables H and E, are independent, Bayes' theorem holds, but it is the basic statement:

$$P\left[\bigwedge_{i=1}^{s} H_i=h_i \middle| \bigwedge_{i=1}^{t} E_i=e_i\right] = P\left[\bigwedge_{i=1}^{s} H_i=h_i\right].$$

That is, if the hypotheses are independent of the evidence, then the evidence is not relevant to the probability of the hypotheses occurring. In other words, the evidence does not play a role in supporting or causing the hypotheses.

Bayes' theorem is sometimes called backward probability. That is, the probability of the hypotheses given the evidence is based on the probability of the evidence given the hypotheses weighted by the ratio:

$$\frac{P\left[\wedge_{i=1}^{s} H_i=h_i\right]}{P\left[\wedge_{i=1}^{t} E_i=e_i\right]}.$$

Pearl developed Bayesian networks. A Bayesian network is built on a directed acyclic graph (DAG). In these DAGs, each vertex is a random variable. Edges from a vertex E_i to a vertex H_i

indicates H_i is dependent or supported by the evidence nodes E_i. That is, edges show dependency and possible causality.

When there are no paths from one node to another, this indicates independence between the nodes. In general, all E_i are H_i's parent nodes while H_i is a child node of each E_i. This equates to the conditional probability.

$$P\left[H_i = h_i \middle| \bigwedge_{i=1}^{t} E_i = e_i\right].$$

In an IoT system, each evidence node may be a sensor. The hypothesis node may be another controlling sensor. This hypothesis node may make a causal estimation based on its inputs from the evidence nodes.

For a vertex V, its set of parent nodes is $pa(V)$, and its set of children nodes is $ch(V)$. In this example, $pa(H_i) = \{E_1, E_2, \ldots E_t\}$ and $ch(E_i) = \{H_i\}$. A critical insight motivating Bayesian networks is the parent nodes may not involve possible situations. Thus many events seem to occur when modest numbers of input values are selected. The process is tailored to align with the natural DAG model. Sparse DAG instances are more tractable when computing probability of hypotheses.

In general, each node has a transition probability matrix $P[A_i|pa(A_i)]$. This transition probability matrix comprises of two factors: the node A_i and its parents $pa(A_i)$. Consequently, the probability computed using the fact can be exhibited in the equation:

$$\mathrm{P}[A_1 \wedge A_2 \wedge \cdots \wedge A_n] = \Pi_{j=1}^{n} P\left[A_j \middle| pa\left(A_j\right)\right].$$

Since Bayesian networks are acyclic, therefore if $B = \{A_1, A_2, \ldots, A_n\} - p(A_j)$ is nonempty, leading to the question of how we compute conditional dependencies of $\{A_1, A_2, \ldots, A_n\}$ with B. This is done by computing conditional independence. The variable A is conditionally independent of B conditioned on C is written as $P[A \perp B|C]$. That is, $P[A \perp B|C] = P[A|C]P[B|C]$. Graphically, C leads to both A and B and there is no path from A to B or from B to A. It may be common for some IoT systems to have such a physical set up.

7.10.1 BAYES PROBABILITY AND DO-CALCULUS

Causality predicts system response to intervention. Pearl's Do-Calculus handles exogenous interveions using Pearl's three rules [10]. For instance, in the Bayesian DAGs, these rules can be defined as interventions. The Bayesian network comprises two edges notably ($E_1 \rightarrow H$ and $E_2 \rightarrow H$) for nodes (E_1, E_2) and H. Supposing these are the only two edges terminating at H, this indicates.

$$P[H] = P\left[H \middle| pa(H)\right] = P[H|E_1 = e_1, E_2 = e_2].$$

If the hypothesis "H" is that a sensor's water sample is not liquid, see Table 7.1, and E_1 is the sensor's temperature. If we hold the environmental temperature at 0°C, then E_2 is the atmospheric pressure, then we can determine the validity of the hypothesis H. This intervention helps understand the causality of the state of the water monitored by the sensor. Fixing a variable to a value is done using Pearl's Do-Operator. In this case, we write $do(E_1 = 0°C)$ and this gives

$$P[H] = P[H|do(E_1 = 0°C), E_2 = e_2].$$

Table 7.1 Experiment Outcomes Displaying States of Water Under Varying Conditions as Detected by the Sensors.

Experiment Outcome	H_2O State	Atmospheres	Temperature
Success: 0.9, failure: 0.1	Solid	1/1000	−50°C
Success: 0.8, failure: 0.2	Gas	1/1000	0°C
Success: 0.9, failure: 0.1	Solid	1	0°C
Success: 0.05, failure: 0.95	Liquid	1	35°C
Success: 0.8, failure: 0.2	Gas	1	110°C

In this section, we follow Pearl's notation. Consider three disjoint sets A, B, and C of nodes in a Bayesian DAG. Pearl writes, and we follow, $G_{(\overline{A})}$ to be the DAG G where all edges going are to node(s)A are deleted. Likewise, $G_{(\underline{A})}$ is the DAG G hence, all edges coming from the node(s)A are deleted. These may be mixed and matched, so $G_{(\underline{A},\overline{B})}$ is the DAG G, where all edges coming from the node or nodes A are deleted, and all edges going to the node or nodes B are deleted.

Pearl's three rules are:

- Ignoring evidence: If B is conditionally independent of C while A is fixed, then B depends only on D and whatever A was fixed to.

$$P[B|do(A),\ C,\ D] = P[B|do(A), D] \text{ if } [B \perp C \mid A, D]_{(G_{\overline{(A)}})}$$

- Evidence exchange: Here, B is conditionally independent of C conditioned on A and D when A is fixed and has no inputs, and C has no outputs and so it cannot influence anything else.

$$P[B\big|do(A), do(C), D] = P[B|do(A), C, D] \text{ if } [B \perp C \mid A, D]_{(G_{(\overline{A},\underline{C})})}$$

- Ignoring actions/interventions: Here, B is conditionally dependent on C conditioned on A and B where A has no incoming edges, and all nodes of C but not parents of D have no incoming edges.

Since A has no incoming edges, it is fixed. Since all nodes of C and not also parents of D have no incoming edges, it is fixed and cannot influence D.

$$P[B\big|do(A), do(C), D] = P[B\big|do(A), D] \text{ if } [B \perp C \mid A, D]_{(G_{(\overline{A},(\overline{C}(\overline{D})))})}$$

Where $C(D)$ represents the nodes in C that are not parents of D. Thus $\overline{C}(\overline{D})$ indicates deleting all edges into nodes of C that are also not parents of D.

7.11 HALPERN–PEARL CAUSALITY

The Halpern–Pearl (HP) model is built on Pearl's CC. The HP casual definitions have evolved by strengthening their applicability. The HP approach includes two views on causality. These views are type causality and actual causality. Halpern [13] describes type causality as a critical goal of causality. Type causality gives general predictions. That is, type causality is for forecasts [13]. Actual causality focuses on specific instances that occurred. For instance, the 1906 earthquake caused the building to tumble in 1911. In theory, there is a parallel between the tragedy of 1906 and the specific actual causes that involve history, courts, and reasoning [13]. The HP model views causality as forward-looking or backward-looking. In either view of the HP model, its representation on IoT devices seems similar to the representation of Pearl's model.

Intelligent sensors rely on different causality processes to collect and process data. Smart sensors help report real causes, then try to understand how to predict the causes involving devices and sensors in the ecosystem. Despite these technological advances, the process of computing practical or type causality using the HP model is an expensive process. The worst-case cost of computing causality is often notable when acquiring and deploying IoT devices and systems. For example, devices and sensors to the IoT environment. In a Boolean expression, a single actual event is supported by the actual causality HP model, such as $X_i = x_i$. Hence, X_i is an arbitrary event, while x_i defines specific occurrence of such an event. In this context, the HP causality includes conjunctions of simple features.

$$X_1 = x_1 \wedge X_2 = x_2 \wedge \cdots \wedge X_k = x_k.$$

These events can be written as $\vec{X} = \vec{x}$ [13]. Whereas in HP causality, the notation of $Y_i \leftarrow y_i$ shows that an intervention has occurred by setting the variable Y_i to y_i. This interventional notation is not hard for an IoT device. The causal variables of $Y_1, Y_2, \ldots, Y_k$ can be described as external to the causal system under consideration. Intervening by setting these variables to $y_1, y_2, \ldots, y_k$, may cause a primitive event φ to occur. This event can be shown as:

$$[Y_1 \leftarrow y_1, Y_2 \leftarrow y_2, \cdots, Y_k \leftarrow y_k]\varphi \text{ or in shorthand } [\vec{Y} \leftarrow \vec{y}]\varphi.$$

The expression $(M, \vec{u}) \vdash [\vec{Y} \leftarrow \vec{y}](\vec{X} = \vec{x})$ means the model M in context $\vec{u}$ with interventions $[\vec{Y} \leftarrow \vec{y}]$ which has a unique solution of $(\vec{X} = \vec{x})$.

The following three expressions can be described as A1, A2, and A3 [13]. In the original context, the HP model consists of variables that are represented in the CMs—for instance, variables outside the model that affect causality. Also, the HP model uses the expression $(M, \vec{u})$ to denote a CM M in the context $\vec{u}$. The model M is made of Boolean expressions of causal variables. Such variations encompass both external and internal causal variables.

The conjunction $\vec{X} = \vec{x}$ is an actual cause of an event φ given the setting $(M, \vec{u})$ assuming the three conditions

A1: $(M, \vec{u}) \vdash (\vec{X} = \vec{x})$ and $(M, \vec{u}) \vdash \varphi$

A2: Necessary and enough causal conditions

A3: The expression $\vec{X} = \vec{x}$ is the minimal actual causal conjunction satisfying the two earlier steps

Plausibly*A2* is the most interesting causal condition, while the HP actual causality offers notions of blame and responsibility (Chockler & Halpern, 2004). An example of a measure of responsibility is as follows,

$$dr\big((M, \vec{u}), (\vec{X} \leftarrow \vec{x}), \varphi\big).$$

That is the degree of responsibility is computed from the model M with context $\vec{u}$ and interventions $\vec{X} \leftarrow \vec{x}$ it caused φ. For example, $dr((M, \vec{u}), (\vec{X} \leftarrow \vec{x}), \varphi) = 0$ if the model and interventions do not cause φ. If k is the smallest number of interventions causing φ, then

$$dr((M, \vec{u}), (\vec{X} \leftarrow \vec{x}), \varphi) = \frac{1}{k+1}.$$

7.12 SHAFER'S PROBABILITY TREES

Probability theory is about predicting events in closed systems. Probability theory uses events to understand experiments. An event is a possible outcome for an experiment. A sample space is all possible event outcomes for an experiment.

Shafer discusses causal probability trees. A probability tree is a series of events and their relations [14]. Building on probability trees, Shafer mentions probability trees go back to the mid 1600s. A probability tree's root is at the top of the trees, and the leaves are below. These trees are useful to describe logical causal derivations. These derivations unravel from their roots and going down [14].

A level of a probability tree is made from all branches coming from a node. Each level of a probability tree is a sample space. That is, the probability of all possible next steps are immediately below a node has total probability 1. Each node and its respective children form a probability space. Fig. 7.1 shows the node A with children B, C, and D. The values 0.1, 0.3, and 0.6 are probabilities of going from the event A to its children B, C, and D, respectively. Suppose the treeing Fig. 7.2 provides the subtree of a larger tree. Shafer [14] states that individual events, at the same level, cannot represent an entire causal system. On their own, these individual events have no context with other events. Shafer defines simple Humean events (after David Hume) as local events transitioning from one level to the next [12]. Humean events do not describe causality; they just describe an experiment and all of its possible outcomes.

Shafer defines concomitants to determine which step is taken down a probability tree. A concomitant is something that occurs together with an event. A concomitant may be a correlation. In particular, if A causes D, then A correlates with D. If A does not cause, then A may or not correspond with D occurring.

Fig. 7.3 shows the basics of concomitants. The notation **1** ATM describes the atmospheric pressure at sea level on earth. In addition, in Fig. 7.3, the expression ATM/1000 is 1/1000th of the atmospheric pressure on the earth at sea level. In nodes 4 and 6 water is in a liquid state. In nodes 5 and 8 is water in a gas state. In node 7, water is in a liquid state. Due to different atmospheric pressures and temperatures, this tree shows causation from nature. In particular, water's phases can be determined scientifically. The temperatures and atmospheric pressures cause the states of water illustrated. Fig. 7.2 describes a probability tree representing some phase transitions of water.

Moivrean events (Shafer named after Abraham De Moivre) are event sequences form the root of a probability tree to a leaf. Moivrean paths may split into several separate subtrees. With concomitant information, Moivrean paths maybe causal. The sum of the probabilities coming out of each simple Humean event node is 1. This is illustrated by the probabilities in Fig. 7.2 where

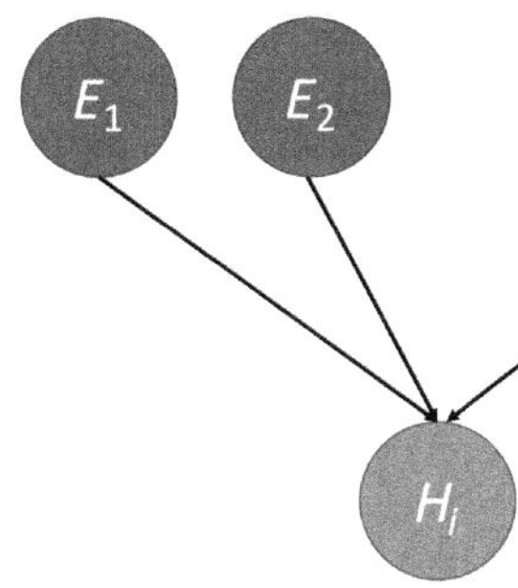

FIGURE 7.1

A simple Bayesian network.

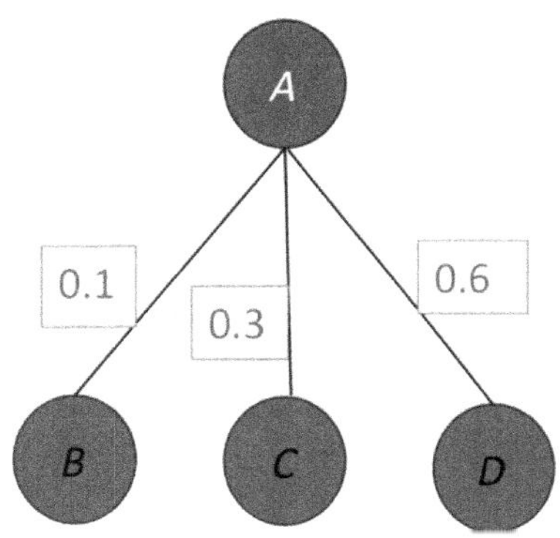

FIGURE 7.2

Three Humean events in a subtree of a more significant probability tree.

1* = *0.1* + *0.3* + *0.6. This is also for the subtree of nodes *1*, *2*, and *3* in Fig. 7.3. Likewise, the two subtrees rooted at nodes *2* and *3* also have sums of 1.

The probability of a Moivrean event or path is computed by multiplying probabilities down the entire path. Since the sum of all Humean events at each level add to 1 at each level. So the products of the probabilities down all paths of a Moivrean event also sum to 1.

The next example illustrates Moivrean paths—it further describes some sensors that are monitoring experiments in extremely different environments. These experiments are sensitive to the state of water in these environments. These environments differ with temperature and pressure. The different temperatures and forces cause water to be solid, liquid, or gas. These states of water, in turn, affect the probability of success of the experiments. In the case of very slight pressure, such as 1/1000th of atmospheric pressure, water jumps from a solid-state directly to a gas state. When warming up the ice at such low pressure, water skips the solid-state and goes straight to its gas state as

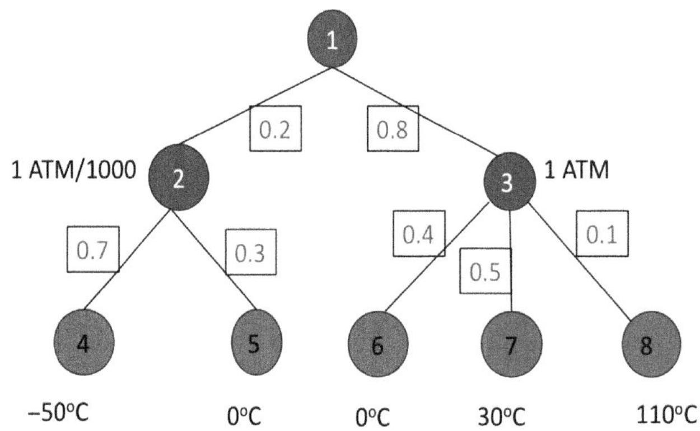

FIGURE 7.3

A probability tree showing water phases for different atmospheric pressures and temperatures.

the temperature rises from (−50°C) up to the low temperature of (0°C). Table 7.1 shows how the liquid state of water dramatically reduces sensor experiment success rates, where solid and gas states have better success rates.

The Moivrean paths to sensors detecting water in liquid states depict the below probability equation:

$$0.32 = 0.8 \times 0.5 = 0.8 \times 0.5(0.05 + 0.95)$$

Summing up, the first column gives a probability of 0.546 for the success of the experiment for all the sensors. We conclude that the likelihood of failure for an arbitrary analysis is 0.454. The below table explains the logic behind such an experiment. The whole Moivrean paths to success, which can be described by each probability given the atmospheric water levels (Table 7.2).

Level 1 down to level 2 is node *1* and its edges going down to nodes 2 and 3 in Fig. 7.4. Subsequently, levels increase gradually. Each level is a simple Humean event. All success states are leaves on the bottom with "*S*" below them.

As depicted above, 1 ATM is the atmospheric pressure at a sea level. Whereas nodes 9, 10, 13, and 14 describe the solid-state. Moreover, nodes 15 and 16 are displayed as a liquid. While nodes 11, 12, 17, and 18 have gas.

7.13 CP-LOGIC FROM PROLOG AND PROBLOG

Causal-probabilistic logic or CP-logic has been implemented in logic programming (LP) augmented by probabilities. It was created to work with causality. CP-logic is causal information and applies causal reasoning. An early application of CP-logic was to implement Shafer's probability trees. At about the same period, it was used to implement Pearl's Do-Calculus.

The CP-logic approach uses ProbLog ([15,16]. ProbLog is an extension of Prolog. ProbLog extends Prolog in two ways. The first way is by adding probability values to facts. The second way

Table 7.2 Probability Atmospheric Water Levels.

Probability	Level 1 Down to Level 2	Level 2 Down to Level 3	Level 3 Down to Level 4
0.126	0.2	0.7	0.9
0.048	0.2	0.3	0.8
0.288	0.8	0.4	0.9
0.02	0.8	0.5	0.05
0.064	0.8	0.1	0.8

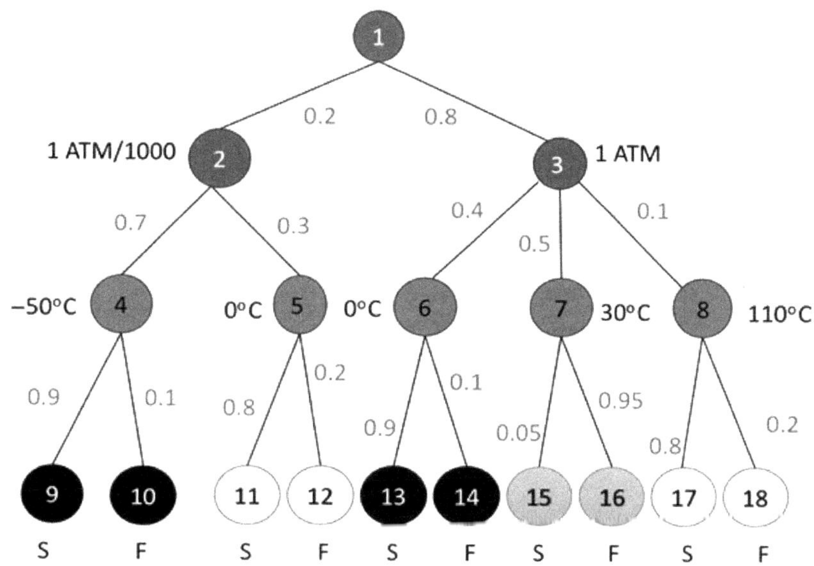

FIGURE 7.4

IoT devices and sensors in different environments. *IoT*, Internet of Things.

ProbLog extends Prolog is by diminishing the possibility of infinite loops by grounding the expressions. This grounding is done using the Herbrand universe.

A LP paradigm is a form of declarative programming [17]. In the declarative programming paradigm, software defines situations, and a computational system determines the solutions. In theory, a theorem prover or generalized solver generates the solutions. The proof is often not produced. However, in some cases, such as causality, how a test is determined may be of interest. LP is very suitable for derivations of expressions found in Shafer's probability trees. Since CP-logic allows probabilities to be assigned to facts, it is very suitable to describe contingency in Shafer's work.

LP starts by standing for knowledge using logical statements and relations. These logic statements are taken as facts. The relationships between events and other links allow the LP systems to find solutions. There are some exacting LP languages. In 1972 Colmerauer and Kowalski created Prolog. A simple example of LP from Prolog is next. Simple KRs in Prolog are given. In Prolog, facts are in lower case. Variables start in the upper case. Each statement ends in a period. Line comments start with "%."

```
%
% The operator ": -" shows a relationship
% so, A: - B can be read "If B then A"
%
%
% This next example is a simple classical infinite loop in prolog.
%
%
sensor_alive(X):- sensor_ alive(X).
sensor_alive(X):- sending_data(X).
sending_data(X).
%
% The variable X will infinitely loop trying to resolve that the
% sensor is alive.
%
ProbLog prevents this infinite looping.
Prolog, and hence ProbLog, uses the closed-world assumption. The closed-world assumption
means if a fact is not explicitly assumed true, then it is false.
%
% These are facts
%
% The first one is read sunlight affects the temperature
%
Impacts (sunlight, temperature).
Impacts (temperature, humidity).
Impacts (humidity, battery).
Impacts (battery, power).
impacts (atmospheric_pressure, freezing_point).
impacts (freezing_point, battery life).
impacts (battery_life, power).
%
% Next is a query about a Humean event
%
query (impacts (sunlight, temperature)).
%
% ProbLog returns true with probability 1. Such a process is due to
% no probabilities that can be defined, so all statements are with
% probability 1
%
```

% The next query finds all X that affect power.
%
query (impacts (X, power)).
%
% ProbLog returns all X that directly affects power. Namely, battery_life and battery
% Each of these is with probability 1
%
The next relationships allow ProbLog and similarly Prolog to compute transitivity.
impacts (X, humidity):- impacts (X, temperature).
impacts (X, battery):- impacts (X, humidity).
impacts (X, power):- impacts (X, battery).
%
%The next query gives
%
% True with probability 1.
%
% However, the closed world assumption makes the next query give True with probability 0.
%
query (impacts (moonlight, power)).
%
% True with probability 0.
%
Fig. 7.5 describes CP-law. Each of the (a_i) terms depicts real values in the [0,1] so that $a_1 + \ldots + a_k = 1$.

The value (A_i) is the probability that the associated fact A_i is true. The equation figure x may be read in two distinctive parts:

- Cause: ϕ
- Potential effects: $A_1, \ldots, A_k$, where effect A_i is caused by Cause: ϕ with A_i.

The expression ϕ is a FOL expression. The expression ϕ is made of the facts A_i, for all $i \in \{1, 2, \ldots k\}$. The $\forall x$ is such that $x = x_1, \ldots x_n$ which are the free variables in the expression ϕ.

As an example, a Problog:
%
% The battery is dead with probability 1.
% The rule, ifbattery_dead then zero (power) with probability 0.5 indicates
% There is an alternate source of energy with a probability of 0.5
%
1:: battery_dead.

$$\forall x \ (A_1 : \alpha_1) \vee \cdots \vee (A_k : \alpha_k) \ \Leftarrow \ \phi$$

FIGURE 7.5

The equation describes the causal-probabilistic law.

0.5:: zeroes (power): - battery_dead.

Query (zero (X)).

ProbLog returns that there is a 0.50 probability the power is zero. In this example, bettery_dead causes the power to be 0 with the given probability. Here the "$\forall x$" in the equation in figure is vacuous.

In the next example,

0.4:: strong_moonlight.

0.25:: zeros (power): - strong_moonlight.

Query (zero (X)).

%

% Shows that the probability of zero power is 0.1.

%

We are considering sensor_1 and sensor_k. If we know that sensor_1 has no power storage, but with probability 0.9, we know that sensor_k has no power storage, and so the following rule expresses the "$\forall x$" so that no_power_storage then it has a 0.3 probability for zero power

1:: no_power_storage (sensor_1).

0.9:: no_power_storage (sensor_k).

0.3::zero(X):- no_power_storage(X).

%

% This returns 0.3 for sensor_1 and 0.27 for sensor_k.

%

%

Instead, in this section, we use ProbLog's query function rather than Prolog is classical "? -" operator. The prologue is used to present an example of CP-logic applied to Shafer's probability trees following.

7.14 PROBABILITIES AND CAUSAL RELATIONSHIPS

Industry has developed systems embedded with publish-subscribe middleware. These schemes help find and disseminate events on the IoT distributed/decentralized IoT networks. IoT intelligent systems played an essential function in aviation traffic management for many years. This practical methodological model provides a step-by-step system validation process aimed at influencing the construction of Bayesian networks (Male & Keyvanpour, 2014). The holistic IoT event prediction and generalized architecture includes event cloud/IoT environment, event capture interface, notably advanced message queuing protocol (AMQP), Hypertext Transfer/Transport Protocol (HTTP), MQ Telemetry Transport (MQTT), Bayesian Event Prediction Model (BPM), and Decision Support System (DSS). Each of these protocols contains solutions that can scale beyond a particular IoT object threshold. The combination of IoT protocols and event prediction models can be enhanced by CMs. The advantage that a generalized architecture for IoT event prediction brings to the environment is that it dynamically computes dependent probabilities. It similarly deals with procedures that can evaluate an event prediction [18].

Probabilities and causal relationships are events that are logically predicted before occurring. Probabilities and causal relationships often guarantee that an effect can be anticipated in advance before allowing proper counteractive or preventive measures to be brought out. In probabilities, the idea remains the process of typical functional structure. This procedure can be supported by representational abstraction together with the advanced programming language. Predictive representation is a method that consists of two properties, that is, productive and graded/guardedness. In the IoT network, causal relationships focus on determining what causes events while triggering remedies [7].

In IoT, this concept provides statistical information about occurring events. This data are generated to aid in assembling a provisional probability occurrence model, also known as the conditional probability event model [18]. Most of all, probabilistic dependency can be determined between events, for example, $e_1 - e_2$. This method allows for each event to display the desired facts/results. In IoT event prediction infrastructure, for instance, a generalized architecture consisting of several events can be exhibited. In this case, IoT objects trigger the events via a network interface named event cloud or IoT environment. This architectural concept applies to machine-to-machine protocols [18]. These protocols, however, are generally stored in the event/big data warehouse repository [7].

Event prediction models continue to provide interoperable services to the aviation or aerospace industry. These systems, on the contrary, can be deployed along with other IoT intelligent devices and sensors to collect data, aimed at predicting flight delays and routes [18]. These networks are designed to support IoT devices and sensors deployed to various industries [18]. Their large-scale generated events and dynamicity to support IoT networks. These underlying variables, for example, large volumes, functions, and dynamicity, can be viewed as terminology that is designed to support object and event interactions [18].

7.15 MATHEMATICAL RULES FOR DO-CALCULUS

Causal discovery may determine the causal effects of variables. Do-Calculus has been used to identify causal effects, based on arbitrary classes, which involve the Semi-Markovian Causal Models (SMCM). In this context, SMCM uses a general logical approach, that is equivalent to that of the class of graphs [19]. A similar approach can be applied to AI and IoT technologies by using causal structure discovery algorithms. In theory, this approach allows for the queried procedures to be implemented in the Do Calculus inference to trigger unexpected effects. In many industries, Do-Calculus has been used more than traditional causal effects [20,21]. The goal is to apply the naïve graph enumeration to a similar class [22].

Do-Calculus is a modular and adaptable method, which is embedded with distinctive features. In data science, both methods, for example, causal effects and Do-Calculus, cannot be effective if there is insufficient data [20,21]. In data sets, a simplified approach is required to confirm the causal structure discovery as well as the desired inference on causal effects [19]. If properly deployed into the desired environment, both Do-Calculus and causal discovery algorithms can interact by sharing data in real time. This technique, however, is described as a method of identifying causal effects, to support arbitrary/equivalence classes in SMCM. DAGs is based on latent variables, despite primary advantages, which do not yield a unique or true causal structure. While partial ancestral graphs (PAGs), can be displayed to make algorithmic outputs to complement the causal effect method [20,21].

The true causal structure is known as a casual graph or Do-Calculus is a method aimed at enabling the entire inference, namely the identification of causal effects. This process occurs when the passive observational distribution/sharing of data is performed through assigned variables. This method further allows for causal effect form to be implemented, for example, $P[y|do(x),w]$ and correctly identified. The method is often performed when a data set is assigned an input to provide a numerical estimate [20,21].

7.16 BAYESIAN NETWORKS

The advent of artificial intelligence (AI) and related applications helps to overcome some of the critical issues are affecting the networks, parents, and child nodes. Pearl has contributed to the progress of Bayesian networks being used in AI. Bayesian networks (BNs) or Bayesian belief networks (BBN) are individually labeled as directed acyclic graphs (DAGs). BNs networks simulate a part of the system with uncertainty. Bayesian networks can be used in medical and mechanical failure diagnoses [7]. High-tech firms have adopted Bayesian networks in support of bots.

Causal dependents are types of nodes that can be denoted by their parent nodes. The node probabilities often follow under a specific state, such as that of its parent. Often the node is referred to as a conditional probability [7]. There is always a dependency between nodes. Such a relationship can be referred to by an edge between two or more event nodes, that it, the probabilistic dependence between events (E_1–E_2). These variables can occur conditionally on the event probabilities. This formula further explains that a phenomenon often occurs in (E_2). Accordingly, (E_1) probability can be labeled as an event [7]. In probability, there are common technical issues involving Bayesian networks and nodes.

Some of these problems require conditional probability functions to establish relationships between systems and nodes [7]. In DAGs, each node is describes a separate random interesting variable. In this type of network setting, there is a demonstrable level of association between a parent and the child nodes [7]. Such a relationship can be established and maintained by the causal direction between these conforming variables. In the BNs environment, some of the child nodes are causal dependents [7]. AI has built-in tools that can be deployed to resolve issues affecting the Bayesian networks and clients. These AI applications can execute large-scale projects to support, Bayesian networks, and clients. AI experts cite that some lingering effects amid the graph structure have been noticed in the last decade [7]. The conditional probability values trigger these technical problems through uncertain probability values. Bayesian networks are generally known as causal probabilistic networks. In manufacture, these nets are sent for "causal nets, graphical probability networks or probable cause-effect models or probabilistic influence diagrams." These types of systems have advanced overtime, given the rise of AI. The Bayesian networks are becoming popular around the world, mainly in the industrialized nations.

7.17 SIMULATION AND EQUIVALENCE CLASS

In the ML and IoT environments, Do-Calculus can be used to perform a sequence of device or object simulations. The model can be implemented in many industries production processes where ML and IoT intelligent devices and sensors are deployed [23]. This method is fundamental to the selected areas, primarily if the causal structure cannot be determined. Such methods are then introduced to solve complex problems involving ML and IoT. Besides, Do-Calculus can be used to ascertain causal effects identifiability in an instance where researchers are unable to determine the causal graph [24,25]. In this context, algorithmic values can be applied using '***R***' language to trigger package provisioning throughout the ML/IoT environment. In the development environment, a variety of techniques can also be applied using a core modern Boolean Satisfiability or, naturally, the SAT solvers [25]. In this context, the Do-Calculus method should be implemented to determine whether causal effects, such as $P[y|do(x),w]$, can be otherwise classified. This process occurs between Do-Calculus inferences and the constrained equivalence class. If accurately determined, this process can determine the causal effect through a joint probability distribution for the variables. In ML and IoT, this approach allows for significant flexibility required to address the credentials issues affecting both environments [24].

Equivalence class (EC) has gained momentum due to the birth of new technologies such as AI, KR, Bayesian Network Structures, and IoT. In AI, EC is used for performing heuristic search algorithms [24,25]. EC comprises two distinctive sections, namely forwarding equivalence class (FEC) and learning equivalence class (LEC). The LEC method may be applied to combine data and scoring metrics within the learning Bayesian networks [24]. In EC, how the privacy of users can be protected is based on the anonymization process. Such a concept, however, entails how records can be displayed or connected quasi-identifier attributes/sensitive attributes are critical for establishing sensitive privacy. Researchers suggest that there is a parallel between FEC, LEC, and ontology classes. In data aggregation, methods such as class Calculus continue to be used in many industries throughout the world [24].

7.18 FUTURE RESEARCH DIRECTION

Causal KR for IoT devices seems to be a promising area. Pearl's Do-Calculus, Shafer's probability trees, and the Halpern-Pearl model can be readily deployed to distributed IoT devices and sensors. IoT devices and sensors can be integrated to support the functioning of causal systems. Designing IoT devices/systems to leverage CC seems promising. This similar process involves proof-systems for declarative languages [17]. It would be interesting to see whether IoT scale devices/sensors can have effective causal calculations. For example, the decision problems for Halpern-Pearl causality are in an expensive complexity class. For instance, looking at variations of HP causality for small, devices/systems whose constraints may allow for less costly solutions would be fundamental to future research. Causality for small non-Turing complete tools is a domain, which requires further research.

7.19 CONCLUSION

Distributed IoT networks or causal systems are a reality. Together IoT devices/sensors and SCM have a great deal of potential. In general, causality is fundamental in reasoning systems. In the end, pushing causal reasoning down to the IoT device-system level may open many new opportunities. The process entails KR for casualty, that is, suited for small, constrained devices and systems. It is not clear how the nature of IoT devices/sensors will be suitable for any of the three systems mentioned here: Pearl's Do-Calculus, Shafer's probability trees, and Halpern-Pearl causality. These systems are based on logic, probability, and are expensive to deploy and perform a range of tasks. Determining causality intuitively seems to be essential and potentially costly. In essence, declarative systems offer the foundations for causal reasoning [17]. How a cause-effect relationship can be computed may not impact on the functions of most causal reasoning systems. Hence, there may be cases when the reasoning is essential to analyze [17].

REFERENCES

[1] J. Malpas, Donald Davidson, in: E.N. Zalta (Ed.), The Stanford Encyclopedia of Philosophy, Winter 2012 Edition, n.d. Available from: <https://plato.stanford.edu/archives/win2012/entries/davidson/>, 2017 (accessed 06.01.17).

[2] G.P. Zarri, 2014 special issue: sentiments analysis at conceptual level making use of the Narrative Knowledge Representation Language, Neural Netw. 58 (2014) 82–97. Available from: <https://sciencedirect.com/science/article/pii/s0893608014001099>, 2019 (accessed 06.01.19).

[3] M. Purdy, P. Daugherty, Why artificial intelligence is the future of growth, Accenture, 2016. Available from: <https://www.accenture.com/us-en/_acnmedia/PDF-33/Accenture-Why-AI-is-the-Future-of-Growth.pdf>, 2019 (accessed 06.01.19).

[4] R. MacGregor, Using a description classifier to enhance knowledge representation, IEEE Expert. 6 (3) (2017) 41–46. Available from: <http://ieeexplore.ieee.org/xpl/login.jsp?tp = &arnumber = 87683&url = http%3A%2F%2Fieeexplore.ieee.org%2Fxpls%2Fabs_all.jsp%3Farnumber%3D87683>, 2019 (accessed 05.01.19).

[5] S. Ahmad, Visit: an efficient computational model of human visual attention (Ph.D. thesis), University of Illinois, 1991.

[6] S.L. Keoh, S.S. Kumar, H. Tschofenig, Securing the Internet of Things: a standardization perspective, IEEE Internet Things J. 1 (3) (2014) 265–275. Available from: <http://ieeexplore.ieee.org/xpl/articleDetails.jsp?arnumber = 6817545>.

[7] P.M. Mishra, Internet of Things and Bayesian Networks, 2014. Available from: <https://www.analyticbridge.com/profiles/blogs/internet-of-things-and-bayesian-networks>, 2019 (accessed 24.02.19).

[8] J. Beal, M. Viroli, D. Pianini, F. Damiani, Self-adaptation to device distribution in the Internet of Things, ACM Trans. Auton. Adapt. Syst. 12 (3) (2017).

[9] M. Viroli, J. Beal, F. Damiani, D. Pianini, Efficient engineering of complex self-organizing systems by self-stabilizing fields, in: Proceedings of the IEEE Conference on Self-Adaptive and Self-Organising Systems (SASO'15), IEEE, Piscataway, NJ, 2015, pp. 81–90.

[10] J. Pearl, Causality: Models, Reasoning, and Inference, second ed., Cambridge University Press, New York, 2009.

[11] J. Pearl, An introduction to causal inference, Int. J. Biostatistics 6 (20) (2010). 10 Article 7.

[12] J. Pearl, M. Glymour, N.P. Jewell, Causal Inference in Statistics – A Primer, Wiley, 2016.

[13] J.Y. Halpern, Actual Causality, MIT Press, 2016.

[14] G. Shafer, The Art of Causal Conjecture, MIT Press, 1996.

[15] L. De Raedt, A. Kimmig, H. Toivonen, ProbLog: a probabilistic Prolog and its application in linkdiscovery, in: R. Sangal, H. Mehata, R.K. Bagga (Eds.), Proceedings of the 20th International Joint Conference on Artificial Intelligence (IJCAI'07), Morgan Kaufmann Publishers Inc., San Francisco, CA, 2007, pp. 2468–2473.

[16] D.A.A.N. Fierens, et al., Inference and learning in probabilistic logic programs using weighted Boolean formulas, Theory Pract. Log. Program. 15 (3) (2015) 358–401.

[17] M. Kifer, Y.A. Liu, Declarative Logic Programming: Theory, Systems, and Applications, ACM Books, 2018.

[18] S.M. Molaei, M.R. Keyvanpour, An analytical review for event prediction system on time series, in: Second International Conference on Pattern Recognition and Image (IPRIA), 2015.

[19] S. Triantafillou, I. Tsamardinos, Constraint-based causal discovery from multiple interventions over overlapping variable sets, arXiv: 1403.2150, 2014.

[20] A. Hyttinen, P.O. Hoyer, F. Eberhardt, M. Jarvisalo, Discovering cyclic casual models with latent variables: a general SAT-based procedure, in: Proceedings of the UAI, AUAI Press, 2013, pp. 301–310.

[21] A. Hyttinen, F. Elberhardt, M. Jarvisalo, Constraint-based causal discovery: conflict resolution with answer set programming, in: Proceedings of the UAI, AUAI Press, 2014, pp. 340–349.

[22] A. Hauser, P. Buhlmann, Characterization and greedy learning of interventional Markov equivalence classes of directed acyclic graphs, J. Mach. Learn. Res. 13 (2012) 2409–2464.

[23] M.H. Maathuis, D. Colombo, A generalized backdoor criterion, Ann. Stat. (2015) 1060–1088.

[24] M. Kalisch, M. Machler, D. Colombo, M.H. Maathuis, P. Buhlmann, Causal inference using graphical models with the R package pcalg, J. Stat. Softw. 47 (11) (2012) 1–26.

[25] S. Tikka, Package "Causal effect," 2014.

FURTHER READING

Abbot, 1990 L. Abbot, Learning in neural networks memories, Network: Computation Neural Syst. 1 (1990) 105–122.

Beckers and Vennekens, 2016 S. Beckers, J. Vennekens, A general framework for defining and extending actual causation using CP-logic, Int. J. Approximate Reasoning 77 (2016) 105–126.

Gelfond and Lifschitz, 1993 M. Gelfond, V. Lifschitz, Standing for action and change by logic programs, J. Log. Program. 17 (2–4) (1993) 301–321.

Koller and Friedman, 2009 D. Koller, N. Friedman, Probabilistic Graphical Models, Principles and Techniques, MIT Press, 2009.

Pearl, 2012 J. Pearl, The Do-Calculus Revisited, Keynote Lecture, UAI, 2012.

Traub et al., 2014 K. Traub, et al, The GS1 EPC global Architecture Framework 2, GS1 Version 1.6, 2014.

CHAPTER

EXAMINING THE INTERNET OF THINGS–BASED ELEGANT CULTIVATION TECHNIQUE IN DIGITAL BHARAT

8

Rajeev Kr. Sharma[1], Rupak Sharma[1] and Navin Ahlawat[2]

[1]*SRM University, NCR Campus, Modinagar, India* [2]*SRM University, NCR Campus, Modinagar, India*

8.1 INTRODUCTION

The cultivating and horticultural industry depends on inventive thoughts and innovative headways to help increment yields and better distribute assets. Because of gigantic development in advances, cultivating has bowed away to be more famous and huge. Distinctive devices and procedures are accessible for advancement of cultivating. As indicated by the UN Food and Agriculture Organization, with a specific end goal to bolster the developing populace of the world, the world should create approximate 71% further nourishment in 2051 than it did in 2006 [1]. To fulfill this need, ranchers and horticultural organizations are swinging to the Internet of Things (IoT) for examination and more prominent generation capacities. IoT can assume enormous job in expanding efficiency, acquiring immense worldwide market, thought regarding late patterns of products.

Today numerous horticultural ventures swung to receive IoT innovation for savvy cultivating to improve proficiency, efficiency, worldwide market, and different highlights, for example, least human intercession, time and cost, and so on. The headway in the innovation guarantees that the sensors are getting littler, advanced, and more monetary. The systems are likewise effortlessly opened universally so savvy cultivating can be accomplished with full promise. Concentrating on empowering development in agribusiness, keen cultivating is the response to the issues that this industry is right now confronting. It can be achieved through the mobile phones and IoT devices. The farmers can get the necessary information for his agriculture area.

8.2 INTERNET OF THINGS

IoT is a system of connected computer devices, operating systems and digital devices, objects, animals, or unique service providers with the ability to transmit data to a network without the need of humans. Communication, network communication, and communication policies used by these web-enabled devices are largely dependent on the specific IoT applications installed.

Artificial Intelligence to Solve Pervasive Internet of Things Issues. DOI: https://doi.org/10.1016/B978-0-12-818576-6.00008-3

In addition the IoT is actually a building block that allows for the integration and exchange of data between the physical world and computer systems over offered network infrastructure.

8.2.1 AN INTEGRAL PART OF THE INTERNET OF THINGS

Many people mistakenly view IoT as an independent technology. Surprisingly, the IoT is enabled for the presence of other independent technologies that create the basic elements of IoT.

The basic features that make the IoT a reality are as follows:

- *Hardware*: Make physical objects are responsive and enable them to retrieve data and respond to commands.
- *Software*: Enables data collection, storage, processing, trick, and instruction.
- *Communication infrastructure*: Most importantly it is a communications infrastructure that contains the principles and technologies that enable two virtual objects to exchange data.

8.2.2 NEED OF INTERNET OF THINGS IN AGRICULTURE

From a UN Food and Agriculture Organization survey, global food production should increase 70% by the year 2050 for the people to grow. Farming is the backbone of the human species as it is the main source of food and plays an important role in the economic growth of the country. It also provides huge employment chance for people. Farmers still use traditional methods of agriculture, which results in low yields of crops and fruits. Therefore crop yields can be improved by using automated equipment. There is a need to make science and technology better for agriculture to increase yields. With IoT, we can expect an increase in low-cost production by monitoring soil performance, monitoring temperatures and humidity, rainfall monitoring, fertilizer efficiency, water-storage capacity, and agricultural availability. The arrangement of habitual methods and modern technologies like the IoT and wireless sensor networks (WSN) can lead to the modernization of agriculture [1].

The data is collected by the WSN through the number of sensors and sends it to the core server using wireless protocol. The productivity can be affected by a number of factors. Features include pest and pest control that can be controlled by spraying the right insects and pests and attacking wildlife and birds when the crop is growing. Crop yields are declining due to unexpected rainfall, water shortages, and poor water use.

If we want to accomplish the integration and interoperability in IoT some minimal set of measures must be contented or satisfied and IoT framework to be reliable and dependable. These frameworks extent across the IoT research communities ranging from academic research to organizational research, which focus on integrating things in IoT.

Here we are proposing set of minimal actions to be contented by IoT frameworks for assimilation as IT standard itself is still in growing state. These are thc following:

- *Contract termination*: The IoT system consists of large devices and different communication channels. The merger framework must be sufficiently competent to properly manage the contract finalization. Termination of the contract is the ability of service buyers and service providers for the IoT architectural framework to independently appear without terminating the convention among them.

- *Scalability*: It gives the emerging environment of IoT forecasting and calculations [1]. An effective integrative framework should be efficient and effective enough to support billions of Internet connectivity.
- *Testing simplicity*: The assimilation framework must support relieve of testing and error handling. It must offer support for failures and repair failures, integration tests, component tests, program tests, compliance tests, installation tests, performance and malfunction tests, performance tests, and safety checks.
- *Ease of development*: An integrated IoT structure must provide a path for easy improvement for the developers. The structure must exclude all complications and offer relevant documentation for nondevelopers and developers with basic programming knowledge so that they can simply recognize the contents of the structure.
- *Fault tolerance*: The IoT system must reliable and robust. A smart integrated framework must successfully handle errors as IoT devices can finally switch between offline and offline areas. The structure must provide remedies to temporary errors (network errors, node-level errors, and so on), unauthorized access error, server failure, omission failure (when the server does not receive incoming requests from the client), session error, and so on.
- *Lightweight implementation*: Integration frameworks should have a lightweight overhead both in its development and deployment stage. It should be lightweight and easy to install, uninstall, activate, deactivate, update, versioning, and adaptable.
- *Service coordination*: Service coordination is the orchestration and choreography of services. Service orchestration is the coordination of multiple services by a mediator acting as a centralized component. Service choreography on the other hand, is the chaining of services together to execute a particular transaction. Integration frameworks should support at least either or both to achieve reliability.
- *Inter domain operability*: The framework should further be extensible to support inter domain communication. For example, in a smart car domain, an integration framework should also provide support for communication and interaction with traffic lights, road closure, and so on belonging to a smart-city domain. Regardless of the research community or disparity in research, they all aim to achieve extensibility, flexibility, scalability, design reuse, and implementation reuse. The next subsections will present an overview of some IoT frameworks.

8.2.2.1 Benefits of Using Internet of Things in Agriculture

As in other industries, application of IoT in agriculture promises previously unavailable efficiency, reduction of resources and cost, automation, and data-driven processes. In agriculture, however, these benefits do not act as improvements, but rather the solutions for the whole industry confronting a range of dangerous problems.

1. *Standout efficiency*: As we know that farming is also in competition. Farmers have to grow more products in deteriorating soil, declining land availability, and increasing weather fluctuation. IoT-enabled agriculture allows farmers to monitor their product and conditions in real time. They get insights fast, can predict issues before they happen and make informed decisions on how to avoid them. In addition, IoT solutions in agriculture introduce automation, for example, irrigation based on demand, fertilizer, and harvesting of robots.

2. *Expansion*: By the time we have 8 billion people in the world, 70% of them will be living in cities. IoT-based greenhouses and hydroponic systems enable short food supply chain and should be able to feed these people with fresh fruits and veggies. Smart closed-cycle agricultural systems allow growing food basically everywhere—in supermarkets, on skyscrapers' walls and rooftops, in shipping containers and, of course, in the comfort of everyone's home.
3. *Reduced services*: Many IoT solutions focus on optimizing the use of resources—water, energy, and land reduced resources. Precision farming using IoT relies on the data collected from diverse sensors in the field, which helps farmers accurately allocate just enough resources to within one plant.
4. *Cleaning process*: The same applies to pesticides and fertilizers. Not only do IoT-based systems for precision farming help producers save water and energy and, thus make farming greener, but also significantly scale down on the use of pesticides and fertilizer. This approach allows getting a cleaner and more organic final product compared to traditional agricultural methods.
5. *Agility*: One of the benefits of using IoT in agriculture is the robustness of processes. Thanks to real-time monitoring and prediction systems, farmers can quickly respond to any significant change in weather, humidity, air quality as well as the health of each crop or soil in the field. In the conditions of extreme weather changes, new capabilities help agriculture professionals save the crops.
6. *Improved product quality*: Data-driven agriculture helps to grow more and better products. Using soil and crop sensors, aerial drone monitoring, and farm mapping, farmers better understand detailed dependencies between the conditions and the quality of the crops. Using connected systems, they can recreate the best conditions and increase the nutritional value of the products.

8.2.3 HOW INTERNET OF THINGS WORKS

The IoT environment is made up of web-enabled smart devices that use embedded processors, sensors, and communication devices to collect, send, and process data they receive in their environments. IoT devices share the sensor data they collect by connecting to the IoT gateway or other device where data is transmitted to the cloud for analysis or analysis locally. Sometimes, these devices communicate with other related devices and do something about the information they receive from one another. Most of the work is done by devices without human intervention, or people can interact with devices, for example, setting up, giving them instructions or accessing information. The communications, networking, and communication policies used with these web-enabled devices depend largely on specific IoT applications (Fig. 8.1).

The issues with the smart farming can be resolved through the IoT as it is the most capable and fundamental structures for these issues. The data can be exchange without human collusion over the network in it. It consist of number of building slab which joins software's, lots of sensors, and other electronic process. Basically it makes data more profitable.

Interfacing the unquestionable contraptions and amassing of the data is the true task of IoT. When we take a gander at the IoT framework then IoT is used to hold and brains with statistics and data.

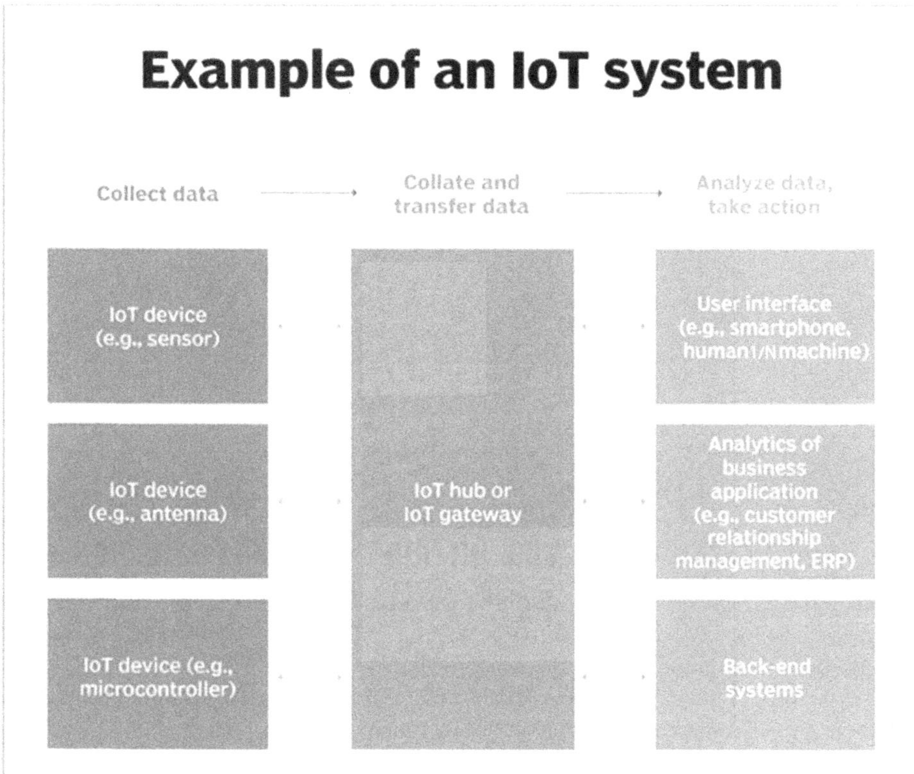

FIGURE 8.1

Example of an IoT system. *IoT*, Internet of Things.

In the structure client can pick their sensors first then data is flow by client so that it can be processed. IoT are giant in a variety of structures of making. Occupations of IoT are town that are smart, elegant atmosphere, neat water, elegant metering, protection and crisis, business control, elegant cultivation, home Automation, e-Health, and so on.

In IoT, we can speak to things with regular way simply like ordinary person, similar to sensor, similar to vehicle driver, and so forth. This thing is doled out an Internet protocol address with the goal that it can exchange information over a system. According to the report created by Garner, toward the finish of 2016 there will be 30% ascent in consider of associated gadgets contrasted with 2015. Garner added that, this check will increment to 26 billion by 2020 [2]. The IoT innovation is further effective because of following grounds:

1. overall connectivity during any devices,
2. least individual undertakings,
3. quicker admittance,
4. time competence, and
5. expert communiqué.

FIGURE 8.2

Smart agriculture component.

8.2.4 FOR WHAT REASON WE REQUIRE INTERNET OF THINGS IN AGRIBUSINESS?

As of examination of UN Food and Cultivation Organizations, the general sustenance age ought to be broadened roughly by 71% of each 2051 for pushing individuals. Development is the clarification behind the individual group as it is the fundamental wellspring of sustenance and it acknowledges basic occupation in the enhancement of nation's wealth. It besides offer enormous bounteous business chances to the comprehensive network. The ranchers are up to this point utilizing normal procedures for developing, that consequences in low acquiescent of things and characteristic things. Thus the immense yield may be enhanced by the use of changed apparatus. Here is need to complete current science and enhancement in the agribusiness for building up the yield.

The enhancement in evolution with immaterial exertion by review the advantage of the earth, temperature, and moistness checking; rainwater fall checking; manures capacity; and watching limit cutoff of water tanks furthermore theft insistence in creating districts.

The combination of standard methods and the latest types of development as IoT and WSN can force the development of innovation. WSN that stores statistics on various sensors and sends them to a primary server using remote culture. There are a number of different factors that influence production at the highest level. The only thing that joins the invasion of the scary little animals and the weeds that can be prevented is by spraying the right prey and pesticides without a strike of wildlife and flying animals as the crop grows. The yield of the item decreases as a result of heavy rainfall, lack of water, and poor water supply (Fig. 8.2).

8.3 ISSUES TO CONSIDER BEFORE DESIGNING YOUR ELEGANT FARMING SOLUTION

8.3.1 WHAT WILL YOUR APPLICATION BE MONITORING?

Farmers, agriculturalists, and modern sustenance makers alike are taking a gander at IoT answers for increment efficiencies and yields and decrease misfortune and burglary. At the end of the day, they are hoping to upgrade assets and lower costs. Whatever the end client will screen ought to be up front as you structure the application.

For instance, a corn farmer might be essentially worried about water use. He would not like to utilize excessively water; however, he likewise should make certain that enough water is getting where it is required. Ongoing observing, then again, can enable a farmer to find a wiped out bovine in the group before it defiles whatever is left of the creatures. Doing this will drastically lessen animals misfortunes, and diminish costs related with acquiring antiinfection agents expected to treat a huge gathering.

8.3.2 THE AMOUNT WIRELESS RANGE IS NECESSARY?

The separation that information needs to travel has an enormous effect on what kind of innovation ought to be utilized. On the off chance that you measure something 10 m away, you would not utilize a similar innovation you would use for something 1500 m away.

For short separations, you may utilize radiofrequency identification (RFID) proof or near field correspondence (NFC), which is normal in mobile phones. NFC or RFID might be utilized in case you are labeling a feedbag and need to know what number of pounds of soybeans are in each pack.

8.3.3 WHERE IS THE POWER SOURCE COMING FROM?

There is an exceptionally close relationship between battery life and range. A sensor that is extremely far away requires more vitality to get data starting with one point then onto the next. To get around that, IoT item makers regularly engineer applications to send considerably less information (or send information less often) to save money on expenses and influence.

Along these lines, you will have to figure out where your sensor application will draw control from. Given that most IoT farming is ordinarily outside or spread over a vast territory, you will have to think about a low power application. The administration, and upkeep of numerous removed sensors will overpower for the end client.

8.3.4 HOW OFTEN DOES THE END USER NEED TO GATHER DATA?

You may believe that the more information parcels a sensor can send the better; however, this is really not the situation. What number of information parcels are fundamental relies upon a wide range of variables, including the end-client application and the neighborhood condition.

For instance, if a farmer has a dampness sensor in a faraway potato field, he likely does not have to assemble data at regular intervals. More than once per day is most likely adequate, which implies the battery life will be far more noteworthy.

This to state: Before you make a machine to machine (M2M) agribusiness application, ensure you think about how much information is excessively.

8.3.5 WHAT TYPE OF SENSORS ARE NECESSARY, AND HOW WILL THEY BE INTERFACED?

Each remote detecting innovation is extraordinary, so you will need to think about which ones you are going to utilize and how you will be interfacing with them before you begin. A few sensors—like dampness sensors—are inserted, and expect microcontrollers to interface. Making the sensor and weatherizing it is a building test that should be met.

8.4 APPLICATIONS OF INTERNET OF THINGS IN AGRICULTURE

1. Agricultural drones

   a. *Drone deploy software*: Drone deploy makes it feasible for anybody to work a little automaton and break down the caught mapping pictures utilizing a PC or cell phone. The organization centers on businesses, for example, horticulture, development, investigation, and protection. With a single tick, clients can dispatch any monetarily accessible automaton on a computerized way to get same-day ethereal maps and 3D field models. The innovation can enable them to see where their products require consideration, gauge yields, and store precise information for correlation after some time.
   b. *SenseFly's eBee*: The little SenseFly eBee ramble is intended to wipe out human blunder from trim exploring. The settled wing unmanned aerial vehicles (UAV) enables agriculturists to review more sections of land quicker, and in addition catch close infrared band information for vegetation assessments. In addition, it is for the most part self-sufficient. Hurl it into the air and it will fly, obtain pictures, and after that land itself. In the wake of flying, it can rapidly produce maps of harvests, distinguish issue regions, redo rural application maps, and make ramble-to-tractor work streams for trim medicines, all around the same time.

2. Precision farming

Precision agriculture, satellite cultivating, or site-specific yield administration is a cultivating administration idea dependent on watching, estimating, and reacting to bury and intrafield fluctuation in crops. It is tied in with making the best choice, in the correct place, in the correct path, at the ideal time. Overseeing crop creation information sources, for example, water, seed, manure, and so on to expand yield, quality, benefit, decrease squander, and progresses toward becoming eco-accommodating.

The determination of access to quick web, phones, and trustworthy, negligible exertion satellites (for imagery and arranging) by the creator are little key advancement depicting the exactness agribusiness incline.

Exactness agribusiness is a hero among the most conspicuous utilizations of IoT in the provincial part and distinctive affiliations are utilizing this structure the world over. Item Measurements is an accuracy agribusiness connection concentrated on ultrapresent-day agronomic strategies while having some ability in the association of exactness water structure.

8.4.1 WATER CONSERVATION AND IRRIGATION

Ongoing advances that can screen soil dampness and water use, and accordingly oversee water costs, are basic. This sort of exactness cultivating can limit waste and deliver brilliant harvests, better profitability, and expanded benefits.

3. Livestock monitoring

As now the world's population increases, there is a more noteworthy interest for nourishment and a flood being developed. With a lessening in the measure of land accessible for homestead utilize and a consistently developing worry about water assets, ranchers should be keen about their yield and domesticated animals' administration on the off chance that they need to diminish waste and cut generally costs. The IoT is making it workable for ranchers and producers to upgrade their harvest yields and advance domesticated animals' wellbeing through remote checking and information driven basic leadership.

The IoT can possibly change cultivating and sustenance generation through enhancing item quality, expanding crop efficiency, supporting in asset protection, and helping rancher's better control costs [3]. Here are only a couple of ways ranchers utilize constant information taken from rural IoT answers for improve trim yields:

- Gather information on soil quality, dampness levels, and climate conditions with the end goal to viably get ready for streamlined gathering.
- Use climate estimates to advance profitability and take safeguard measures to diminish risks of product harm.
- Monitor ecological parameters and plant development to foresee bother conduct and address any pending nuisance issues before they affect crops.
- Analyze and mange water system necessities, and utilize accessible water assets economically to diminish squander [4].

4. Smart greenhouses

The smart greenhouse is an insurgency in agriculture, making an automatic, microclimate reasonable for plant development using sensors, actuators, and checking and control frameworks that streamline development conditions and computerize the developing procedure.

The worldwide ability nursery platform was valued at approximately USD 680.3 million of every 2017 and is necessary to attain around USD 1.31 billion by 2023, developing at a compound annual growth rate (CAGR) of around 14.12% wherever in the range of 2018 and 2023.

The market is relied upon to observe huge development because of expanding populace, environmental change, and urbanization. Keen cultivating is additionally anticipated that would create at a quick rate [4] (Fig. 8.3).

5. Smart cultivating utilizing Internet of Things

Cultivating is an important backbone of India's practical improvement. The most important snag that creates the most common is climate change. Parts of the effects of climate change are extreme weather patterns, a very insane storm and a generous result, a small amount of rain. The most imperative snag that creates in standard creating is change in the weather. The proportion of result of ambiance alters wires tremendous precipitation, most ridiculous tornado and tenderness effect, a lesser amount of precipitation. In context of these the preferred standpoint decays to true

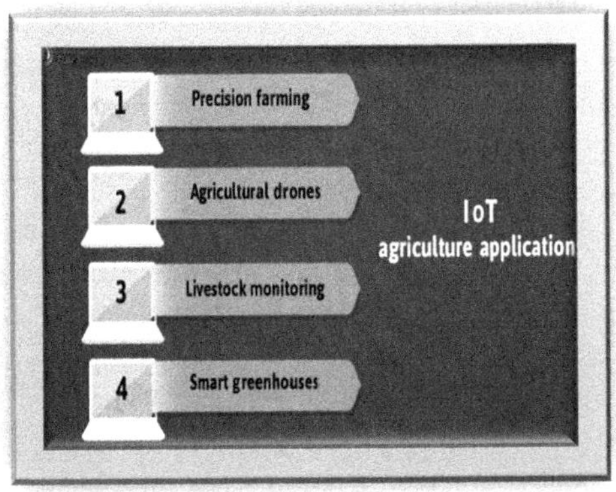

FIGURE 8.3

IoT agriculture applications. *IoT*, Internet of Things.

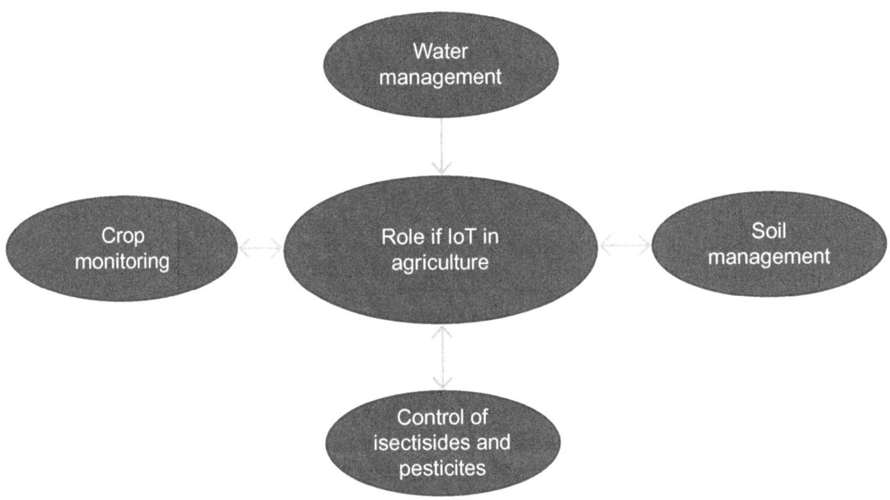

FIGURE 8.4

Roles of IoT in agriculture. *IoT*, Internet of Things.

blue degree. Climatic change other than raises the reliable results, for instance, discontinuous transform in life cycle of plant life. To engage the benefit and most remote point the obstructions in agribusiness field, there is need to use imaginative enhancement and techniques described IoT [3].

These days, the IoT is changing toward improvement diligence and interfacing with agriculturists to measure up to the titanic confront they go up against. Agriculturists can get beast information and data about late points of reference and movement using IoT (Fig. 8.4).

The devices that used IoT can be of awesome assist in improving the generation and capitulate in the agribusiness part since these gadgets can be utilized to screen soil sharpness level, temperature, and different factors. Besides, keen agribusiness will help in checking domesticated animals efficiency and wellbeing too. Farmers can get the information about the crop yields, rain, and bug pervasion through the IoT sensors. So that sufficient action to be taken to handle the same. That information can be utilized to enhance cultivating methods after some time. IoT, with its continuous, precise, and shared attributes, will convey extraordinary changes to the agrarian production network and give a basic innovation to setting up a smooth stream of horticultural coordination [5].

The key central purposes of using IoT in redesigning developing are according to the accompanying:

1. We can save the water through the IoT by using the sensors. This will ensure that no water will be wastage.
2. IoT is working to close the world tightly with the intention that special measures can be taken at the earliest opportunity.
3. Outputs are generated, reduce the hard work of employees, shorten the time, and make the growers more skilled.
4. An item inspection can be completed successfully to check the yield.
5. Soil mixing, for example, PH quality, poor content, and so on can be easily identified by the target that the farmer can plant the seed as indicated by the soil area.
6. Contamination in plants can be seen on the chip as well. RFID names recommend data using and integrated on the web [6]. At that point a technical analyst can obtain this data from a single location and make an important focus. Our crop can therefore be protected from future diseases [7].
7. Collect arrangements will be stretched out in by and large market. Agriculturist can beyond question related with the general market without obstacle of any land zone.

8.5 WHY WILL INTERNET OF THINGS?

- As the telecommunications sector grows more and more efficient, the broadband Internet becomes more widely available. With advances in technology it is now much cheaper to produce the necessary sensors with built-in WiFi capabilities that make connecting devices more expensive.
- Most importantly, smartphone usage has exceeded all predicted limits and the telecommunications industry is already working on their toes to keep their customers satisfied by improving their infrastructure. Since IoT devices do not require different connections than the existing IoT tech is cheaper and more accessible.

8.6 FUTURE OF INTERNET OF THINGS IN AGRICULTURE (SMART OR ELEGANT FARMING)

"Smart or elegant farming" is a promising thought that focuses on farm management using technologies such as IoT, robots, drones, and AI to maximize the value and quality of products while performing labor-demanding tasks. The following are the technologies available for farmers:

- *Sensors*: The things that comes are quality of earth or soil, water, daylight, moisture, and temperature of the area.
- *Software*: Some specific software resolution that focus on the specific farm types.
- *Communication*: Cell phones and others.
- *Location*: Through GPS, satellite, and so on.
- *Robots*: Independent, functional tractors, and so on.
- *Data analytics*: It is used for the integration of the various machines together for the use of technology so that the relevant information can be driven. The condition of the field can be managed without going to the physical location. The decision can be taken about the field at single place.

8.6.1 THE INTERNET OF THINGS–BASED SMART FARMING CYCLE

The heart of IoT is the data that can be drawn from objects or things (T) and put out to the Internet (I). To extend the agricultural process, the collected data must be processed through the technical devices. The recursive cycle that gives farmers the chance to quickly come out on emerging issues and changes in changing conditions. Cycle that is followed by elegant farming is as follows:

1. Recognition
   Sensors evidence sensitive information from plants, livestock, soil, or the atmosphere.
2. Diagnosis
   Sensor values are provided on a cloud-managed IoT platform with predetermined rules and decisions—also called "business logic"—to verify the status of the object being examined and identify any deficiencies or requirements.
3. Decisions
   After the disclosure of the problems, the user, and/or components operated by the IoT machine learning equipment decide if a site-specific treatment is needed and if so, what.
4. Action
   After checking and using the end user, the cycle repeats from the beginning.

8.7 CONCLUSION

Consequently, the IoT provincial applications are making it serviceable for agriculturists and farmers to assemble noteworthy data. Immense landowners and little farmers must understand the capacity of IoT exhibit for cultivating by acquainting splendid headways with extends force and viability in their manifestations. The enthusiasm for creating masses can be successfully met if the agriculturists and little farmers complete plant IoT courses of action productively.

Cultivating will accept basic employment in next couple of years in country. Thusly we require the splendid Cultivating. IoT will update sharp developing. IoT effort in different spaces of cultivating to enhance time capability, water organization, alter checking, soil organization, management of bug splashes, insect repellent, and so on. It in like manner limits human undertakings, adjusts techniques of developing, and expands shrewd developing. Close by these features bright cultivating can develop the bazaar for agriculturists with solitary touch and least endeavors.

REFERENCES

[1] <http://www.businessinsider.com/webof-things-shrewdfarming2016-10?IR = T>.

[2] C. Jim, The evolution of the Internet of Things, White Paper, Texas Instruments, September 2013.

[3] N. Gondchawar, R.S. Kawitkar, IOT based smart agriculture, Int. J. Adv. Res. Comp. Commun. Eng. 5 (6) (2016) 2278–1021.

[4] P. Rajalakshmi, S. Devi Mahalakshmi, IOT based crop-field monitoring and irrigation automation, Tenth International Meeting on Intelligent Frameworks and Control (ISCO), January 7–8, 2016 distributed in IEEE Xplore, November 2016.

[5] X. Wang, N. Liu, The use of web of things in agrarian methods for creation store network administration, J. Chem. Pharm. Res. 6 (7) (2014) 2304–2310. ISSN: 0975-7384, 2014.

[6] T. Baranwal, N.P.K. Pateriya, Improvement of IOT based smart security and monitoring devices for agriculture, Sixth International Conference - Cloud System and Big Data Engineering, IEEE, 2016, 978-1-4673–8203-8/16.

[7] D. Jain, P. Venkata Krishna, V. Saritha, A study on Internet of Things based applications, 2012. <https://www.researchgate.net/publication/227172798_A_Study_on_Internet_of_Things_based_Applications/link/5721a9e508ae0926eb45c4f9/download>.

CHAPTER

MACHINE LEARNING AND INTERNET OF THINGS FOR SMART PROCESSING

9

Ramgopal Kashyap

Amity School of Engineering and Technology, Amity University Chhattisgarh, Raipur, India

9.1 INTRODUCTION

Examination of the actual fluctuation test was performed on F1 scores to research the impacts of these three components. The outcomes show that recurrence groups and spatial channels rely upon one another. The blends straightforwardly influence the F1 scores, so they must be picked cautiously. The consequences can utilize as rules for brain-computer interface (BCI) specialists to viably plan a preprocessing technique for a progressed nonconcurrent BCI framework, which can help the stroke restoration. Over the previous decade, numerous calculations have proposed to distinguish the development goal before actual development execution; in any case, it is as yet indistinct about how to channel or clean the procured flags before taking care of them to a classifier. For instance, strategies using coordinated channel and territory affectability discriminant investigation to recognize development goals. However, they experienced troublesome decisions in choosing both a fitting recurrence band and a spatial sifting procedure, rather than finding a proper arrangement, the decisions made subjectively.

Albeit a few specialists knew about these issues and endeavored to investigate a fitting setup or mix of calculation strategies for the development aim or the development of creative mind discovery in the BCI framework, none of them studied these issues thoroughly. The most efficient investigation where mixes among three variables of the spatial channel, transient channel, and classifier were broke down. In any case, the work just spotlights on development goal grouping among left and right hand, not on development goal recognition. Additionally, electrocardiogram (EEG) information in time-space yet besides information utilize as it very well may be a deterrent for constant use. The creators considered the impacts of recurrence and spatial channels on the unexpected harmful variety. Even though their investigations were orderly and concerned numerous viewpoints, they did not concentrate on self-guided examinations.

The impacts of classifiers were likewise not referenced in their work except direct segregation examination. Besides, their outcomes are constrained to just disconnected utilization because of the deferral of the recurrence channel. For different tasks, tests under specific requirements (e.g., a preset classifier, a preset recurrence band, or a preset spatial channel). These circumstances could cause issues; for instance, transforming one factor may influence different factors as will be appeared in our investigation. Likewise, the information not procured in a self-guided way, nor the

Artificial Intelligence to Solve Pervasive Internet of Things Issues. DOI: https://doi.org/10.1016/B978-0-12-818576-6.00009-5

recognition cannot utilize progressively, which makes the outcome unusable for development expectation identification issues [1].

Right now, make the primary endeavor to contemplate impacts and relationship among the three elements recurrence groups, spatial sifting strategies, and classifiers, explicitly on EEG information obtained in a self-guided way for development aim identification. In detail, we recorded 19-channel EEG information from nine subjects while playing out a progression of self-managed lower leg dorsiflexions. This investigation fundamentally centers around discovering the relationship and impacts of different variables and on deciding great blends among them. The outcomes give significant bits of knowledge toward advancement in a compelling development expectation identification calculation for offbeat BCI restoration frameworks. Moreover, a novel shape-based layout coordinating count is added to our examination to research a chance of using a state of utilizing time arrangement mining methods in BCI applications. The analysis gives an understanding that can utilize to help improve execution and productivity of development aim recognition in offbeat BCI frameworks. Commitments and effect of this work can abridge as follows: (1) this work explores the relationship and impacts of the three primary factors in nonconcurrent BCI frameworks for stroke recovery. The outcomes exhibit that these elements are very delicate. Various blends of the ingredients can considerably impact the framework's performance. (2) This work can be utilized as crude rules to process self-guided EEG information in an offbeat BCI framework, which can help stroke rehabilitation. (3) This work investigates a chance of using states of time arrangement signs to identify self-guided EEG information utilizing unusual time arrangement mining procedures.

9.2 INFORMATION LABELING AND INFORMATION SEGMENTATION

After the EEG information securing, every time of the development execution must name as a standard for grouping. Electromyogram (EMG) information used right now; was outwardly assessed to find development onsets and balances. Just preliminaries whose EMG development is beginning at any rate 4 seconds separated from the past balance considered. Recorded crude information and their mark made freely accessible, as expressed in the data availability area with the goal that the trial right now repeated. The report physically fragmented for grouping. For development expectation, fragments removed a gathering beginning before each positive development and consummation after the construction. Machine learning, web on items, and massive data are dependent on each other while finishing the first movement, as found in Fig. 9.1.

Further clarifications and instances of divided details can download through the connection portrayed in the data availability area [2]. EEG is an electrophysiological strategy to record voltage vacillations inside the neurons of the mind. In any case, these voltage variances are little to such an extent that the EEG signal is effectively defiled by commotion or even electric fields from other cerebrum districts, bringing about a low sign to clamor proportion of the procured EEG. Numerous spatial sifting methods have acquainted with emphasizing confined EEG exercises to augment the sign to clamor proportion. Right now, we depict a few subtleties of the generally utilized spatial separating methods that have been applied in our analyses to decrease clamors or to upgrade the nature of EEG signals. The strategy used for building another portrayal from high-dimensional information to lower-dimensional information whose difference expanded between two classes of

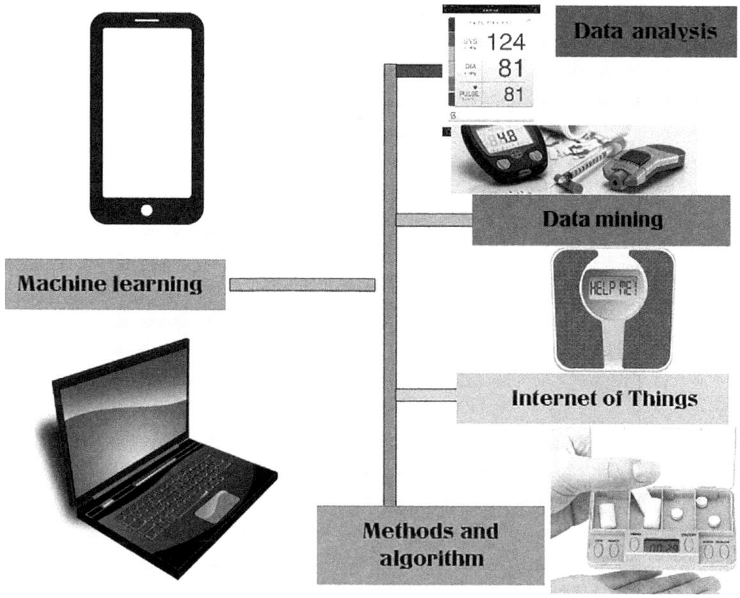

FIGURE 9.1

Machine learning interoperability.

data. In EEG space, contributions of two types of multichannel EEG information computes a projection network that utilized to extend multichannel knowledge into a low-dimensional spatial subspace by a straight change. The first and the last column of the projection network are generally reasonable for separating two unique classes. They give a move that expands a fluctuation in one type and limits a difference in another. Right now, the first line of the projection lattice was used as a load for multichannel EEG to make another portrayal as acted.

9.3 MACHINE LEARNING TOOLS

9.3.1 PRINCIPAL COMPONENT ANALYSIS

Principal component analysis (PCA) has utilized as a dimensionality decrease procedure for most information excavators. In any case, PCA can likewise be used in information pressure as it can bring out solid examples from complex information while protecting inconstancy. PCA ventures the first information onto another arrangement of tomahawks using its eigenvector, which is called the ruling segment. PCA can likewise utilize to disintegrate unique information into another method of decorrelated information by anticipating the information onto symmetrical tomahawks. Subtleties of PCA and machine learning results shown in Fig. 9.2. Its standard, computation, and

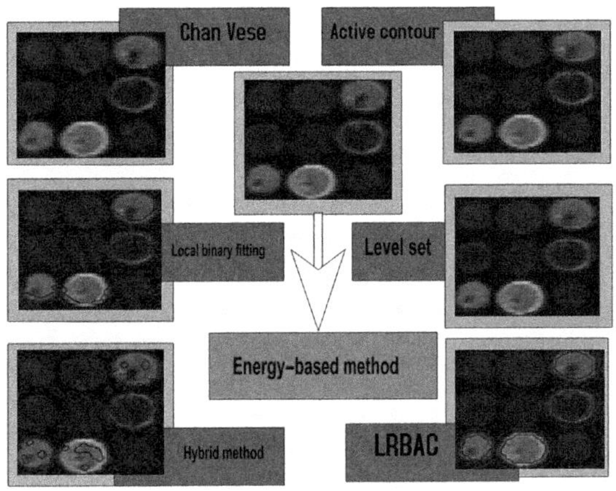

FIGURE 9.2

Machine learning methods case intelligent result comparison.

application all around depicted. Right now, disintegrated multichannel EEG information into free segments and afterward utilized.

9.3.2 CLASSIFICATION AND CLUSTERING

The most generally used order calculations in BCI frameworks because of its straightforwardness and effectiveness. It depends on the utilization of hyperplanes to isolate various classes of information. A hyperplane can be controlled by limiting the intraclass difference and expanding the separation between the methods for two categories of data. Support vector machine (SVM) depends on the thought of hyperplane development, which gives the most significant edge that can utilize to isolate the information. Something else, a mapping of highlights to a higher-dimensional space, is done using a part work before a hyperplane development. Right now, the SVM works in MATLAB and performed improvement on all parameters using a framework search with default settings. A separation measure used to quantify similitude between double cross arrangements, right now, ages of EEG information. A straightforward separation measure is a Euclidean separation, and a progressively perplexing one is dynamic time traveling. The comprehensive limitation set to forestalling nonsensical twist an extensive factual examination on the execution of various recurrence groups, spatial channels, and ordering strategies to uncover impacts and relationships among these elements. The exhibition was broke down dependent on F1 scores. The outcomes indicated that the classifier is an autonomous factor, while the recurrence and the spatial channel are needy components. In this way, the repetition and the spatial channel must consider at the same time. These outcomes can be utilized as rules to look into identification, mainly on lower appendage development. A part space semisupervised fluffy C-implies bunching division calculation joining using neighborhood spatial dark data with fuzzy participation data proposed right now. The mean force data of the neighborhood window inserted into the target capacity of the current semisupervised fluffy C-implies bunching, and the Lagrange multiplier technique to acquire its iterative articulation is comparing to the iterative arrangement of the streamlining issue. In the meantime, the neighborhood Gaussian bit work utilized to outline pixel tests from the Euclidean space to the high-dimensional component space with the goal that the group versatility to various kinds of picture division upgraded [3]. Test results performed on several types of boisterous pictures demonstrate that the proposed division calculation can accomplish preferable division execution over the current regular powerful fluffy bunching calculations and altogether improve the antinoise performance shown in Fig. 9.3.

Fuzzy C means (FCM) is an unaided grouping strategy, which broadly utilized in various applications. Indeed, it is a significant device in multiple fields, including information mining, AI, and picture investigation. FCM has a primary iterative usage method, quick union rate, and low stockpiling necessities. Be that as it may, it is a test to take care of the picture division issue viably through direct use of the current FCM calculation and bunching the pixels of the full pictures. Such an impediment is, for the most part, because the current FCM calculation does not think about the high connection between the present pixel and its local pixels. To take care of this issue, initially proposed a hearty FCM calculation with spatial data limitation for clinical picture division, yet it has high time cost. Afterward, numerous scientists proposed a progression of improved quick, vigorous FCM calculations utilizing nearby and nonlocal sifted data, insufficient recreation data of neighborhood window. The deficiency of these hearty FCM calculations is that they cannot naturally decide spatial data requirement parameters.

Furthermore, the limitation parameter of vigorous fluffy bunching with spatial imperative built by fluctuation or deviation of neighborhood pixels in writing gave a versatile memetic fluffy

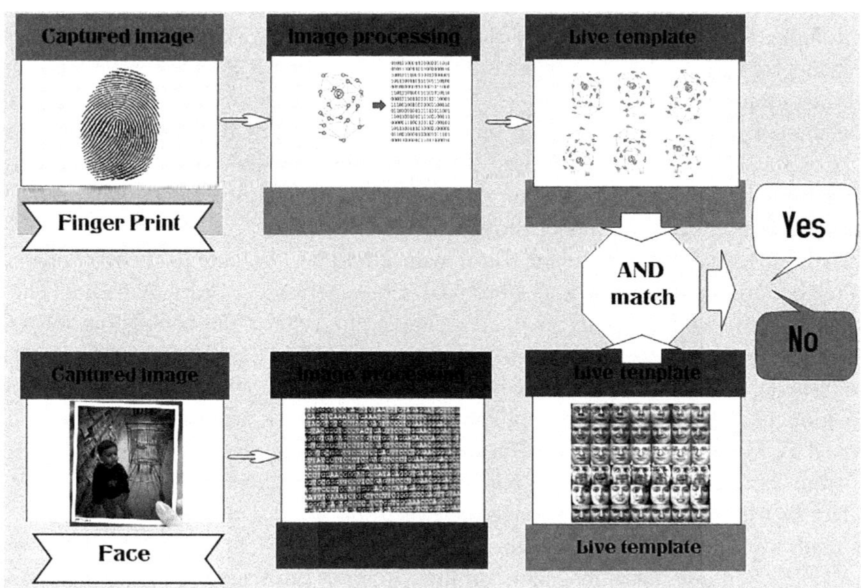

FIGURE 9.3

Machine learning model.

bunching with segment entropy for remote detecting picture division. These handy fluffy grouping calculations with spatial imperatives extraordinarily advance the quick improvement of the strong fluffy bunching division hypothesis, yet their capacity to smother commotion is frail. Taking into account that hearty fluffy bunching calculation with spatial dim data imperatives has a high computational time cost, a robust fluffy bunching calculation with spatial fluffy enrollment degree data limitations. Additionally, proposed a contingent spatial fluffy bunching for clinical picture division. In numerous hearty fluffy grouping calculations with fluffy data limitations, Pham right off the bat offered a vigorous regularized fluffy bunching division calculation, and its grouping target work implanted with neighborhood enrollment data. Afterward, additionally improved a robust regularized fluffy bunching calculation with neighborhood enrollment requirements. A kernelized fluffy weighted nearby data grouping calculation, and it has more grounded commotion concealment capacity than the vigorous division calculation. Afterward, numerous researchers have talked about a progression of improved fluffy weighted nearby data grouping calculations, and their primary contrasts reflected in the weighting coefficient development strategies for fuzzy neighborhood data. These explore significantly advance the improvement of fluffy nearby data grouping calculation [4]. Yet, their calculations require high time cost and are not so much appropriate for the continuous division of enormous scope pictures.

In this way, a gathering of researchers has proposed the utilization of bionic likelihood improvement techniques, including the hereditary calculation, to acquire rough ideal grouping results. In addition, other researchers have proposed a semisupervised fluffy grouping calculation that utilizes incomplete earlier data for bunching information tests to improve the presentation of the solo fluffy bunching. Semisupervised improvement of the example enrollment degree requirement dependent on the current FCM calculation. The use of grouping earlier information in the FCM calculation commonly founded on the adjustment of the bunching target work or the bunching procedure. Given the change of target work, a semisupervised fluffy grouping calculation with the enrollment deviation. In his proposed count, imperatives are limited by presenting the bunching procedure, and a fitting punishment work is applied to advance the participation of the managed data limitation to the known classification. The semisupervised fluffy grouping calculation has effectively used in the information examination, shape explanation, remote detecting picture division, and picture change location, and Big Data processing model shown in Fig. 9.4.

Moreover, they acquainted a standard factor with a bargain the extent of the regulated and the solo pieces of the target work and got an improved semisupervised FCM calculation. The common term factor as the believability of the marked examples to assume the controlling job of the case earlier data on the grouping. In addition, a semisupervised FCM calculation that can successfully manage the issue that the quantity of tests is littler than the number of bunches. A semisupervised FCM calculation is dependent on the replicating piece Hilbert space. The previously mentioned semisupervised FCM calculations depend on the alteration of target work. They give new plans to the change from the old-style solo mode bunching to the semisupervised mode grouping. In addition to the fact that these transitions improve the execution of thc current FCM grouping, yet additionally, it tends to apply in various fields, for example, the picture division, remote detecting picture change recognition, and the shortcoming finding. Embraced the semisupervised fluffy bunching for entangled clinical dental picture division and made extraordinary progress right now.

Notwithstanding, the calculation is convoluted and requires false parameter choice when it applied in the intuitive picture division in complex circumstances. The strength and adequacy of

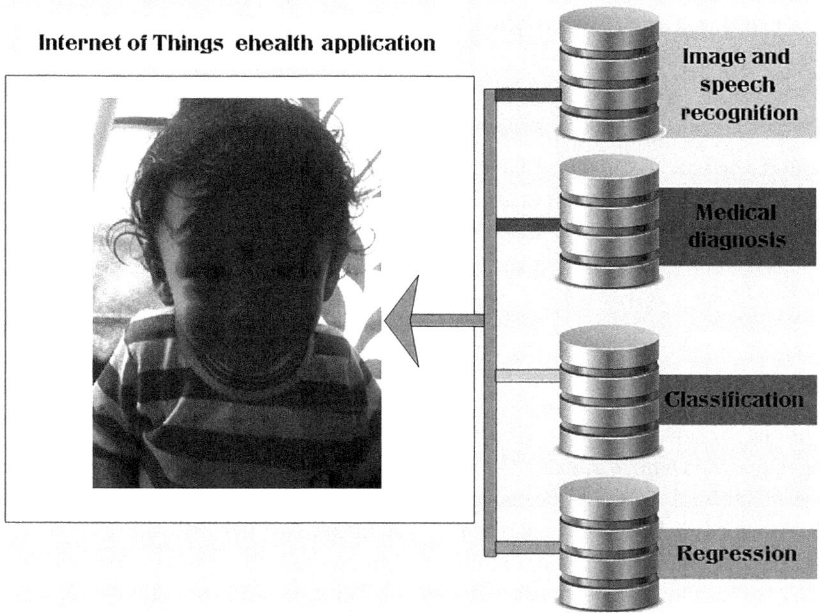

FIGURE 9.4

Big Data processing model.

the fluffy grouping to take care of complex picture division issues, a few researchers consolidated the spatial dark and the fuzzy enrollment data. In any case, the acquaintance of the uniqueness leads to the point that there is power activity existing in the computation of participation, which brings about high time cost. The difference-based semisupervised fluffy bunching is hard to pick the customary parameters of disparity thing successfully. In a couple of words, these endeavors enormously advanced the quick improvement of the semisupervised fluffy bunching division hypothesis. The procedure moves toward the ideal grouping arrangement by iterative strategy. In the interim, the standard requirement term of neighborhood data implanted in the grouping target capacity of semisupervised fluffy bunching, which is relied upon to improve the ability of the fluffy grouping calculation to smother clamor [5]. In addition, consider using the neighborhood Gaussian piece capacity of the replicating bit Hilbert space to delineate examples acquired from the European area to the high-dimensional component space. At that point, a semisupervised fluffy neighborhood C implies grouping calculation for the bit space will propose to take care of the issue of productive characterization of straightly indistinguishable example sets in high-dimensional component space. A division trial of various kinds of methods shown in Table 9.1 those are standard, clinical, and remote detecting pictures perform.

Limit pixel obscure and class irregularity are underlying issues that happen during the semantic division of urban remote detecting pictures. Roused by DenseU-Net, this paper proposes another start to finish organize SiameseDenseU-Net. To start with, the system all the while utilizes both genuine orthophoto pictures and their relating standardized computerized surface model as the

Table 9.1 Overview of Time-Utilized Machine Learning Algorithms for Shrewd Information Examination.

S.N.	Machine learning algorithm	Data processing tasks	ROC curve	Accuracy	Training time (s)
1.	K-nearest neighbors	Classification	0.912	78.6	0.38
2.	Naive Bayes	Classification	0.965	90.7	0.04
3.	Support vector machine	Classification	0.953	86.2	0.13
6.	Classification and regression trees	Classification/Regression	0.911	90.9	0.02
7.	Random forests	Classification/Regression	0.989	91.5	0.48
8.	Bagging	Classification/Regression	0.875	89.4	0.35

contribution of the system structure. The profound picture highlights are separated in equal by downsampling squares. Data, for example, shallow surfaces and elevated-level unique semantic highlights, are melded all through the associated channels. The highlights separated by the two equal handling chains. At long last, a softmax layer utilized to perform the expectation to create thick mark maps. Examinations on the Vaihingen dataset show that SiameseDenseU-Net improves the F1-score by 8.2% and 7.63% contrasted and the Hourglass-ShapeNetwork (HSN) model and with the U-Net model. As to limit pixels, when utilizing a similar spotlight shortfall work dependent on middle recurrence balanced weighting, contrasted and the first DenseU-Net, the little objective "vehicle" class F1-score of SiameseDenseU-Net improved by 0.92%. The general precision and the normal F1-score likewise improved to shifting degrees. The proposed SiameseDenseU-Net is better at recognizing little objective classes and limit pixels, and it is numerically and outwardly better than the differentiation model.

9.4 COMPUTER VISION AND NEURAL NETWORK

In the computer vision field, the semantic division is a significant issue in the previous hardly any decades, numerous exemplary customary division calculations have risen, including locale-based strategies, watershed calculations, edge techniques, and group-based division techniques. In functional applications, high-goals pictures are hard to computerize for two reasons: first, their spatial goals are higher. However, their otherworldly goals are lower; second, the surface highlights of molecular targets become noticeable. These two variables lead to an expansion in intraclass inconstancy in the picture, while the contrasts between classes decline. Picture semantic division means to decide the most proposed class name for every pixel in an image drawn from a lot of predefined restricted marks. In 2012, the AlexNet organized caused another upsurge in imaging applications in the field of profound learning. A convolutional neural system (CNN) with an utterly associated contingent irregular field way to deal with getting familiar with the lost earlier data. In the information combination rivalry in 2015, a CNN model as a component extractor to order land spread utilized genuine orthophoto pictures, a comparing advanced surface model picture, and a standardized computerized surface model picture to prepare a moderately little arrangement of CNN models. At long last, the outcomes additionally advanced utilizing contingent irregular field, a completely convolutional organize to arrange images at the pixel level. Dissimilar to the exemplary CNN, FCN can acknowledge an info picture of any estimate and reestablish it to a similar size as the information picture, in this way creating an expectation for every pixel while holding the spatial data in the first information picture. A CNN-based framework depends on downsample-then-upsample engineering so that CNN figures out how to thickly mark each pixel at the primary goals of the image. A design for a semantic pixel-wise division named SegNet that wipes out the need to figure out how to upsample. The upsampled maps are convolved with trainable channels to deliver thick element maps. In 2018, weighted harmony work and a neural system dependent on a multifeature pyramid structure. Three new personality mapping secure route associations between the even encoder-decoder sets. This methodology guarantees that the angle data can be passed legitimately to the upper layers of the system. The novel regulated profound CNN-based structure to manage fragmented superpixels and presented two veil approaches for arranging models. DeepLabv3 + applied depth-wise separable convolution to both the decoder modules, and included Atrous Spatial Pyramid Pooling, bringing about a quicker and more grounded encoder-decoder organize. The TreeUNet model was the first to utilize both a ready order and profound neural systems in a bound together intelligent learning structure. TreeSegNet embraces a versatile method to build the characterization rate at the pixel-wise level.

Through broad research on satellite remote detecting pictures, scientists have discovered that high-goals remote detecting pictures have lower otherworldly goals than low-goals remote detecting pictures. The three RGB channels are accessible, and class data completely caught high-goals remote detecting pictures, examining surface and the spatial setting is especially significant. Numerous examinations have concentrated on separating highlights from pixel spatial neighborhoods [6]. The semantic division task for high-goals remote detecting pictures intended to foresee every pixel as classification from a predefined set of semantic classifications, for example, structures, low vegetation, trees, or vehicles. Opportune access to precise division results necessary for undertakings, for example, urban arranging, ecological checking, and financial anticipating. As showed up in Fig. 9.5, how to show base and case base models are accepting an essential part in machine learning.

In the previous barely any decades, countless factual strategies dependent on ghostly highlights, including the most extreme probability technique and the K-implies strategy, just as machine learning-based strategies, for example, neural systems, the SVM, object-arranged characterization, and meager portrayal, have been broadly utilized in remote detecting picture division assignments. In any case, these shallow system techniques regularly neglect to think about the interrelationships among worldwide and neighborhood tests satisfactorily. Lately, profound learning techniques, particularly convolutional neural systems, have performed well on visual learning undertakings. An intelligent system accepts the first picture as information. It changes the chart through numerous handling layers by collecting the highlights to the continuously expanding setting neighborhood, and the data turns out to be increasingly express along these lines accomplishing a differentiation between various article classes. The parameter set for the whole system model found out from the first information and labels, including the underlying layer containing the first highlights, the center layer containing explicit errand setting data, and the elevated level that plays out the natural grouping. The remote detecting picture semantic division undertaking can portray as follows: given a lot of marked preparing informational collections, the classifier takes in prescient contingent likelihood

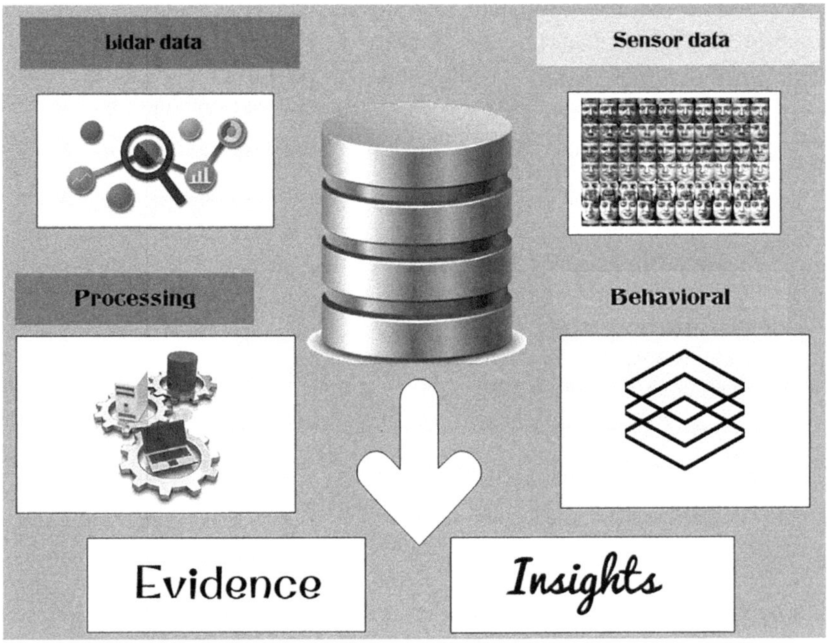

FIGURE 9.5

Model-based machine learning.

number crunching from the ghostly highlights. The first-pixel force, an essential blend of crude qualities, and different sorts of factual data depicting the nearby picture surface are average decisions for input highlights. Another typical strategy is to precalculate countless excess capabilities for preparing and afterward let the classifier select the ideal subset. Right now, essential data can overlook during the element encoding process. It utilizes beginning and lingering modules [7]. The initiation module empowers the system to extricate data from multiscale responsive regions. Leftover modules are used together with the skip association, taking care of data forward from the encoder legitimately to the decoder to utilize the spatial data.

9.5 CHALLENGES AND SOLUTIONS

The Internet of Things (IoT) thought rises out of the need to manage, motorize, and examine all weird-looking machines, tools, or objects used to do work or measure something, also, sensors on the planet. Altogether, to make smart decisions both for people and for the things in IoT, data mining developments are joined together with IoT developments or increases over time for significant authority support and structure change. Data mining incorporates discovering novel, exciting, and possibly productive vital things that happened before from data, what's more, applying to figure's to the extraction of covered information. The information mining in three particular viewpoints:

data, method, and application. In data, the review arranges packing, connection examination, a time course of action examination, and a weird-quality examination. In use, survey the keep running of the process data mining application, including electronic business, industry, richness care, and open organization. The method is discussed with a learning point of view, also, the app. Now massive data is a new issue for data mining and IoT, Fig. 9.6 showing the problems with machine learning.

In usual done information act of asking questions and trying to find the truth about something, a model would be based on past information and master feeling to set up a connection between the factors. Be that as it may, machine learning begins with the result factors and after that, as a result using indicator factors and their connections. A famous model is Google's happening now use of machine learning on their server farm cooling invention of new things to keep up related to surrounding conditions reasonable for their server's job. To expand energy effectiveness, Google connected machine learning and cut its general energy use by 15%. It speaks to a vast number of dollars in investment money for Google in the coming years. Smart about the future abilities to hold or do something is, to a high degree, valuable in a modern setting. By illustration information from different sensors in or on machines, machine learning calculations can "understand" what's familiar and weird and unexpected conduct for a computer, sense soil dampness, and adds to additions in farming, oversee smart homes, control wearable's, change social insurance, and et cetera. The billions of sensors and gadgets that will keep on connected with the Internet in the up and this huge small step forward/upward in the report will control amazing upgrades in machine picking up; opening cannot be counted or figured out chances for us to receive the rewards.

Improving IoT with AI can also honestly make new items and groups of managers; natural language processing is showing signs of improvement and better at giving people a chance to talk

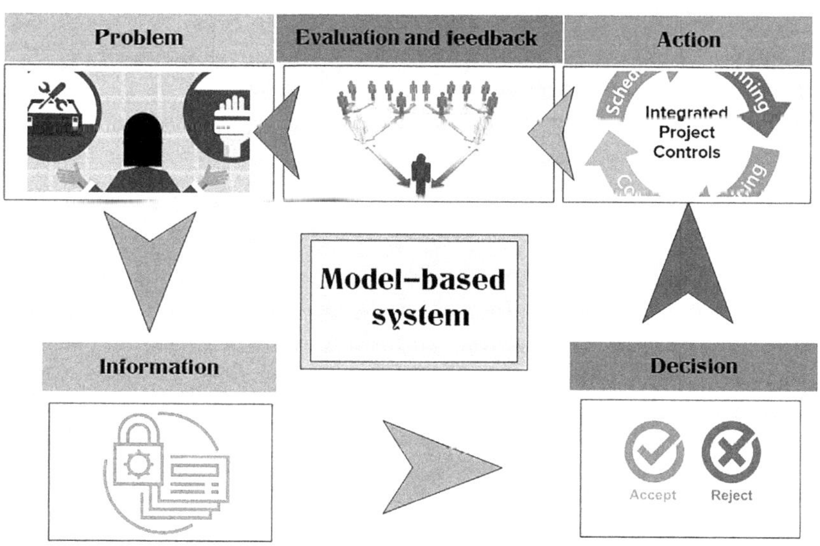

FIGURE 9.6

Challenges with machine learning.

with machines, instead of needing a human manager. Artificial intelligence (AI) controlled robots and robots can go where people cannot bring every single new open door for watching and examination that just did not exist before. Cloudera claims its ship-related management AI has cut downtime for ship-related vehicles, followed by Navistar gadgets up to 40%. Different applications matching IoT with AI are helping associations better understand and predict a mixed group of dangers and also mechanize for quick reaction, give power to them to all the more likely oversee laborer security, money-related lousy luck, and digital risks.

The information collected over time could be produced from different out into different changes to flow on a different path to point to, or focus on, something else of information sources in both organized and without rules, schedules, etc. arrangements. For human services associations, the information may ask of now be easy to get to, use, or understand in information lakes or information distribution centers. Anyway, it needs information extraction and stacking from source setup to target organizes as a part of an information-gathering stage [8]. The information still may not be a wrong organization for building the smart about future models. In such a case, information getting can act to construct the news all the more precisely. It is the essential stage to build the expectation models by choosing and applying a clearly stated machine learning calculation by making as mart about the future model. The datasets separated into preparing and testing datasets. The preparation information used to make the model; the other part of the new data used to check to decide the execution test of the smart about the future model. The testing models can be repeating gone through different importance and focus given to things with outfit machine learning calculation to keep away from underfitting and overfitting and wiping out exceptions and evaluate the machine learning calculation that fits perfectly to put together the expectation display. The model can send once the best fit for the forecast model and execution evaluation finish. Not prevented by part of the issue, there is reusability of the forecast shows or proves over many divisions of medicinal services associations or other airplane business associations. Such reusability of forecast display needs arrangement through web management and computer file full of information over the association all through the country or over the globe.

The IoT has gotten advancements in the everyday lives of a client; the eagerness and receptiveness of shoppers have fueled the producers to dole out new gadgets with more highlights and better style. While trying to stay aware of the challenge, the producers are not giving enough consideration to the digital security of these savvy gadgets. The gravity of security vulnerabilities is additionally bothered because of their associated nature. Subsequently, an undermined device would quit giving the planned help, yet could likewise go about as a host for malware presented by an assailant. The examined the security issues which have brought up before and the correspondence conventions utilized by gadgets made by these brands. It discovered that while security vulnerabilities could be acquainted due to the absence of consideration regarding subtleties while planning an IoT gadget, they could likewise get presented by the convention stack and insufficient framework design. Scientists have emphasized that conventions like transmission control protocol (TCP), user datagram protocol (UDP), and domain name systems (DNS) have natural security inadequacies, and producers should be aware of the reality.

Moreover, if assemblies like EAPOL or Zigbee have been utilized, at that point, gadget engineers should know about shielding the keys and other validation systems. It additionally broke down the parcels caught during the arrangement of 23 gadgets by the previously mentioned makers. The examination gave us an understanding of the fundamental convention stack favored by the makers. This investigation planned to decide whether a producer could be recognized dependent on the tokenized conventions. The demonstrated classifier could then utilize to drive a calculation to the plan against conceivable security vulnerabilities, which are typical for the meetings and the maker history. Such a robotized framework will be instrumental in the customary diagnostics of a savvy structure proposing a few estimates a client can take to ensure their neighborhood organizes and associated gadgets.

The motivation behind why the savvy machine is turning into an open field for programmers is because the makers progressively focused on conveying new items at a quicker pace. While the new things gloat of better proficiency and more highlights, the makers may wind up giving less consideration to the security part of the items. Additionally, with customary firmware refreshes, it is conceivable that some new code option could present a bug which was absent previously. It ought to likewise notice that while the significant onus is on the makers, it is additionally critical to upgrade the security of correspondence conventions utilized by the gadgets. The correspondence conventions are used either in their freely distributed structure or adjusted by the producers to suit their necessities. Hence, it is essential to thoroughly test the gadgets to check for any weakness which could get abused by somebody sniffing and examining the gadget transmissions. Right now, security bargains which hit the features and how the makers dealt with them. The basic pattern followed by the producers is to build up a healing patch and offer it with the current clients through a firmware update. While this methodology lets the makers free, the component of firmware refreshes is not stable. Since various gadgets not arranged for programmed refreshes, they will, in general, pass up the security fixes and make them exposed targets for programmers. Likewise, there are various conditionals for a fruitful firmware update, for example, power supply necessity during the update, or a nonstop web network. In this manner, they are making the client liable for verifying their gadgets and the home system.

The experiment has likewise broken down the transmissions of 23 gadgets during their introduction to the system [9]. The dataset is openly accessible on kaggle.com. The dataset comprises parcel catch documents that were recorded with the gadgets killed and on multiple times. The devices utilized for the investigation produced by notable brands like D-Link, Fitbit, Philips Hue, EdiMax, Smarter, etc. The examination of the pop records for the correspondence conventions utilized by the gadgets, their appropriation over the beginning up traffic, and what vulnerabilities plague the particular protocols [10]. In the accompanying segments, we have dissected the exploration addresses which coordinated this examination. Further sections examine the ongoing defenselessness abuses on the gadgets under observation, trailed by powerlessness investigation of the significant correspondence conventions utilized by brilliant machines. This examination with the outcomes found from our research of the records and propose a few practices for a nonspecialized client to defend against potential trade-offs.

9.6 RESEARCH QUESTIONS

While completing this examination, the scientists had the accompanying inquiries at the top of the priority list

1. What have been significant security assaults on intelligent machines in later past?

2. How do makers handle announced security vulnerabilities?
3. Are security vulnerabilities a factor of the fundamental convention pile of a keen apparatus?
4. Are security vulnerabilities uniform over gadgets made by a similar firm?
5. Could a prescient model be utilized to schedule a recently introduced savvy apparatus against potential reasons for vulnerabilities and propose countermeasures for the equivalent?
6. What preventive measures can a client take to keep their imported savvy machines from getting bargained?

The dataset gave by kaggle.com [9], which contained bundle catches of 23 gadgets. The dataset was framed by killing the devices and on multiple times. Fitbit Aria Wi-Fi scale made news for inappropriate reasons in the year 2016, as security helplessness was accounted for by an analyst working for Google Project Zero [11]. The security issue emerged from the nearness of a static exchange identifier in DNS demands created by the intelligent scale. The keen range utilized DNS demands during arrangement or synchronization to find Fitbit servers. Be that as it may, the static exchange identifier could be sniffed by a potential assailant, empowering the individual to send a spurious reaction and synchronize the keen scale with their server. Since a client would log individual data like body insights, exercises, and rest information, an assault like this would render the personal client information accessible to programmers. Even though Fitbit agents affirmed that no security episodes accounted for because of this issue, the advancement group at Fitbit immediately settled the matter and discharged it as a firmware update. It should likewise notice that for a programmed firmware update to complete, it necessitated that the keen scale would have a functioning Fitbit account, had been synchronized with Fitbit servers, and have working batteries set up. The issue and goals imparted to open by Fitbit, and clients were given help with instances of any problems with the firmware refreshes. To date, no further security issues have accounted for the Aria Wi-Fi scale, and another model to be specific, Aria 2, was propelled in 2017. No security vulnerabilities have considered for Fitbit Aria 2 scale at present.

With regards to arranging gadgets, D-Link Systems has directed a significant piece of the pie. Tragically, D-Link organizes video controllers engaged with a long-standing controversy brought about by different security defects found in the gadgets [12]. The debate began in 2013, when scientists at firm Qualys found numerous blemishes in D-Link network video recorders (NVR) gadgets, to be specific DNR-322L and DNR-326 system video controllers. NVRs are brilliant stockpiling gadgets for IP cameras. They are customized to store the video bolsters for review through approved logins. They additionally utilized for remote observing of continuous video feeds, and additional reinforcement of chronicles to remote FTP servers. The columnists found some significant security defects. Initially, any potential aggressor could make a client account on the gadget by sending a solicitation parcel. No validation instrument set up for the making of an extra client. Moreover, the secret word for the chairman record could reset by the additional client. An aggressor could hence make an extra client with framework benefits, and afterward, find and change the secret head phrase to seize the gadget.

Another significant structure blemish in NVRs would let an aggressor transfer a noxious firmware to the gadget. As there were no prerequisites for confirmation of firmware refreshes, the information on gadget uniform resource locator was sufficient to get to the online user interface and transfer the firmware. The gadgets were additionally helpless to refusal of administration assaults since an unapproved client could close down or reboot the device. They could likewise reset the device to its default plant settings. D-Link also fabricates network attached storage (NAS) gadgets. NAS gadgets are outside hard drives that are associated with a system rather than a PC. Such system-associated memory encourages simple sharing, booked reinforcements, and more straightforward information to the executives. It discovered that the above-detailed defects applied to D-Link NAS gadgets also. Gadgets like NVRs and NAS are utilized generally by organizations, libraries, emergency clinics, etc., and imperfections like these can render touchy information presented to burglary and spying. D-Link delegates acknowledged the specific security issues and expressed that the problems had fixed and the arrangements shared over the introduced gadgets.

In any case, it was not the finish of troubles for D-Link. In the year 2016, analysts at Senrio detailed that more than 120 items sold by D-Link were tormented by a product issue, which made them defenseless against remote assaults [13]. The powerless items included D-Link Connected home items, for example, IP cameras, arrange capacity, switches, etc. The particular issue empowered aggressors to complete code infusion assaults. Code infusion assaults are the most well-known weapons store utilized by programmers. They are helped out primarily through four techniques, viz. XSS assaults, SQL infusion assaults, shell infusion assaults, and document consideration assaults [12]. Such charges permit assailants to get benefits to organize assets. In this manner, a real client would, at present, be helpless regardless of whether they have made an unbroken secret key for their certifications. An assailant could utilize this weakness to make another client, award it heads benefits and complete tasks with no prevention. Such undermined gadgets would then be able to additionally use to access different devices associated with a similar system. In light of the raised concerns, D-Link discharged DCS-930L firmware to impart answers for the clients.

While D-Link delegates continued promising answers for security issues and resulting firmware refreshes, the producer was prosecuted in the year 2017 by The United States Federal Trade Commission (FTC) over their lacking corrections. The claim documented by FTC expressed that D-Link had neglected to verify their items and therefore uncovered they are a great many clients to digital security assaults [14]. D-Link had not just ignored to contain the significant issues, for example, hard-coded logins and unapproved access to arrange assets, yet had likewise dishonestly advanced the idea of security of their gadgets. As a measure to settle these claims, in the year 2019, D-Link Systems Inc. reported a product security program to guarantee the redressal of past issues and superior anticipating imperfections found in the future. The proposed game plan would incorporate strides to warrant the security of web-empowered cameras and other system gadgets, extensive predischarge testing for dangers, predictable checking, and consideration regarding blemishes announced by clients and security analysts. Likewise, for the following decade, the D-Link security programme and its usage will be surveyed by an unbiased outsider substance on a biennial premise.

While internetworking gadgets like switches and NVRs are clear powerless connections because of their situation in the systems administration chain of command, programmers have found the probability of misusing vulnerabilities of savvy apparatuses, for example, a keen fitting. Savvy plugs are gadgets that go about as an interface between electrical machines and a UI as a versatile application. These gadgets are quick getting mainstream as they give superior control to the client to deal with their premises remotely. A shrewd fitting may assist clients with controlling their vitality utilization, remotely work different apparatuses, and screen necessary hardware, for example, a heart screen in a clinic. A security bargain of a savvy fitting may not just cause the client to lose control of their system-associated gadgets; however, it may likewise uncover their touchy system certifications [15].

The scientists recommended that such imperfections can recognize to a significant degree using a four-overlap approach. Right off the bat, gadget filtering can utilize to decide whether the brilliant attachment is revealing merchant related data in a decoded structure and whether the client has verified the gadget arrangement. Besides, the quality of the client secret phrase can try utilizing an animal power calculation. The beast power calculation would utilize to attempt to log in using all generally used secret phrase mixes. Thirdly, the framework satirize gadget to check if the remote server would interface with the phony device and open client certifications to an aggressor who possesses the mock device. Fourthly, the fitting framework can try to look at the firmware update process, since an unapproved malignant firmware can utilize to execute an assault.

Another keen attachment framework that was accounted for to have security issues was Belkin Wemo Insight Smart Plug. In 2018, this brilliant attachment considered to have a support flood bug in one of the libraries of gadget programming. This defect would permit vindictive hypertext transfer protocol parcels, made by aggressors, to sidestep the neighborhood to arrange security channels. While the subsequent littlest level trade-off could be an assailant having the option to kill an associated electrical gadget, the associated idea of the deep attachment could cause a significant occasion of an aggressor picking-up section to the home system. The way amplified the gravity of the imperfection that while the switch does not store any client data itself, it might give indirect access to a programmer through other associated gadgets. Purchasers have additionally promptly acknowledged shrewd lighting arrangements, which can make a customized vibe dependent on the client's decision. Well, known lighting arrangements by Philips Hue and Osram Lithify have seen as inclined to specific helplessness because of the utilization of the Zigbee convention in their validation strategies. While the Zigbee convention and the Hue and Lightify items are secure naturally, the issue had emerged from the way Zigbee's meeting took care of the keys utilized for the gadget verification system. Despite considering this issue as a minor hazard, engineers at Philips lighting arrangements fixed the problem and imparted it to the clients through a firmware update. Osram additionally paid heed to the blemish, the firm likewise empowered security testing of its items by scientists at Rapid7 and remediation of the found imperfections.

The well-known shrewd espresso creator brand Smarter additionally got into high temp water when specialists had the option to bargain its keen espresso machine. In 2017, FortiGuard Labs educated Smarter that their gadgets were inclined to the disavowal of administration assault, because of the suspicion by the advancement group that the device will associate with a protected wi-fi. The issue quickly fixed around the same time. Once more, in the year 2019, Avast detailed that misuse of the utilization of default settings of Smarter espresso-making machine implied that a programmer could transfer vindictive programming on the computer. A potential assailant could even hack and change over the gadget into ransomware.

9.7 EXAMINATION AND RESULTS

The experiment broke down the dataset containing bundle catches of 23 devices during their introduction to the system. With the end goal of analysis, we utilized Python libraries of Scapy to parse the parcel catch documents and Matplotlib to plot the perceptions. Scapy is an incredible library given as a piece of Python stage to complete various systems administration activities. The place of the center was the conventions these gadgets used to speak with the system components. Above all else did a superficial investigation of convention dissemination in the parcel transmissions of devices made by the ten makers to be specific, Fitbit, D-Link, Ednet, Edimax, Osram, Homematic, Philips Hue, Belkin, Smarter, and Withings. By considering the conventions which make up most of the bundle trade, we observed that the conventions Extensible Authentication Protocol over LAN (EAPOL), TCP, UDP, and DNS hold the highest rate. EAP utilizing MD5 hash calculation is a mainstream validation strategy used by IoT gadgets. While it is difficult to snoop on an EAP trade, a caught EAPOL parcel can be used by a programmer to decide the test and the reaction key. When an aggressor has the test and the reaction key, they can use animal power to separate the client's secret word, in this way, making the gadget helpless against a lexicon assault. Since the confirmation restricted to the gadget side, EAP-based correspondence is additionally powerless to man-in-the-middle (MITM) assaults. The second significant convention TCP is one of the essential vehicle layer conventions of the Internet protocol suite. It depends on a three-path handshake to frame an association, and the originators did not consider a lot of security estimations for the customer server association. Since a TCP server acknowledges customer demands with no verification, a potential aggressor can parody the IP address of a signed-in client and access the server. Another part of defenselessness because of TCP is the way that information trustworthiness exclusively actualized through checksum checks. An individual with pernicious purpose can present a malware parcel by virtually guaranteeing that the checksum esteem is the same as the normal one.

Late years have seen various distributed denial of service (DDoS) assaults a notable example being the assault executed by Mirai botnet in the year 2016. The Mirai botnet assault had abused the default setup of web gadgets, for instance, IP cameras to hack into them and like this, make them into a host for additional DDoS assault. UDP convention has been instrumental in driving flood assaults since a UDP association is datagram-based and does not have a committed customer server association. A potential assailant would send various UDP parcels to shut ports of an objective machine; this will bring about the accepting device to send a blunder reaction. It renders the accurate tool overwhelmed and the dominant part of figuring power spent in taking care of the food parcels. The fourth convention Multicast DNS (mDNS) is utilized by gadgets to find different devices and administrations on a system. In certain occurrences, reaction to DNS inquiry gotten from outside the nearby system. This reaction may uncover more data about the system and discover different gadgets on the nearby network. In addition, since the size of the response is more significant than a question, it could be abused to intensify DDoS assaults. To dishearten such attacks, a client may square mDNS parcels coming into neighborhood organize from an external system, or to obstruct any mDNS bundles from leaving the nearby system. As a piece of this examination, we likewise dissected if a grouping of convention trade could be utilized as a component to distinguish a maker. The conventions that are trademark to a maker could use to show an AI-based answer for a drive a calculation to check against conceivable security vulnerabilities [8].

It would, subsequently, computerize testing of gadgets at whatever point they reconnect with the system—the extricated convention groupings through a CountVectorizer. The sklearn library offers various modules to extricate highlights from the content or a picture; in an organization reasonable for use by AI calculations, CountVectorizer of the sklearn library offers both tokenization and event check of removed words. While this module has generally used to assemble a jargon from an examination of a book record, we used it to tokenize conventions and use them as highlights to bolster into an AI model.

9.8 WEB AND IMAGE MINING

Everybody who needs any sort of data goes to the World Wide Web today computerized pictures can be gotten of any scene around the globe from the web or the World Wide Web. Be that as it may, grouping the different real views into legitimate requests is inconceivable right now because there are vast amounts of data that should get to every day. An answer has been, and it is as of now being worked on, this includes utilizing naturally produced pictures to list the conventional images as opposed to using hard pictures of this present reality scenes. The traditional order of picture instrument has portrayed as information securing from the immense database of the web. There are three stages in the whole procedure of picture arrangement. The first is the social event organize where similar pictures of a particular watchword assembled from the web utilizing the procurement component. The subsequent stage is known as the learning stage, and right now, data or the significant substance separated from each picture. This stage additionally helps in finding out about the class in which it has a place [16]. The third stage, where the pictures are separated dependent on the particular watchwords and the classes. Estimating "Visualness" is another considerable idea, which was the visual substance of the images is as significant as the data it holds, and finding the imagery of the pictures dependent on their web positions will be useful in clarifying them. In web picture recovery, various things can utilize, and one such significant idea is the different modalities between the different pictures. It has dived deep in his investigation of the patterns between two meaningful ideas, which are watchwords and groups utilizing the multimodel affiliation rule. Picture arrangement in picture mining has become a significant and muddled subject. Numerous analysts have chipped away at this point. They are finding the issues in picture characterization with concocting the problem as the separation between the preparation complex and the test complex.

Getting the necessary data from any picture from the web is a troublesome errand, as these pictures are typically not clarified. A compelling technique has been proposed, which uses picture mining to draw out the necessary data. This arrangement of picture recovery can likewise be known as the inquiry picture. The fundamental idea of driving this methodology is to achieve the required content data that related to each image. At that point, this data can utilize to portray other low-quality pictures on the web using the comparability appraisal procedure. There are a lot of picture mining methods, yet the genuine test is to separate the semantically right images from the internet. It incorporates finding a significant picture with the assistance of Google Image Searcher. At that point, the yield then utilized to find out about the photo, particularly its spatial coevent on the calculation of the touchy Markov stationary component. The sacks of word portrayal and the idea data are then appropriately incorporated to get the best possible spatial event of the picture on the web. The order finished with the assistance of picture mining business, or e-business has gotten significant in today's world. To have an online nearness, each organization needs to have an official site, and the data fragmented without legitimate pictures. Finding a single image from these picture sources can be extreme; however, two techniques help in it, and these are the arrangement-based recovery or the shading-based recovery of the pictures [17]. It utilizes the web e-inventory picture recovery framework, which utilizes client log and metadata. Right now, different strategies that have to get the client use designs like the surface and shading-based picture grouping and the ordering methods like the bit vector list. Each client question followed, and inclinations altered or included based that to the client picture database. Another technique gathers and stores the pictures dependent on the people who use it. Age estimation done depends on the responses to the inquiries, and afterward, the images are index as needs Flickr, the picture sharing site, is utilized for the equivalent. For distinguishing the pictures of the faces, the active shape model utilized. This model additionally dispenses with any nonface pictures. Biorosued highlights used, which perceive the age of the face, and afterward, the calculations label the images as needs be on Flickr.

9.9 IMAGE MINING FOR MEDICAL DIAGNOSIS

Attractive reverberation imaging is a significant method of clinical sciences. In any case, another considerable methodology that is critical is the division of Corpus Callosum in midsagittal segments. This piece of the cerebrum has applications in both nervous system science and neurocognitive examinations and aides in controlling the conduct and feelings of the individual in question. Hence, the remarkable piece of the picture mining process is a division of the Corpus Callosum as it helps in learning a great deal about the individual. The measurable attributes centered around first, and a picture mining calculation has created the high-power zones of the corpus callosum are then dissected graphically. Integrated retinal information system was designed for clinical experts, and this strategy helps in examining the eyes for any bother, which may be looked at by diabetic patients. Recognition of tumors in the cerebrum is another significant piece of the CT explore, and the calculation to help right now the prepreparing strategy is the initial segment, which helps in dispensing with conflicting information from the examined CT Images [18]. Novel fuzzy association rule mining is the second step that contains all the necessary data from the CT output and aids in giving a legal conclusion to the patient. The whole data of the mind examine then dissected to come to a result in regards to the filtered pictures and the state of the person cerebrum.

The identification of the tumors in the cerebrums is a significant piece of the checking. Since it is the subject of human life, it is essential that the picture mining calculations which created built up a calculation that arranges the hazardous regions of the mind as ordinary or irregular. The unfamiliar territories are then additionally breaking down to discover any cerebrum-related ailment. To order the tumors in mind, a strategy [19]. The five phases of this strategy are: prehandling, include extraction, affiliation rule mining, and half-breed classifier. The principal arranges accomplished by separating the middle—the watchful edge location method utilized to get a reasonable picture. Two picture mining methods used here to get the best possible images from the CT filter. The two techniques are the choice tree strategy and the crossover technique. The picture mining method has additionally been applied to bosom mammography to get away from any underlying dangerous tissues. Here the half-breed approach is better utilized for include choice of the pictures. The crossbreed hereditary calculations decrease the highlights by 75% and give the image more excellent clearness and accuracy.

9.10 CONCLUSION

The most fitting wellspring of discovering information is the advanced database. Digitization has occurred in pretty much every area, and this has prompted enormous development in uncomplicated access to gigantic information. The report does not need to be just as writings. It tends to be in the types of sound, recordings, and even pictures. Individuals who need to settle on choices every day are the ones who need to get to the prior databases for better data. The way toward discovering pertinent data from previously existing pictures in the understood and accessible databases is called picture mining. This strategy utilized in pretty much every division, including agribusiness, remote detecting, clinical analysis, space research, and others. The focal point of this proposition will be on the utilizations of picture mining, and it will likewise consider the existent material on this theme to comprehend the idea of picture mining better. Capacity innovation and gaining pictures have become simpler today in light of the enormous advancement in the change that supplements these two capacities. It is the reason for separating data and information that will be valuable for individuals who have additionally gotten simpler. Picture mining centers around the necessary information or designs and other picture relationship, which will not be noticeable to the layman from the start look. While a few people guarantee that picture mining is essentially a subset of information mining. Image mining is disciplinary in itself and uses a lot of mechanical advancements like human-made consciousness, computer vision, picture extraction, computerized picture handling, picture learning, and information mining too. There are issues in picture mining and its applications. The center around the picture mining system and have built up specific rules for what the eventual fate of picture mining holds.

REFERENCES

[1] S. Rodpongpun, T. Janyalikit, C. Ratanamahatana, Influential factors of an asynchronous BCI for movement intention detection, Comput. Math. Methods Med. 2020 (2020) 1–12. Available from: https://doi.org/10.1155/2020/8573754.

[2] R. Kashyap, V. Tiwari, Energy-based active contour method for image segmentation, Int. J. Electron. Healthc. 9 (2–3) (2017) 210–225.

[3] S. Tiwari, R.K. Gupta, R. Kashyap, To enhance web response time using agglomerative clustering technique for web navigation recommendation, in: H. Behera, J. Nayak, B. Naik, A. Abraham (Eds.), Computational Intelligence in Data Mining. Advances in Intelligent Systems and Computing, vol. 711, Springer, Singapore, 2019.

[4] R. Shukla, R.K. Gupta, R. Kashyap, A multiphase pre-copy strategy for the virtual machine migration in the cloud, in: S. Satapathy, V. Bhateja, S. Das (Eds.), Smart Intelligent Computing and Applications. Smart Innovation, Systems and Technologies, vol. 104, Springer, Singapore, 2019.

[5] R. Kashyap, V. Tiwari, Active contours using global models for medical image segmentation, Int. J. Computational Syst. Eng. 4 (2/3) (2018) 195. Available from: http://dx.doi.org/10.1504/ijcsyse.2018.091404>.

[6] R. Kashyap, S. Rahamatkar, Medical image segmentation: an advanced approach, in: S. Paul, P. Bhattacharya, A. Bit (Eds.), Early Detection of Neurological Disorders Using Machine Learning Systems, IGI Global, Hershey, PA, 2019, pp. 292–321. Available from: http://dx.doi.org/10.4018/978-1-5225-8567-1.ch015.

[7] F. Xiao, Y. Sun, D. Du, X. Li, M. Luo, A novel malware classification method based on crucial behavior, Math. Probl. Eng. 2020 (2020) 1–12. Available from: https://doi.org/10.1155/2020/6804290.

[8] N. Waoo, R. Kashyap, A. Jaiswal, DNA nano array analysis using hierarchical quality threshold clustering, in: Proceedings of 2010 Second IEEE International Conference on Information Management and Engineering, IEEE, Chengdu, 2010, pp. 81–85.

[9] Kaggle.com, IoT Device Captures. [online]. Available from: <https://www.kaggle.com/drwardog/iot-device-captures>, 2020 (accessed 01.09.19).

[10] S. Marchal, M. Miettinen, T. Nguyen, A. Sadeghi, N. Asokan, AuDI: toward autonomous IoT device-type identification using periodic communication, IEEE J. Sel. AreasCommun. 37 (6) (2019) 1402–1412.

[11] Software Integrity Blog, Application Security & Software Security Blog | Synopsys. [online]. Available from: <https://www.synopsys.com/blogs/software-security/fitbit-iot-bathroom-scale-vulnerabilities>, 2019 (accessed 02.09.19).

[12] Threatpost.com, D-Link wi-fi camera flaw extends to 120 products. [online]. Available from: <https://threatpost.com/d-link-wi-fi-camera-flaw-extends-to-120-products/119097/>, 2019 (accessed 05.09.19).

[13] S. Ragan, D-Link vulnerability impacts 400,000 devices. [online]. CSO Online. Available from: <https://www.csoonline.com/article/3092497/d-link-vulnerability-impacts-400-000-devices.html>, 2020 (accessed 01.01.20).

[14] Federal Trade Commission, D-Link agrees to make security enhancements to settle FTC litigation. [online]. Available from: <https://www.ftc.gov/news-events/press-releases/2019/07/d-link-agrees-make-security-enhancements-settle-ftc-litigation>, 2020 (accessed 02.01.20).

[15] Z. Ling, J. Luo, Y. Xu, C. Gao, K. Wu, X. Fu, Security vulnerabilities of Internet of Things: a case study of the smart plug system, IEEE Internet Things J. 4 (6) (2017) 1899–1909.

[16] R. Kashyap, P. Gautam, Fast level set method for segmentation of medical images, in: Proceedings of the International Conference on Informatics and Analytics (ICIA-16), ACM, New York, NY, Article 20, 2016, 7 pages.

[17] R. Kashyap, Advanced diagnosis techniques in medical imaging, in: R. Kashyap, A. Kumar (Eds.), Challenges and Applications for Implementing Machine Learning in Computer Vision, IGI Global, Hershey, PA, 2020, pp. 64–91. Available from: http://dx.doi.org/10.4018/978-1-7998-0182-5.ch003.

[18] G. Quellec, K. Charrière, Y. Boudi, B. Cochener, M. Lamard, Deep image mining for diabetic retinopathy screening, Med. Image Anal. 39 (2017) 178–193. Available from: https://doi.org/10.1016/j.media.2017.04.012.

[19] Y. Meshcheryakov, R. Meshcheryakov, Automated system of monitoring and positioning of functional units of mining technological machines for coal-mining enterprises, MATEC Web of Conf. 161 (2018) 03017. Available from: https://doi.org/10.1051/matecconf/201816103017.

CHAPTER

INTELLIGENT SMART HOME ENERGY EFFICIENCY MODEL USING ARTIFICIAL INTELLIGENCE AND INTERNET OF THINGS 10

Harpreet Kaur[1], Simar Preet Singh[2], Supreet Bhatnagar[3] and Arun Solanki[4]

[1]Computer Science and Engineering Department, Chandigarh University, Mohali, India [2]Department of Computer Science and Engineering, Chandigarh Engineering College (CEC), Landran, Mohali, India [3]Electronics and Communication Engineering Department, Chandigarh University, Mohali, India [4]School of Information and Communication Technology, Gautam Buddha University, Greater Noida, India

10.1 INTRODUCTION

Today, humans are introducing more technological advancements to make human life more comfortable and secure. To follow the same pattern, there is a need to devise the next-generation house that will provide the next-generation living to the people. These houses will change the traditional standard of living. In the starting phase, houses are just treated as the place to have shelter, but as time passes, people started thinking and working on it [1,2]. More and more advanced versions of these houses are coming in the market nowadays. But the problem with currently constructed houses is that technology and construction techniques are not working hand in hand. Thus, in this work, we found a way to devise an intelligent energy efficient smart home, which can concurrently work with technology and can provide the best living place. It is a house that will have electrical devices that can think for themselves and provide feedback to the user. In this the user does not need to worry about small things such as security systems, etc. These will self-monitor the house entrance and the people who enter the house. These systems will monitor the emergency system to recover from every possible worst-case situation in which the present-time houses cannot meet. These houses will provide a platform for sustainable development as these houses will use renewable sources of energy [2–4]. These houses will monitor the usage of water and will try every possible way to reduce water usage in the house. The proposed model will be an intelligent energy-efficient smart home model, that will be using artificial intelligence and Internet of Things (IoT) technologies. This model will devise the responsibility for saving the smart home system and their owners (from energy) in the worst situations. These homes can give feedback on various issues such as how to reduce water consumption today, such as the day's weather updates, perception prediction, humidity, wind speed, and reports of daily electricity and water consumption. This creates a very healthy and friendly interaction with machines and humans [5–7].

Artificial Intelligence to Solve Pervasive Internet of Things Issues. DOI: https://doi.org/10.1016/B978-0-12-818576-6.00010-1

This system is equipped with voice commanding and speech recognition, which allows the user to interact with machines very easily as well as it is capable of understanding the various languages that also allow the user to interact in their own language without any extra effort. Through speech recognition that allows the owner of the house to set the priority to follow commands on that hierarchy, it secures the system in aspects that no other person can command the system [8,9]. This system also saves the wastage of food by monitoring the daily consumption and predicting the quantity of food or other eatables required by using the machine-learning supervised library that can do so [10].

The proposed model will be programmed to deal with the various unconditional situations occurring in a day, like the live calling of emergency services by saying SOS or pressing an SOS switch through the phone or switch present in the home, starting of water sprinklers in case of fire, and alarming in case of an earthquake.

To do all these activities we are going to use various present-time technology as well as the equipment that is modified according to the need in the smart homes [11–15]. But to ensure the reliability we are using monitoring systems on each system so that the working and accuracy of the system can be measured and in case of error, maintenance message will be displayed to the owner.

10.2 LITERATURE REVIEW

This is about controlling all the systems in the home using a digital remote. This remote will control all the devices that are synchronized through this digital remote. The biggest advantage of this remote is that it can be used for a large range and for many devices, from switching on the air conditioner to opening the latch of the garage. This ensures the conservation of electricity and comfort to the owner [16]. In this research paper authors are focused on controlling the lights and other systems through the various communication protocol that come under the term IoT. We have two ways to do this: first, by controlling the appliances through the app that is equipped with the virtual switches, and second, by voice command [17].

The work done in the research paper is that introducing an idea that will store the electric energy for cars and home, as well as in plug-in electric vehicles (PEV). The various application of PEV is that it can convert grid power for the household usage as well as for the charging of the e-vehicle, second for giving power to the house from the power of e-vehicle as well as the intelligence system that works on the similar circuit of PEV array battery models are capable of identifying or predicting the model of traveling time, speed, and traveling length to transform the PEV to a smart home energy management system [18].

This is all about the advancement in the smart home to make them smarter, and this is done by linking IoT and electricity efficient systems with each other, giving a user-friendly system that is easy to use and understand, and including renewable resources in it such as solar lights, etc. [19].

A risk analysis is being conducted for the development of the safest and most acceptable smart homes. A total of 32 tests are made: nine were classified as low risks and four as a high risk, that is, most of the identified risks are termed as moderate. Risk that is in high classification is either related to human factors or the software systems. The results indicate that with the implementation of standard security features, new as well as current risks can be minimized to acceptable levels and the most serious risks [20].

Automatic speech recognition is a computer-based software that processes the audio signals recorded by a microphone and converts them into machine-understandable language or code this code commands the other machines their works and tasks. To process the speech to a command requires a separate processor or cloud computing with an internet link to it for connection. The "Alexa," "Cortana," and "OK Google" start working as the machine recognizes the audio "OK Google," "Alexa," or "Hello Cortana," the uplink starts over the internet starts. Voice analysis is not usually suitable for all the computer because this requires the high-speed processing data rather than the old low-power processors. To achieve this in the real world, this paper presents special techniques and details an efficiency voice command method compatible with an embedded IoT low-power device [21]. To save energy we are switching ourselves toward smart energy management systems that require not only the way energy is transformed and transmitted but also requires two types of ZigBee networks that will emit low-power digital radiowaves for getting control over the devices and energy management and conservation. These will also be able to connect to the nearby area networks in search of energy, using the ZigBee network for subenergy present in the home or nearby houses [22].

This research is about the advancement of in-home energy management. We control our energy consumption by managing it critically and analytically. In this research paper, it is done by installing a thermostat. The values at the thermostat have to perform the activities that can be done by using the communication protocol called Wi-Fi. That is, user friendly, as the dependency on the remote is removed and control is over the smartphone. So, according to the situations, either system will set the values automatically or either by the command given by the user through Wi-Fi. In this way, we can reduce our usage of electricity very efficiently as well as easily [23].

In this paper, we discovered how to decrease the amount of electric bills without any human interference. This is done by the smart-charging algorithm that uses a preinstalled array of batteries that will store low-cost energy for use when the cost of electricity is very high. This smart-charging algorithm reduces power consumption by switching between the grid power supply and the preinstalled battery power supply. The algorithm enables the prediction model that develops by obtaining daily data sets, which predict the future demand using statistics that have been analyzed in the past through machine-learning techniques. The authors of this paper examined Smart Charge installed in real homes and using that data obtained from real homes to quantify its potential to lower bills. After installing this system in real-life smart homes, a linear price of electricity can be seen which means there is no impact of increased pricing on the billing amount which is possible by the machine-learning algorithms [24].

The literature based on the smart grid provides a way to use the grid power supply in the most efficient and useful way. This literature also gives the idea of charging the electric vehicle using a grid power supply that can provide power to your home as well. This system will include the interlinking part that is capable of charging your e-vehicle in a regulated manner that gives your battery a great life and in case of power loss this system will also help provide power to your house in case of power loss from the grid power supply the name to this interlinkage part is given as multiagent system [25].

In the way of home automation, this research paper is about the automation of the home appliances such as lights and air conditioners by a text message that would be sent by a global standards for mobile communication (GSM) protocol. In this research paper the author emphasizes the working of the GSM protocol over various conditions and commands that can be used for controlling the devices, which allows the user to control the desired system that is far away from home using various frequencies present in bandwidth. The latest technology of serial communication and its codes have been used to develop the smart global system for mobile communication-based automation systems. Homeowners will be capable of receiving the feedback status of any appliance that is connected remotely from their mobile phones. PIC18f452 is the integrated part of GSM that provides the smart automated house system with the desired baud rate that is required for various devices for sending and receiving the data, which is up to 9600 bps [26].

Naito et al. [27] is implementing the remote access to the smart device without making use of the server, provided by the service provider of the internet. The author is using mobile that is technology to achieve NAT traversal and end-to-end encrypted communication that increases the security of the system. With recommended systems, the operating system of Android devices can control the smart applications installed in the house from far distances safely and securely without installing the remote control servers that are designed by manufacturing companies. Another paper by Sidduqi et al. [28] discusses an accelerometer, which is a sensor that follows the principle of gyroscope as well as the orientation of the object, so using the same principle, and to increase the accuracy, we are using the statistical data to generate required output. In this we can achieve satisfactory accuracy of 79.58% with random forest.

Mao et al. [29] focus on the lightweight human wearable devices that can sense the human gestures of eyes and facial gestures with the help of physiological signal measurements, since presently, we interact with each other through the facial expression or by the gestures of our eye in a similar way if we create the algorithm for machine vision and store these expressions in it. With the help of an algorithm, the machine works according to the gestures passed by the user. This will help create the bots or applications that will be controlled through your facial or eye gestures. There will be hands-free interaction with the devices. Hazard detection systems according to realization described are functional to enable a user to interface with the hazard detection system by performing gestures from a far distance. The touchless gesture can be performed during an emergency from far away. The distance detection system does not require any physical connection to the hazard discovery system [30].

Krichen et al. [15] research is about power management, using various layers. The first layer control shows demand response algorithm that shifts the load from hours when the load is very high to the hours when the load is very low; the second layer is used to optimize the electric vehicle power management when it is plugged in is alsotested. The main contribution of this study is to achieve constant or lower daily load power for smart homes.

Istre et al. [31] research are about the experiment done by designing a smart home energy management system designed to automatically identify and differentiate between different incoming signals to keep a record of power consumption each device per hour on the basis by monitoring the current (and voltage) flowing through a device that is connected to that outlet The main objective is to calculate and manage the power consumption as effectively as possible.

Hussain et al. [32] presented the current rising aims of IoT systems in various industries as well as discuss the key privacy challenges about to increase the growth of IoT to reach its potential in the smart home concept. The majority of the present literature on IoT smart home platforms focuses on applications provided by smarter devices connected to it.

Ni [33] mapped interface for control of smart home appliances as more and more people use and install these smart devices identifying and differentiating these devices. For example, an average house is consisting of lots of appliances such as lights, fans, computers, televisions, home theater, hot air blowers, air conditioners, etc. The approach is to designate a unique ID or name to it to become mortal and shows errors. These revel techniques that visually situated creating a virtual map of the device installed in the house and to control the function of the devices can be set through that virtual map.

The demand-side management (DSM) aim is to manage the power supply and load over the peak time; it is one of the most important aspects of the power grid system. The power is balanced between the demand of energy and production of energy, which is done through a real-time pricing scheme and the calculations from the correlation of power consumption among users. Maximization model is a suggestion based on Markov decision process (MDP) in the research work. A probabilistic transfer to characterize the elastic devices present in the system matrix has been introduced. Many state transfer functions have been used also to shoe the operation of various semielastic appliances. In this way, the specific characteristics of SHA can be developed [34].

Purwanda et al. [35] designed, developed, and evaluated a user interface prototype for the smart home system. The designed mobile app is named by MINDS-apps V1, and this system is capable of performing three tasks. The first mode is soft mode, that is, controlling the ambient lighting of the house that is done by RGB. Second mode is hard mode. This mode will control the heavy devices such as ACs, heaters, etc. Third mode is the monitoring mode. This mode will monitor the humidity, temperature, sunlight, etc. to control the other devices that require such data.

A system-level model of computing in the thread mesh network is presented by Eliassen et al. [36]. Fixed computing consists of various kinds of delay from the application layer to the physical layer of the network. The problem is solved by a Markov chain model that is used to derive the probability distribution of the medium access control service time. The system-level model is experimentally examined in a multiple network thread mesh network. The outcomes show that the system model results match well with the experimental results.

Hosseinian and Damghani [37] showed another stage that provides power to the imaginative evaluation of IoT information from eager homes. We propose the utilization of cloud frameworks to allow information controlled by administrations and address the difficulties of complexities and requests required by the web and in case of disconnection information handling, storing the unsend data, and arrangement investigation. Along these lines, we can perform energy inspection, the execution, and productivity are the most vital part of that is performed by the power division of the economy. So to provide the interrupted power supply to the clients, the power supply from the grid is measured and analyzed at each point.

Li et al. [38] proposed an idea that is based on an approach made by the multiagent that is based on knowledge and mainly consists of a layered structure, consisting of multiagents and the ontology branch that deals with metaphysics, technologies to automatically gather systematic knowledge, and support a variety of information and have the ability to exchange the information. In such a type of architecture, a normal inference algorithm is presented that takes the input from unordered action and temporary actions for inferring both personal services and composite activity in real-time. The idea is to introduce agents that will understand the human activities and knowledge to that into language that will be understood by the agents to act in the same way a human performs.

Ueyama et al. [39] observed the functionality of the application in an event (1) collect relevant data from the working environment, (2) send data to an actuator, that is done by multichip communication; (3) process the obtained data in or outside the network; (4) make decisions based on the event; and (5) take appropriate actions. As a result, a wireless sensor and actuator network (WSAN) becomes an important part of various systems smart home applications, which allow information exchange in bidirectional, as well as provide them the power to take decisions.

E-key technique of smart industrial process and household applications, IoT connects devices via the internet to realize information sharing, intelligent control, remote monitoring, data statistics, etc., which significantly improve the intelligence, flexibility, and convenience for our industrial production and daily life. Along with more and more movable electric-driving devices to join the IoT, the energy supply is an increasingly serious technique issue for smart industrial and household applications [40].

In research by Stolarik et al. [41], the research is conducted in two-parts. The first part of the research is detecting the human presence in the room by making use of fiber Bragg grating. In fiber Bragg grating, a particular type of wavelength is reflected, and the rest of the wavelengths travel normally. The second part of this paper is focusing on detecting and continuously monitoring the room occupancy by measuring the change in the concentration of CO_2 with the help of an artificial neural network. The results of this experiment are verified for long- and short-term experiments and obtained an accuracy of 90%. Usage of fiber Bragg grating in SH was proven on the experiment based on the results obtained from CO_2 for room's occupancy monitoring in SH.

Alt et al. [42] studied interaction with IoT devices in users logging the device, sometimes becoming difficult due to authentication issues or due to some natural issues. Hence, an innovative technology has been introduced to overcome such logging issues. We can connect devices by smartphone technology simply by scanning the tags that are placed along the devices. We tested our method with two flat shares in two cities and provided preliminary insights concerning the strengths and weaknesses of our study approach.

Shir et al. [43] took the help of IoT to provide connectivity between the sensors installed on the patient and the medical systems, that gives optimal solution based on the inputs provided by the sensors. This study aims to establish IoT-based smart home security solutions for real-time health monitoring technologies in telemedicine architecture. A multilayer taxonomy, drive, and practicality is conducted in this research. The first layer consists of the analysis of telemedicine, which focuses on the client and server sides and shows that other studies associated with IoT-based smart home applications have several limitations that remain unaddressed. Monitoring the condition of the patent remotely is a promising application of IoT smart home technology in the health care sector without reducing the security stem requirement. Detailed research is conducted to handle all the security risks in such a kind of application where the devices are dealing with someone's life. Reviewed and a coherent taxonomy for these articles is established.

10.3 ENERGY EFFICIENCY MODEL

This section will define what do we mean by the terms "energy efficiency" and "energy-efficient model." This section will also explain why there is a need for energy-efficient models. Efficiency is not just whether a device works efficiently or not, it also includes how wisely it uses the power that is provided to the system, and even more than that [44,45]. Currently we are conducting more research into introducing efficient devices that are powerful, fast, accurate, and energy-aware, that is, the reduction in power consumption does not affect the aspects that are required for a device to fulfill the requirement for which it is designed. Thus, to design a model that is energy efficient and can fulfill all the desired aspects for which it is designed is the main motto of this work. The model that is saving energy of the resource proves to be the efficient one.

In mathematical terms, the efficiency of any device, machine, or model cannot be 100%. If we assume that a device is giving that much efficiency, it means that output of that device is equal to the input, that is, not practically possible as some amount of energy waste will be there in form of heat or friction or, depending upon the environment it is being used, to satisfy the law of energy conservation, represented in Eq. 10.1.

$$Input = Output + Loss \tag{10.1}$$

Hence, to avoid these types of loss, we are focusing on building the model that can enroll themselves with less loss state and to return more output (Fig. 10.1).

FIGURE 10.1

Saving feature of smart home.

10.4 NEED FOR ENERGY-EFFICIENT INTELLIGENT SMART HOME MODEL

This section will explain why there is a need for the proposed model. We are using the available resources very extensively. We are just extracting and processing them, and finally selling them to earn money. No one is interested in conserving these precious resources. If these resources go away, to an extent, it will take millions of years to form these resources again, or even longer.

Hence, by understanding the importance of these resources, we must start working to locate the best alternative methods for replacing these nonrenewable resources, else it will give birth to the very worst situations, leading to the shortage of products or materials that will affect the economic crisis [46]. To avoid such a situation, many scientists and engineers are working to replace the currently used resources from the upcoming renewable resources.

Renewable resources are those resources that will not extend easily as these resources can be replenished naturally, for example, solar-powered cars and lights, wind energy, etc. They produce electricity at very low operating costs and they do not harm the environment.

Presently, more than 45% of the electricity we use is produced in thermal power plants that use coal for producing electricity. These plants generate energy but at the cost of emitting pollutants in the environment. Emitting these pollutants consists of harmful toxic gases that mix in the atmosphere. The toxic gases that are emitted are serious concerns about the health of the animals and the environment but also for an environment [47,48]. Every year, millions of humans and other living organisms die due to inhaling these toxic gases into their respiratory organs. That is why air pollution is one of the most serious issues that we face in today's scenario. Mainly, homes use two variants of energy the most.

1. *Water*: This the most important resource that we obtain either from the ground or from nearby rivers. But in the house, we either obtain it by using jet pumps or from nearby government water tanks. The houses use this water for performing different tasks. These tasks involve cooking, watering plants, bathing, and various other uncountable activities. First, to decrease the loss of water, our smart system will be installed at various locations. These systems monitor the flow of water and stop the flow of water when the water tank gets full (water is flowing uselessly) [49]. Second, part of this system will act as a small water treatment plant that will take care of the water quality, which needs to be sent through pipes. It also cleans the water according to its use. For drinking purposes, it adds minerals to it; in case of washing utensils purpose, it either warms the water or softens the water so that detergent cleansing action can take place easily. This will use less detergent that indirectly leads to conserving nature.
2. *Electricity*: This smart home system is going to use the advanced microprocessors and microcontrollers that have an inbuilt battery-saving/power-saving feature that will not only switch off the appliances automatically but also the devices can be controlled through it. The selection of turning on and off the device will be done by the algorithm that is programmed in it, which uses deep learning technology [50,51].

Deep learning allows us to remove the errors using its predefined library that will work on the decision taken by the device and turning off the appliances when they are not required, like closed-circuit television cameras, fire alarm systems, etc. Thus, using this system there is no worry about the power saving, as this system will manage its own. This system will also give the most convenient method one can adopt to save energy [52]. For example, during sunny days it will take the permission to unfold the sunblind that will give us the natural light in our area where we are going to switch on the lights. It also switches automatically to the natural source to produce electricity.

The smart home can produce electricity by wind and solar medium. It is equipped with wind turbines that can move themselves according to the direction of the wind to optimize the power. Similarly, in the case of solar panels, the smart home is equipped with vacuum pumps and washers that automatically clean the surface of the solar panels to regain the power-producing capacity that was degraded due to sand and dust particles. Also, this system will keep monitoring the power output given by these devices and stores it in lithium-ion batteries. These batteries can be further used for obtaining power when these systems are not producing power.

10.5 BASIC TERMINOLOGY USED

This section will represent some basic terms and definitions that will be used throughout the research work. Some of them are as shown in the following:

1. *Smart home*: Home that takes care of itself, from the saving of electricity to water. The home is the basic requirement for all the people, and from various years we are making efforts to improve our comfort level and security in a place where we are living. We are now installing CCTV cameras. But they cannot be tracked 24-7 [53], so they are used to keep the record of all in and from the house. So that in case of any issues, we can see the recording. And other new technologies are introduced to transform a simple house to smart-like energy-saving auto on-off lights, use the air conditioner that is circulating air by using the human presence or switch between turbo cooling and moderate cooling according to the number of people in the specific region to save the electricity but these systems at some part required human interference [54,55]. So there is a need to introduce the next level of intelligence system to make the existing smart home smarter and with less human interference.
2. *Artificial intelligence (AI)*: It is the ability of a computer program or a machine to learn and think. It also includes a study that tries to make computers "smart," as now they can learn from past experiences. AI, as the word specifies, is made up of two words: "artificial," which signifies that it is not real, it is human-made, and "intelligence" signifies the power to think, apply, and use your memory or learn from past experiences that will give the best possible result [56]. We have created and worked with various algorithms that are going to work according to our need and save the previous results and path that is being followed.

For example, in image processing we use various algorithms to do that. The image that we obtain is broken into three parts. These three parts are as follows:

1. To tell us about the size and dimensions of the object that is coming to the image. This part is going to give us the physical significance of the image. This part of the image can be used in the CCTV image to detect the presence of the object. This presence of the object then can be processed and so to detect which object is where we can match the image with the database that is storing the relevant data according to the need.

After scanning our database, it will give the output in the form of information of that image like in case the camera is going to detect the presence of the dog or other animal then first the image is being processed and then the physical outline of the image will be processed so that we can detect that the processed image is either an animal or it is another object.

2. The image color that is being used; this is done by breaking the image in small pixel sizes and these pixels can be processed with the help of MatLab [57]. This MatLab software is used to convert the image pixel into matrix form this matrix we can process the data according to the algorithms.
3. IoT is the interconnection of the various electrical devices over the internet. We believe in connecting people to people. This term is coming to be known as the people of things [58] created in various platforms that perform like what we create in IoT but concatenated with the concept of remotely connecting and monitoring real-world things through the internet. This can be used to make it smarter, safer, and automated as well as reliable. Our focus is on building a smart wireless home security system that sends feedback to the owner by using the internet. This is done by a smartphone application that is connected through data cloud that can be accessed from anywhere in the world. Besides, it can also be used for home automation by making use of the same set of sensors [59]. This can be done by inserting or creating a virtual switch in the application in your smartphones. These applications will command the microcontroller that is connected over Wi-Fi. This enables the system to received data by the user on his phone from far distances without taking the care of connecting smartphones over the internet [60] (Fig. 10.2).

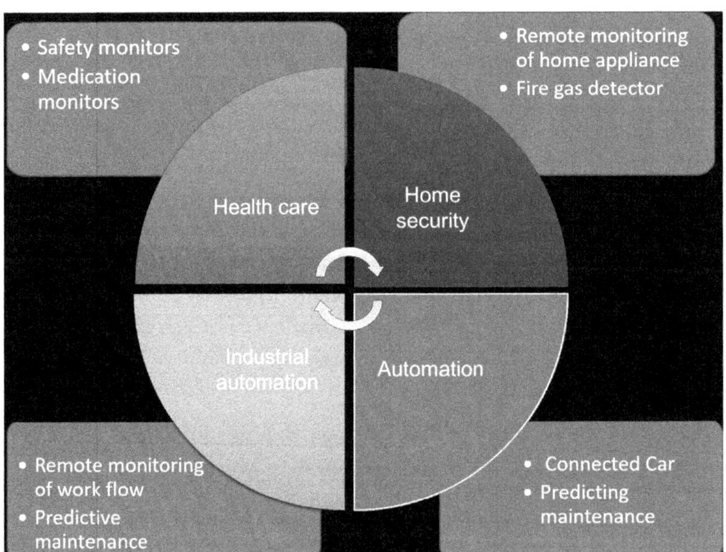

FIGURE 10.2

Working process of IoT.

IoT makes a data cloud that is connected with the user, and a microcontroller that is connected to the device, which are synchronized. The microcontroller connected to the device receives the internet connection, and it starts sending data to the cloud, which stores the data received from the microcontroller [61,62]. This could not only save the data but also allow the user to see the output given by the devices. For example, if I connect the temperature sensor to the controller that is connected with the cloud, users can see the instantaneous time of that location.

1. *Data Cloud:* It is the plac, outside of your computer, to store the data, and this location can be accessed by you at any time and at any place. The data cloud is a place for storing your data provided by various companies and firms. Some provide it free, while some charge for it [63–65]. It is believed that it is more revival and safe as compared to the physical or local storage mediums as there is a probability of losing your data due to physical or technical damage. This is not the case with the data cloud [66].
2. *Sensors:* Sensors are something that can sense and respond to physical or chemical change [67]. In the future, this change will then be processed electronically according to their working principle. For example, with the case of heat sensors, it is the makeup of the semiconductor that increases its conductivity after a certain temperature, and this increase in conductivity increases the flow of current that is measured using an ammeter, which enables users to sense the temperature.

10.6 COMPONENTS OF PROPOSED MODEL

This section is consisting of various components in the proposed model along with the framework. The following various components of the model are discussed.

1. This model will consist of a moisture sensor that will look after the water or moisture in the soil. If it is below the threshold value then it will command the water pumps to water the plants.
2. Servomotors that will allow the camera to move 360 degrees to monitor all of the area carefully and accurately.
3. Solar panels and windmills to generate electricity in an ecofriendly way.
4. Temperature monitoring sensors that will monitor the temperature of the outside and inside, and according to that, switches on the heater, AC or windows to provide a comfortable environment.
5. PH sensor that will continuously monitor the pH level of water either it is drinking or for other purposes.

10.7 WORKING OF PROPOSED MODEL

The working of the proposed model will be divided into four components.

1. Monitor the source for providing power to the electrical equipment from the various sources available (solar, battery, or from a power station).

2. Decide the priority of the device that must have interrupted the power supply.
3. Handle the operations over the IoT system. This unit is also responsible for controlling smoke sensors and water spray jets that are operated in the case of fire.
4. Take care of maintenance for various devices that need to be periodically maintained. So by this means, we can have a smart system for saving electricity as well as use it efficiently.

In this model, we will program the microcontroller to switch between the three available power sources, that is, power from the solar panel, power from the electric board, and power that is stored in available LI-ion batteries.

The proposed model also takes care of storing the adequate power supply of the batteries so that in the worst-case scenarios, emergency systems work efficiently. Switching is done on the uninterrupted power supply (UPS) mode so that a power trip does not occur and ups provide stability while switching. The proposed model also deals with the regular maintenance of the devices as the devices need periodical maintenance. This system in the proposed model will diagnose them, and if needed, it will send the diagnostic report to the manufacturer. It simply works like the computer diagnostics system. It will periodically check every component with various means of checking.

The proposed model is represented in Fig. 10.3. Sen13322 is a moisture sensor. It will sense the moisture of the soil, this will be done by the metal contractors that will be buried inside the soil. As the moisture in the soil increases, the conductivity of the metal contractors increases, so this output is sent to the sensor driver, and this will maintain humidity in the soil and saves water.

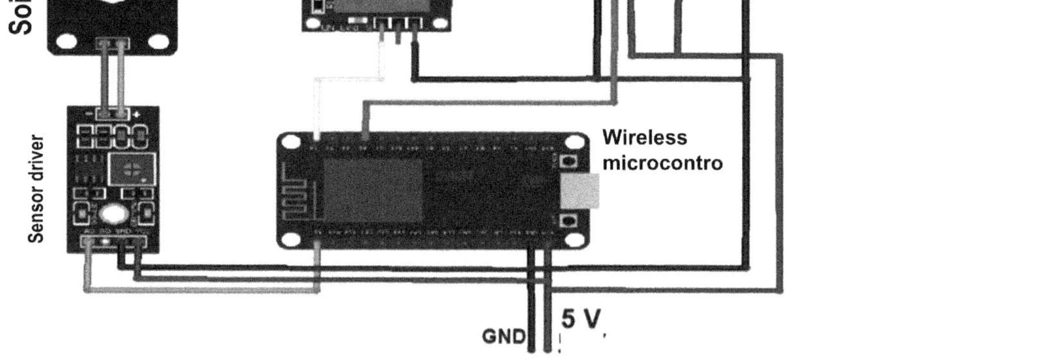

FIGURE 10.3

Proposed intelligent smart home energy-efficient model.

LM393 is a moisture-sensor driver. It converts the conductivity into the electrically readable form for the microcontroller.

Motor driver l293D-IT is responsible for controlling the motor speed or switching on or off by obtaining command from microprocessor DC motor. This motor will act as a pump that will be going to provide water to the plants, and this will become active when the moisture in the soil will be less.

Node MCU wireless controller is the brain of the whole system. It will manage all the processes. It takes input from the humidity sensor as well as from the moisture sensor and accordingly manage the motor function. This NODE MCU can upload all the data on the cloud that can be accessed anywhere in the world. You can change the moisture of the soil according to the plant.

Humidity sensor DHT-11 will sense the moisture in the environment and give this data to the microprocessor. This sensor works on the principle of the capacitance, as the capacitor has dielectric between them. In this case, we have pores that will sense the environment, and as the moisture increases, there will be a change in the capacitance that will be decoded by the microprocessor.

Home must be the safest place to live. We build homes to protect us from various types of dangers that can harm us. Considering this, we need to constantly focus on the safety of the house. We introduce the latest safety measures to do this work more efficiently. To avoid system failure, we have introduced a series of hi-tech devices that not only monitor the four boundaries of a house but also keep the unwanted interferences away from the house. It keeps a check on each in and out in the house with respect to time. This house is not only smart in terms of safety but also in terms of equipment, which provides next-generation living environment like auto-dim window mirrors that control the incoming sunlight just with a small-gesture touch in the smartphone. The systems equipped in the home sends the daily report to the owner's phone.

This report will be consisting of water usage, recycled water, power usage, and other important stuff. The main objective of designing this house is to provide the next generation of living to more and more people in the world so that we can use the most powerful tool available to mankind. Using AI, the smart home will work and gain experience and then apply this experience in future applications just like humans, thus providing the platform for sustainable development.

To meet the 21st century needs, we have introduced IoT in the chain of making hi-tech systems that allow the user to have control over the electrical system from faraway places

10.8 TECHNOLOGY USED IN MAKING THE PROPOSED MODEL

This section will explain various technologies that are used for making the proposed model. We use the help of various technologies to make this smart home come alive. The use of technology is working together, and they are designed to give a favorable outcome. Fig. 10.4 represents all the controls of the fire alarming system that are controlled by the microcontroller PIC18f452. The microcontroller is the device that can control the specific task. In this case, it is doing only one job: to monitor the heating element if the heating elements give the greater input than the threshold value that we specified in the microcontroller (shown in Fig. 10.5) than it has to perform the following process it firstly operates the 12 v relay that will turn on the buzzer that is deployed at various sections of the houses, which will warn all the members of the house, and this system will also

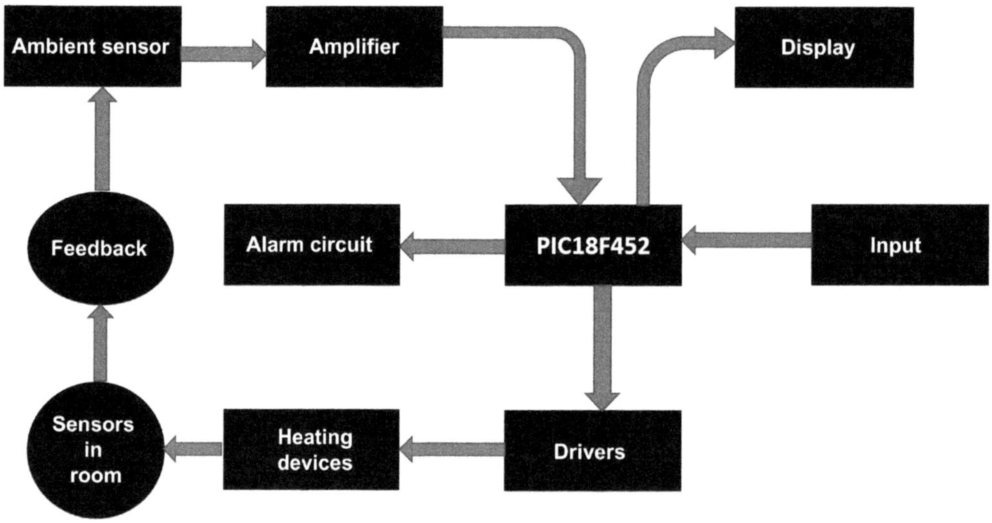

FIGURE 10.4

Working of fire alarm.

operate the world-class function that will turn on the save zone path to all the members, and automatically operate the water and form sprinkler that will off the fire as soon as possible in case of lost amount off error.

This system will always monitor the amount of current. Our circuit is withdrawing the amount of current. As soon as circuits withdraw more amount of current than required, it will switch off the circuit so that it will not short circuit [70]. The notification and path of that damaged zone will be displayed on your phone. This work will be done by the Python library that will create the path of that zone. This microcontroller is having two timmers and 28 pins that is transmitting and receiving ports, which will accomplish our task. Fig. 10.7 shows the circuit diagram.

We are going to use the NodeMCU, which is a microcontroller as well as connect the devices through the data cloud through Wi-Fi that is provided by the user at the same time it provides Bluetooth so that wireless devices can be connected. These wireless devices are connected to the pairing code by the device and the pairing code by the NodeMCU when these codes are matched to the device gets authentication to receive and transmit data [71,72]. Fig. 10.6 shows the NodeMCU appliance. We are going to operate the RGB lighting in the house so that it will give the lighting color according to need, and this will be controlled by this microprocessor.

We are going to produce the R-256, G-256, and B-256. By shifting the values we will get our desired color result. This will be shifted by using WS2811 that will be controlled by data output that will be produced by an addressing code given by the controller to this IC WS2811 [75]. By connecting the power source of the light and one data wire, we can produce all the colors. We need only one data wire for 512 lights because we are going to use 512 demultiplexer that will shift us between various lights and their colors (Fig. 10.7).

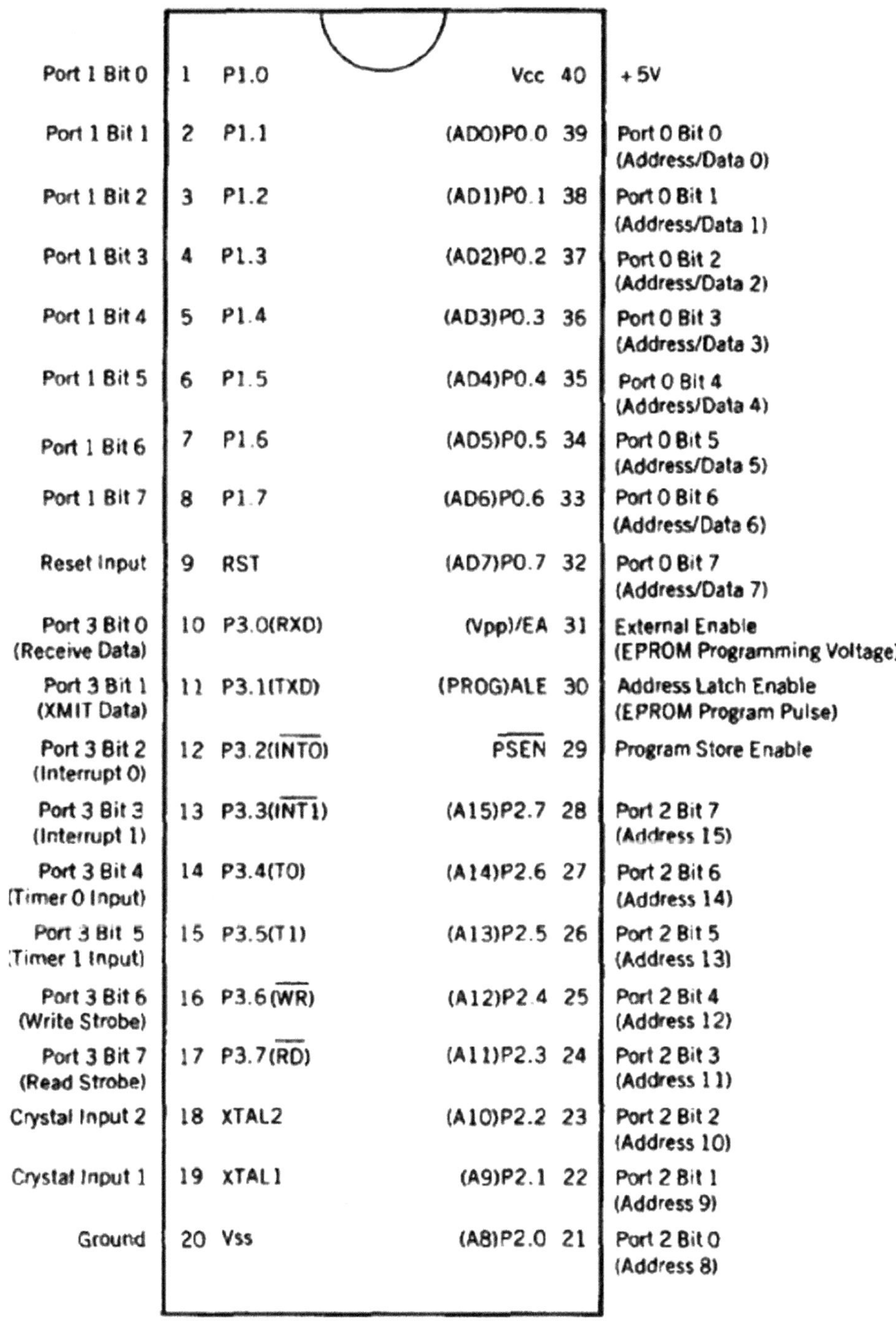

FIGURE 10.5

Microcontroller chip PIC18f452 [68,69].

FIGURE 10.6

NodeMCU [73,74].

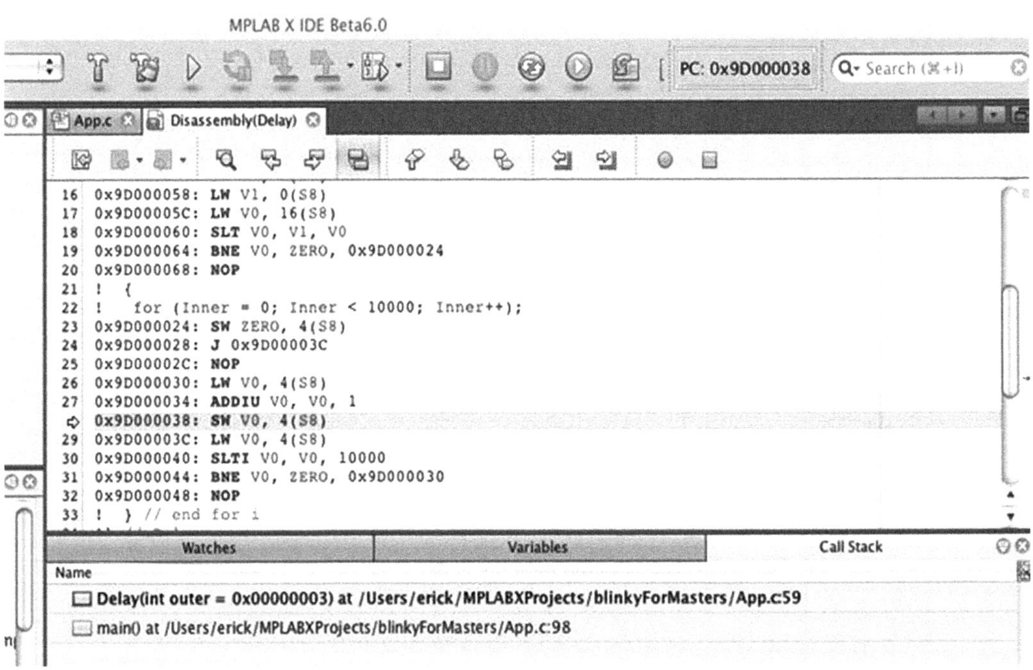

FIGURE 10.7

Coding in PIC18f452.

Table 10.1 Comparison between Proposed Model and Existing Models.

Ref No.	Existing Approach	Our Proposed Approach
[28]	Motion recognition by accelerometer	We used image processing
[35]	Devices controlled by Bluetooth	Devices are controlled by the smartphone connected to Wi-Fi
[41]	Detection of human presence in a room by the change in the concentration of CO_2	Here, human detection is done by the processing of images captured by the cameras capable of identifying the ages
[43]	IoT is used to serve telemedicine	Linking of various sensors attached to the body are sending their data to the cloud, which can be accessed by anyone including doctors and generates alert in case of emergency

10.9 COMPARISON BETWEEN MODELS

The comparison between our proposed model and existing models is shown in Table 10.1.

10.10 ADVANTAGES OF PROPOSED MODEL

This section will describe the advantages of our proposed model. The advantage of the proposed model is that it will make use of present technology that is often used by us. It is not only used by us, but we are surrounded by them (Table 10.1).

For controlling microcontroller and other equipment by mobile phone with the use of a mobile phone-based application, that is connected to the microcontroller over Bluetooth or WiFi, is more easy to use as it is having an easy application portal and user interface. This makes a friendly environment between technology and its user. Worst-case scenario is if the smart home system does not work, the user either can troubleshoot it or cut off the system and switch to manual control mode.

This system will also include various sensors that are connected by various microcontrollers such as NodeMCU that will allow all the sensors to send the data to the cloud created by the user so that the cloud is saving the data of all sensors and can be seen by the user. Fig. 10.8 represents the water-saving process.

10.11 APPLICATIONS OF PROPOSED MODEL

Various applications of the proposed model will be covered in this section. Application of the proposed model is as follows: We can make this proposed model compatible to every household by making small changes to the systems that are used. We will use these smart homes in various applications from saving electricity to saving water, and make use of various techniques that are used by us. These will decrease our efforts.

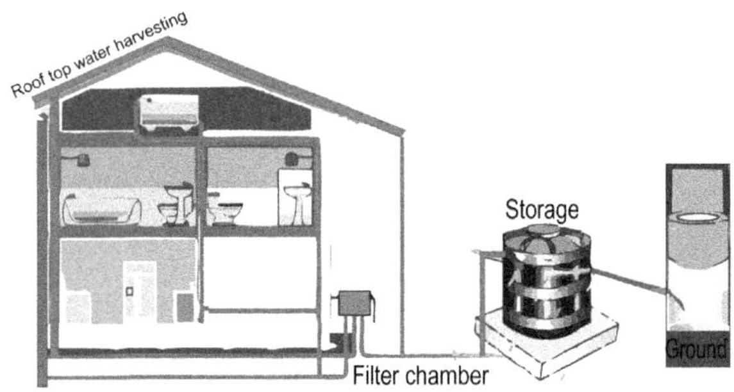

FIGURE 10.8

Water-saving process.

According to need, we can increase or decrease the capability of the system. The system consists of various sensors that can sense the humidity of the air to the intensity of the light coming in the room.

10.11.1 SYSTEM THAT WILL SAVE WATER

We can use this system to purify the water. To purify the water, we use various types of filters to get the most nutrient and distilled water. These filters will include sedimentation, carbon filter, UV, and RO filters. We are also making use of the rain water collected during rain. This water is automatically collected in a separate tank, and extra water is seeped into the ground after treating it [76]. This rainwater system does not start storing the water as the rain starts, rather it firstly allows some water to pass into the toilet flush water tank. This water is used for flushing the toilet.

After some time, when the dust and other contamination in the water are passed or moved into a flash tank and clean water comes, it will be automatically transferred into the other tank that stores the water that can be cleaned and purified according to the need. After storing the water, the system automatically transfers water into the ground by using the pipe that is borewell to the groundwater level.

By using the intelligence system it not only uses rainwater efficiently but also replenishes the groundwater [57]. This system is loaded with hundreds of sensors that are located inside and outside the tanks and pipes and continuously monitor the quality of water. This whole setup for water management is connected through a microcontroller that then sends all the readings to the database, which is further connected to the user's phone. Now the user has access to monitors, the readings, and in case of bad readings, it can command the system for improving the filtration level in case the system is itself not working well [77]. The system also gives the service reminders for the maintenance of the filters that are choking or giving poor quality.

This system is a complete package for monitoring recycling, saving the water, and providing fresh and purified water.

10.11.2 SYSTEM THAT IS RESPONSIBLE FOR TAKING EFFECTIVE DESIGNS

This will take its own decision with the help of the various algorithms that are been used in this system. We are going to take the help of a machine language. Machine language allows the connected devices to make their own decisions. This decision is given to the machine in the form of their inputs. If we connect a door that opens automatically with the help of using infrared sensors or using ultrasonic sensors and motors [78]. In this case, an infrared sensor or ultrasonic sensor will give output to the controller. This controller then processed the inputs given.

This process will include the various algorithms that are preinstalled in the controller. This enables the controller to think and then process. The thinking will include taking the experience from the past or to use the memory that is storing the various outcomes. These various programs in microcontrollers are in the Python language. Python was first introduced in the 1990s, and this language is mostly used to write maching programs. We use the following algorithms in machine learning to obtain various results according to our need. These algorithms are capable of processing images and are capable of playing chess.

One thing that makes Python the most useful and reliable language is its library. The library is nothing but a set of codes that are designed to perform a specific task. When we need to perform a task, we call on that library and make the code according to the pattern to form the code, which is decided by the aforementioned algorithms.

1. *Numpy*: By using this very popular Python library, we can create large and multidimensional arrays. It also reduces the processing time for obtaining results from these large arrays [79]. We can solve a linear equation with the help of this library. It is used to convert the most important form to the other, that is, by using this library we can perform Fourier and Laplace series conversions that are easily used in communication. It converts the time-to-frequency domain that in this smart home project we will perform this operation in various places where we want to send the data from one place to another. In that case we use frequency to carry our original signal that is called carrier frequency [80].
2. *SciPy:* This library is used for performing various mathematics operations, such as linear equations, integration, differential, and statistics. So this provides an enthusiasm in the programmer to perform such types of calculation very quickly and without any extra efforts. There is a difference between the SciPy library and the SciPy stack. SciPy is a core package that makes the SciPy stack. SciPy is also used in image processing.
3. *Scikit-learn:* This library is made up by combining two Python libraries: NumPy and SciPy. This library is used for performing two types of machine-learning operations that are supervised, and the other is an unsupervised learning operation. Supervised learning is the idea of the result, and we calculate the values according to it, while in the case of unsupervised learning we are not knowing the result. We would calculate the result by applying various operations on it. This library is also very useful in performing data-mining and data-analyzing operations [81].
4. *Theano:* This library is meant for mathematical and statisticcs calculations. These calculations may include differentiation, integration, and calculation based on mathematical equations, and consist of a multidimensional array. This library can perform calculations in a very efficient manner by optimizing CPU and GPU. It is also a powerful library for creating a project that involves large-scale computerization that usually takes a long time and complexity by approaching Theano, which performs that operation with less time and simplicity [82].
5. *TensorFlow:* This is a very popular library used for various artificial learning applications. The main advantage of this library is that it can process the raw data of the image easily. This raw

data can be provided by octave or by the MatLab, which gives the data in the form of a multidimensional array. This library firstly uses an AI in the separation of vegetables according to their sizes and shapes [83,84].

6. *Keras:* Keras is a very popular machine-learning library for Python. Keras can run tensor flow programs and run interrupts on CPU and GPU. It is very useful for machine-learning beginners as it can create a prototype easily and quickly. It is also capable of creating a neural network [85].
7. *PyTorch:* This machine-learning library works with C-language along with Lua. It is consisting of various tools that are capable of running on various natural language, computer vision, and different types of machine-learning programs. [86–88].
8. *Pandas:* This library is used for the data analysis work. This will not directly help or play a role in machine learning, but it does the work that makes this library very important. This library works on extracting and processing of the data. It also provides high-level data structures that manage the time and data complexity [89]. It also gives inbuilt functions for groping, combining, and filtering data. Pandas is a popular Python library for data analysis. It is not directly related to machine learning. As we know, the dataset must be prepared before training. In this case, Pandas comes in handy as it was developed specifically for data extraction and preparation. It provides high-level data structures and a wide variety of tools for data analysis. It provides many inbuilt methods for groping, combining, and filtering data.
9. *Matplotlib:* It is not directly used by the programmer in machine learning. It becomes useful when there is a need for two-dimensional designs or layout, such as various types of graphs and plots for data interpretation and visualization. A module in Matplotlib known as Pyplot makes the programming easy to control with line styles, properties, font, axis, etc. [90,91].

Through this, we can understand the basics of machine learning. As in Fig. 10.9 we can see that it is giving the amount of rain an area received in the specific years in inches. So we are given the data of the two alternative years' data, and if one is to find the data of the year 1997 or the year 1999 or 2001, we have to draw the intersection between the x- and y-axis of the specific values that we have to solve. But there are other ways to do that. First, we can form the polynomial equation with two independent variables. This is the example of multiple linear regression. Or we can form the equation with one dependent variable equation. This equation can be formed by making use of approximation. We can assume that either in that year, rain is assumed above 50 or 60 inches, and this dependence or assumption will be formed on the x-axis. If that following year coincides with the rain assumption the following result is obtained. This is the perfect example of linear regression; in this, only one dependent variable is involved.

In similar ways, there are various algorithms to find out the relevant result in various situations. Similarly, our system is going to make use of various assumptions and predictions based upon the past experiences. The artificial Intelligent machine learns from their past experiences; the machine uses its storage for storing the error and uses the same when any similar situation arises in the future also.

10.11.3 SYSTEM THAT IS GOING TO MONITOR IMAGE PROCESSING

In image processing, we use various algorithms. The image we obtain is broken into three parts. These three parts are as follows: part of the image is going to tell us about the size and dimensions

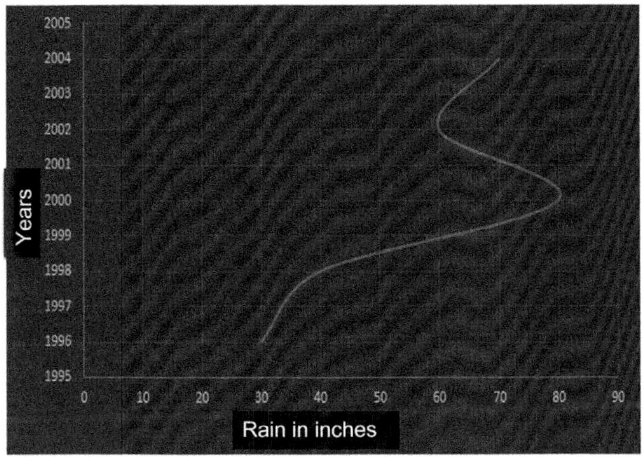

FIGURE 10.9

Prediction diagram.

of the object that is coming to the image. This part is going to give us the physical significance of the image. This part of the image can be used in the CCTV image to detect the presence of the object. Fig. 10.10 shows the processing of an image.

This presence of the object then can be processed to detect which object is where. We can match the image with the database that is storing the relevant data according to need [92]. After scanning our database, it will give the output in the form of information of that image, like in case the camera is going to detect the presence of the dog or other animal. Then the image is being processed and the physical outline of the image will be processed so that we can detect that the processed image is either an animal or it is another object. Fig. 10.11 shows the process of image processing.

10.11.4 DEEP LEARNING: SYSTEM THAT WILL REEVALUATE THE RESULT

Deep learning is the subset or part of machine learning, and as we know, both deep learning and machine learning are the wheels of AI. Both of these languages give the ability to a device to think, store this store input and output, and will work as an experience in the future to store and to use that same output. This is known as intelligence; since it is performed by a machine, it is called AI [93–95].

Plotly is a popular library that allows you to build complex graphics easily. This package is made to work in interactive web applications. Rather, this library has the outstanding ability to create graphics, ternary plots, and 3D charts [96,97].

The continuous enhancements of the library with the latest graphics and features introduced the support for the multiple views that are linked as well as animation and complex integration [98]. It is capable of creating interactive visuals and the animations in web browsers where size or scale can be

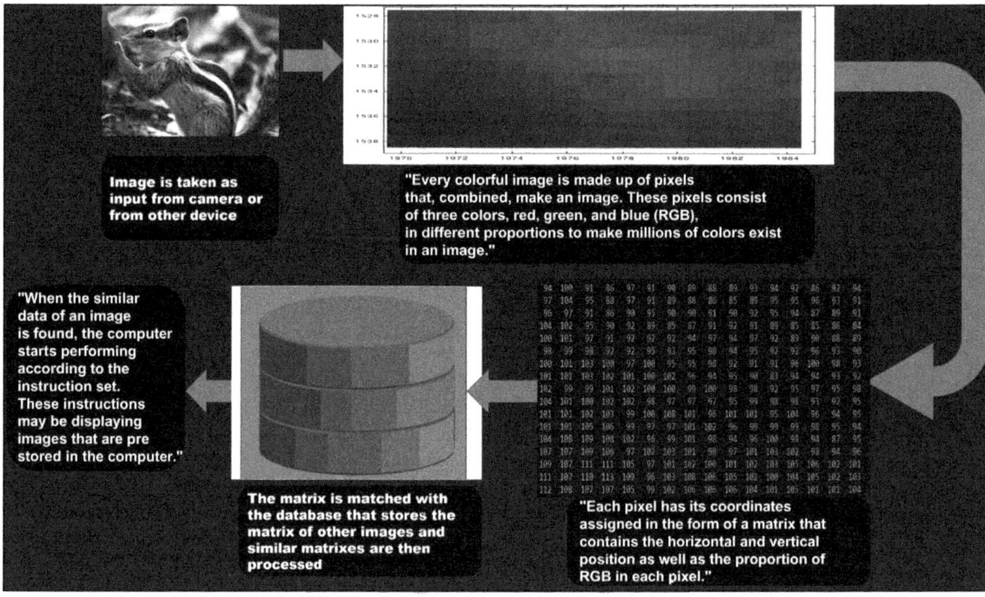

FIGURE 10.10

Image processing.

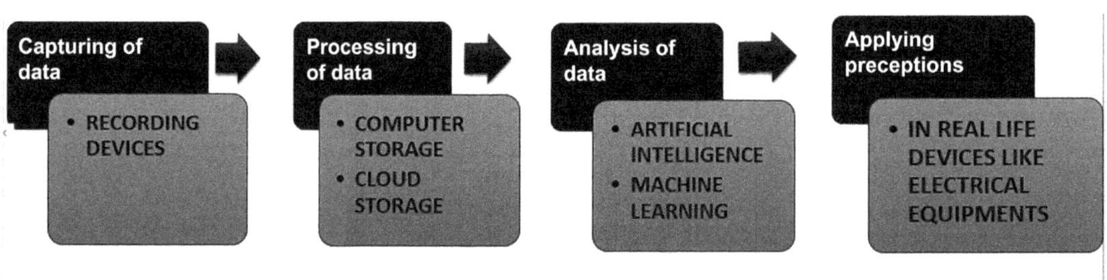

FIGURE 10.11

Image processing process.

changed with the help of well-known JavaScript widgets. This library has preloaded graphs, linking lists, styling, etc. that boosted its capability for giving preferences in the web deployment [99].

10.12 CONCLUSION AND FUTURE SCOPE

Conclusion and future aspects of the proposed model will be covered in this section. The conclusion we can deliver is that this type of house can deliver more convenience to the owners and will provide more sustainable development. As we know, sustainable development is development that will not cost resources for future generations, and we must conserve our resources for future generations as these smart homes use all the resources in a well-defined and planned manner. This will not allow violations of the principles of sustainable development. These types of houses also make it possible to follow the three Rs: reduce, reuse, and recycle.

In the case of water, these houses firstly monitor the consumption of water and try to reduce it, and second, reuse water—that is, it does not draw groundwater unless and until it finds the best resource of it, and last but not least, it recycles the water and uses it for other activities such as gardening and in cooling systems. This not only protects nature it also protects the owners of these houses as it monitors the cameras installed at various places around the house 24-7, and also responds to itself if it finds something dangerous. It also keeps the house well equipped to overcome the natural and unnatural calamities such as long-time electricity cut off, fire, etc.

The future scope of these types of houses is very strong and bright. We are increasing our grip over something as intelligent as a human that has the capability to think and respond according to its needs. It will take no time to absorb the interest of the people to use this very innovative and important technology, which will not only protect nature but also your family at very low maintenance costs. It is simply a one-time investment project that will have the capability to learn itself without any replacement or installation.

REFERENCES

[1] Q. Ni, I. Cleland, C. Nugent, A.B.G. Hernando, I.P. de la Cruz, Design and assessment of the data analysis process for a wrist-worn smart object to detect atomic activities in the smart home, Pervasive Mob. Comput. 56 (2019) 57–70.

[2] D. Sloo, J.W. Boettcher, M. Veron, S. Le Guen, Smart-home hazard detection system providing context-based user notifications, US Patent App. 10/262,507 (Apr. 16 2019).

[3] B. Reeder, J. Chung, K. Lyden, J. Winters, C.M. Jankowski, Older women's perceptions of wearable and smart home activity sensors, Inform. Health Soc. Care (2019) 1–14.

[4] A.M. Fadell, Y. Matsuoka, D. Sloo, M. Veron, Smart-home automation system that suggests or automatically implements selected household policies based on sensed observations, US Patent App. 16/153,147 (Mar. 14 2019).

[5] M. Zouai, O. Kazar, B. Haba, G. Ortiz, N. Kabachi, New approach using an iot robot to oversight the smart home environment, 9 ISTE OpenScience – Published by ISTE Ltd. London, UK, pp. 1–6, 2019.

[6] M. Amadeo, A. Giordano, C. Mastroianni, A. Molinaro, On the integration of information centric networking and fog computing for smart home services, The Internet of Things for Smart Urban Ecosystems, Springer, 2019, pp. 75–93.

[7] D. Sloo, A.M. Fadell, M.L. Rogers, M. Veron, S. Le Guen, Y. Matsuoka, et al., Status indication triggering and user interfacing in a smart-home device, US Patent App. 16/148,404 (Jan. 31 2019).

[8] A. Procopiou, N. Komninos, C. Douligeris, Forchaos: real time application ddos detection using forecasting and chaos theory in smart home iot network, Wirel. Commun. Mob. Comput. 2019 (2019).

[9] G. Wilson, C. Pereyda, N. Raghunath, G. de la Cruz, S. Goel, S. Nesaei, et al., Robot-enabled support of daily activities in smart home environments, Cognit. Syst. Res. 54 (2019) 258–272.

[10] R. Zhang, S. He, X. Yang, X. Wang, K. Li, Q. Huang, et al., Aneog-based human–machine interface to control a smart home environment for patients with severe spinal cord injuries, IEEE Trans. Biomed. Eng. 66 (1) (2019) 89–100.
[11] D. Marikyan, S. Papagiannidis, E. Alamanos, A systematic review of the smart home literature: a user perspective, Technol. Forecast. Soc. Change 138 (2019) 139–154.
[12] N. Panwar, S. Sharma, S. Mehrotra, L-. Krzywiecki, N. Venkatasubramanian, Smart home survey on security and privacy, arXiv preprint arXiv:1904.05476 (2019).
[13] R. Ranjith, N.K. Prakash, D.P. Vadana, A.S. Pillai, Smart home energy management system—a multicore approach, in: International Conference on Advanced Computing Networking and Informatics, Springer, 2019, pp. 363–370.
[14] A. Berger, A. Bischof, S. Totzauer, M. Storz, K. Lefeuvre, A. Kurze, Sensing home: participatory exploration of smart sensors in the home, Social Internet of Things, Springer, 2019, pp. 123–142.
[15] S. Khemakhem, M. Rekik, L. Krichen, Double layer home energy supervision strategies based on demand response and plug-in electric vehicle control for flattening power load curves in a smart grid, Energy 167 (2019) 312–324.
[16] T. Hargreaves, C. Wilson, R. Hauxwell-Baldwin, Learning to live in a smart home, Build. Res. Inf. 46 (1) (2018) 127–139.
[17] O. Hamdan, H. Shanableh, I. Zaki, A. Al-Ali, T. Shanableh, Iot-based interactive dual mode smart home automation, in: 2019 IEEE International Conference on Consumer Electronics (ICCE), IEEE, 2019, pp. 1–2.
[18] X. Wu, X. Hu, X. Yin, S.J. Moura, Stochastic optimal energy management of smart home with pev energy storage, IEEE Trans. Smart Grid 9 (3) (2018) 2065–2075.
[19] S.J. Darby, Smart technology in the home: time for more clarity, Build. Res. Inf. 46 (1) (2018) 140–147.
[20] A. Jacobsson, M. Boldt, B. Carlsson, A risk analysis of a smart home automation system, Future Gener. Comp. Syst. 56 (2016) 719–733.
[21] J.-Y. Fourniols, N. Nasreddine, C. Escriba, P. Acco, J. Roux, G. Soto-Romero, An overview of basics speech recognition and autonomous approach for smart home iot low power devices, J. Signal. Inf. Process. 9 (4) (2018) 239.
[22] D.-M. Han, J.-H. Lim, Smart home energy management system using ieee 802.15. 4 and zigbee, IEEE Trans. Consum. Electron. 56 (3) (2010) 1403–1410.
[23] A. Saha, M. Kuzlu, M. Pipattanasomporn, Demonstration of a home energy management system with smart thermostat control, in: 2013 IEEE PES Innovative Smart Grid Technologies Conference (ISGT), IEEE, 2013, pp. 1–6.
[24] A. Mishra, D. Irwin, P. Shenoy, J. Kurose, T. Zhu, Smartcharge: Cutting the electricity bill in smart homes with energy storage, in: Proceedings of the 3rd International Conference on Future Energy Systems: Where Energy, Computing and Communication Meet, ACM, 2012, p. 29.
[25] W. Li, T. Logenthiran, W.L. Woo, Intelligent multi-agent system for smart home energy management, 2015 IEEE Innovative Smart Grid Technologies-Asia (ISGT ASIA), IEEE, 2015, pp. 1–6.
[26] R. Teymourzadeh, S.A. Ahmed, K.W. Chan, M.V. Hoong, Smart gsm based home automation system, in: 2013 IEEE Conference on Systems, Process & Control (ICSPC), IEEE, 2013, pp. 306–309.
[27] H. Tanaka, H. Suzuki, A. Watanabe, K. Naito, Implementation of secure end-to-end remote control system for smart home appliances on android, in: 2019 IEEE International Conference on Consumer Electronics (ICCE), IEEE, 2019, pp. 1–6.
[28] A. Jalal, M.A. Quaid, M. Sidduqi, Atriaxial acceleration-based human motion detection for ambient smart home system, in: 2019 16th International Bhurban Conference on Applied Sciences and Technology (IBCAST), IEEE, 2019, pp. 353–358.

[29] K.-J. Wang, C.Y. Zheng, Z.-H. Mao, Human-centered, ergonomic wearable device with computer vision augmented intelligence for VR multimodal human-smart home object interaction, in: 2019 14th ACM/IEEE International Conference on Human-Robot Interaction (HRI), IEEE, 2019, pp. 767–768.

[30] M. Dixon, Y. Matsuoka, A.J. Minich, Y. Modi, Systems and methods for detecting gesture events in a smart home system, US Patent App. 10/186,140 (Jan. 22, 2019).

[31] Conaway, James E. Mr. and Istre, Kristina R., "Pico grid-smart home energy management system (2019). Interdisciplinary STEM Teaching & Learning Conference. 23. https://digitalcommons.georgiasouthern.edu/stem/2019/2019/23.

[32] A. Dasgupta, A.Q. Gill, F. Hussain, Privacy of iot-enabled smart home systems, IoT and Smart Home Automation, IntechOpen, 2019.

[33] Ni, Yuzhao, "Map interface for control of smart home appliances", Technical Disclosure Commons, (April 15, 2019) https://www.tdcommons.org/dpubs_series/2143.

[34] H. Zhu, Y. Gao, Y. Hou, Z. Wang, X. Feng, Real-time pricing considering different type of smart home appliances based on markov decision process, Int. J. Electr. Power Energy Syst. 107 (2019) 486–495.

[35] T. Adiono, S.F. Anindya, S. Fuada, K. Afifah, I.G. Purwanda, Efficient android software development using mit app inventor 2 for bluetooth-based smart home, Wirel. Personal. Commun. 105 (1) (2019) 233–256.

[36] D. Lan, Z. Pang, C. Fischione, Y. Liu, A. Taherkordi, F. Eliassen, Latency analysis of wireless networks for proximity services in smart home and building automation: the case of thread, IEEE Access. 7 (2019) 4856–4867.

[37] H. Hosseinian, H. Damghani, Smart home energy management, using iot system, Database 8 (2019) 9.

[38] Zhang, Y., Tian, G., Zhang, S., & Li, C. (2019). A knowledge-based approach for multiagent collaboration in smart home: From activity recognition to guidance service. IEEE T. Instrum. Meas., 69(2), 317–329.

[39] P. Geraldo Filho, L.A. Villas, V.P. Gonçalves, G. Pessin, A.A. Loureiro, J. Ueyama, Energy-efficient smart home systems: infrastructure and decision-making process, Internet Things 5 (2019) 153–167.

[40] Z. Zhang, A. Georgiadis, C. Cecati, Wireless power transfer for smart industrial and home applications, IEEE Trans. Ind. Electron. 66 (5) (2019) 3959–3962.

[41] J. Vanus, M. Fajkus, R. Martinek, S. Zabka, M. Stolarik, Smart home room's occupancy monitoring using fiber bragg grating sensor, Int. Soc. Opt. Photon. (2019) 110281F.

[42] S. Prange, C. Tiefenau, E. von Zezschwitz, F. Alt, Towards understanding user interaction in future smart homes, CHI'19 Workshop, pp. 1–5, May 5, 2019, Glasgow, Scotland UK (2019).

[43] M. Talal, A. Zaidan, B. Zaidan, A. Albahri, A. Alamoodi, O. Albahri, et al., Smart home-based iot for real-time and secure remote health monitoring of triage and priority system using body sensors: multi-driven systematic review, J. Med. Syst. 43 (3) (2019) 42.

[44] A.M. Fadell, M.L. Rogers, Y. Matsuoka, D. Sloo, S. Honjo, S.A. McGaraghan, et al., Environmental sensing with a doorbell at a smart-home, US Patent 9,960,929 (May 1 2018).

[45] D. Sloo, N.U. Webb, E.J. Fisher, Y. Matsuoka, A. Fadell, M. Rogers, Smart-home control system providing hvac system dependent responses to hazard detection events, US Patent 9,905,122 (Feb. 27 2018).

[46] A.M. Fadell, Y. Matsuoka, D. Sloo, M. Veron, Monitoring and reporting household activities in the smart home according to a household policy, US Patent 9,872,088 (Jan. 16 2018).

[47] V. Pilloni, A. Floris, A. Meloni, L. Atzori, Smart home energy management including renewable sources: a qoe-driven approach, IEEE Trans. Smart Grid 9 (3) (2018) 2006–2018.

[48] J. Shen, C. Wang, T. Li, X. Chen, X. Huang, Z.-H. Zhan, Secure data uploading scheme for a smart home system, Inf. Sci. 453 (2018) 186–197.

[49] J. Palanca, E. del Val, A. Garcia-Fornes, H. Billhardt, J.M. Corchado, V. Julián, Designing a goal- oriented smart-home environment, Inf. Syst. Front. 20 (1) (2018) 125–142.

[50] M. Javadi, M. Marzband, M. FunshoAkorede, R. Godina, A. Saad Al-Sumaiti, E. Pouresmaeil, A centralized smart decision-making hierarchical interactive architecture for multiple home microgrids in retail electricity market, Energies 11 (11) (2018) 3144.

[51] C. Keerthisinghe, G. Verbič, A.C. Chapman, A fast technique for smart home management: ADP with temporal difference learning, IEEE Trans. Smart Grid 9 (4) (2018) 3291–3303.

[52] M. Li, W. Gu, W. Chen, Y. He, Y. Wu, Y. Zhang, Smart home: architecture, technologies and systems, Procedia Comp. Sci. 131 (2018) 393–400.

[53] A. Najafi-Ghalelou, S. Nojavan, K. Zare, Robust thermal and electrical management of smart home using information gap decision theory, Appl. Therm. Eng. 132 (2018) 221–232.

[54] U. Datta, A. Kalam, J. Shi, Electric vehicle (EV) in home energy management to reduce daily electricity costs of residential customer (2018).

[55] S.P. Singh, A. Nayyar, R. Kumar, A. Sharma, Fog computing: from architecture to edge computing and big data processing, J. Supercomput. 75 (4) (2019) 2070–2105.

[56] H. Lu, Y. Li, M. Chen, H. Kim, S. Serikawa, Brain intelligence: go beyond artificial intelligence, Mob. Netw. Appl. 23 (2) (2018) 368–375.

[57] R.Y.M. Li, H. Li, C. Mak, T. Tang, Sustainable smart home and home automation: big data analytics approach, Int. J. Smart Home 10 (8) (2016) 177–187.

[58] A.-R. Al-Ali, I.A. Zualkernan, M. Rashid, R. Gupta, M. Alikarar, A smart home energy management system using iot and big data analytics approach, IEEE Trans. Consum. Electron. 63 (4) (2017) 426–434.

[59] A. Dorri, S.S. Kanhere, R. Jurdak, P. Gauravaram, Blockchain for iot security and privacy: The case study of a smart home, in: 2017 IEEE International Conference on Pervasive Computing and Communications Workshops (PerCom Workshops), IEEE, 2017, pp. 618–623.

[60] O. Hamdan, H. Shanableh, I. Zaki, A. Al-Ali, T. Shanableh, Iot-based interactive dual mode smart home automation, in: 2019 IEEE International Conference on Consumer Electronics (ICCE), IEEE, 2019, pp. 1–2.

[61] H. Lin, N. Bergmann, Iot privacy and security challenges for smart home environments, Information 7 (3) (2016) 44.

[62] S.P. Singh, R. Maini, Comparison of data encryption algorithms, Int. J. Comp. Sci. Commun. 2 (1) (2011) 125–127.

[63] M.E. Porter, J.E. Heppelmann, How smart, connected products are transforming companies, Harv. Bus. Rev. 93 (10) (2015) 96–114.

[64] F.Y. Okay, S. Ozdemir, A fog computing based smart grid model, 2016 International Symposium on Networks, Computers and Communications (ISNCC), IEEE, 2016, pp. 1–6.

[65] S.P. Singh, A. Sharma, R. Kumar, Analysis of load balancing algorithms using cloud analyst, Inter. Natl. J. Grid Distrib. Comput. 9 (9) (2016) 11–24.

[66] E.A. Pansotra, E.S.P. Singh, Cloud security algorithms, Int. J. Secur. Appl. 9 (10) (2015) 353–360.

[67] K. Xu, X. Wang, W. Wei, H. Song, B. Mao, Toward software defined smart home, IEEE Commun. Mag. 54 (5) (2016) 116–122.

[68] L.A. Ajao, O.M. Olaniyi, J.G. Kolo, A.O. Ajao, Project-based microcontroller system laboratory using bk300 development board with pic16f887 chip, Int. J. Embedded Syst. Appl. 5 (3) (2015) 15–28.

[69] L.A. Ajao, M.A. Adegboye, E.M. Dogo, S.O. Aliyu, D. Maliki, Development and implementation of microcontroller-based improved digital timer and alarm system, in: International Conference on Information and Communication Technology and Its Applications (ICTA 2016), 2016, pp. 184–190.

[70] Vanaja, K. J., Suresh, A., Srilatha, S., Vijay Kumar, K., & Bharath, M. (2018). IOT based agriculture system using node MCU. Int. j. sci. eng. technol., 5(3), 3025–3028.
[71] S.L. Siregar, M. Rivai, Monitoring dankontrolsistempenyemprotan air untukbudidayaaeroponikmenggunakannodemcu esp8266, JurnalTeknik ITS 7 (2) (2019) 380–385.
[72] M. Kashyap, V. Sharma, N. Gupta, Taking mqtt and nodemcu to iot: communication in internet of things, Procedia Comp. Sci. 132 (2018) 1611–1618.
[73] A.C. Bento, Iot: nodemcu 12e x arduinouno, results of an experimental and comparative survey, Int. J. 6 (1) (2018).
[74] B.R. Babu, B. Anudeep, M. Yugma, M. Meghana, S. Swami, Real time iot based office automation system using nodemcu esp8266 module, Int. J. Res. 6 (4) (2019) 250–254.
[75] M. Edward, K. Karyono, H. Meidia, Smart fridge design using nodemcu and home server based on raspberry pi 3, in: 2017 4th International Conference on New Media Studies (CONMEDIA), IEEE, 2017, pp. 148–151.
[76] M.J. Mudumbe, A.M. Abu-Mahfouz, Smart water meter system for user-centric consumption mea- surement, in: 2015 IEEE 13th International Conference on Industrial Informatics (INDIN), IEEE, 2015, pp. 993–998.
[77] J. Singer, M. Simmons, S. Jansen, "Smart control for home water heater saving" (2016). Mechanical Engineering Senior Theses. 53. https://scholarcommons.scu.edu/mech_senior/53.
[78] M.C. Hemalatha, M.R. Nagarajan, P. Suresh, G.G. Shankar, A. Vijay, Brushless dc motor controlled by using internet of things, IJSTE-Int. J. Sci. Technol. Eng. 3 (09) (2017) 373–377.
[79] L. Drude, J. Heymann, C. Boeddeker, et al., Nara-wpe: A python package for weighted prediction error dereverberation in numpy and tensorflow for online and offline processing, in: Speech Communication; 13th ITG-Symposium, VDE, 2018, pp. 1–5.
[80] J.A. Snyman, D.N. Wilke, Practical computational optimization using python, Practical Mathematical Optimization, Springer, 2018, pp. 311–340.
[81] J. Hao, T.K. Ho, Machine learning made easy: A review of scikit-learn package in python programming language, J. Edu. Behav. Statist. (2019) 1076998619832248.
[82] R. Al-Rfou, G. Alain, A. Almahairi, C. Angermueller, D. Bahdanau, N. Ballas, et al., Theano: a python framework for fast computation of mathematical expressions, arXiv preprint arXiv.1605.02688 (2016).
[83] D.G. Matthews, G. Alexander, M. Van Der Wilk, T. Nickson, K. Fujii, A. Boukouvalas, et al., Gpflow: a gaussian process library using tensorflow, J. Mach. Learn. Res. 18 (1) (2017) 1299–1304.
[84] J. Malmaud, L. White, Tensorflow. JL: an idiomatic julia front end for tensorflow, J. Open. Source Softw. 3 (31) (2018) 1002.
[85] F. Chollet, Deep Learning mit Python und Keras: Das Praxis-HandbuchvomEntwickler der Keras-Bibliothek, MITP-Verlags GmbH & Co., KG, 2018.
[86] N. Ketkar, Introduction to pytorch, Deep Learning with Python, Springer, 2017, pp. 195–208.
[87] M. Ravanelli, T. Parcollet, Y. Bengio, The pytorch-kaldi speech recognition toolkit, in: ICASSP 2019–2019 IEEE International Conference on Acoustics, Speech and Signal Processing (ICASSP), IEEE, 2019, pp. 6465–6469.
[88] P. Mishra, Introduction to pytorch, tensors, and tensor operations, PyTorch Recipes, Springer, 2019, pp. 1–27.
[89] W. McKinney, Pandas: a python data analysis library, see http://pandas.pydata.org (2015).
[90] A. Yim, C. Chung, A. Yu, Matplotlib for Python Developers: Effective Techniques for Data Visualization with Python, Packt Publishing Ltd, 2018.
[91] J. VanderPlas, Python Data Science Handbook: Essential Tools for Working With Data, O'Reilly Media, Inc., 2016.

[92] F. Erden, S. Velipasalar, A.Z. Alkar, A.E. Cetin, Sensors in assisted living: a survey of signal and image processing methods, IEEE Signal. Process. Mag. 33 (2) (2016) 36–44.
[93] J. Wang, Y. Chen, S. Hao, X. Peng, L. Hu, Deep learning for sensor-based activity recognition: a survey, Pattern Recognit. Lett. 119 (2019) 3–11.
[94] M.M. Hassan, M.Z. Uddin, A. Mohamed, A. Almogren, A robust human activity recognition system using smartphone sensors and deep learning, Future Gener. Comp. Syst. 81 (2018) 307–313.
[95] L. Zhao, J. Wang, J. Liu, N. Kato, Routing for crowd management in smart cities: a deep reinforcement learning perspective, IEEE Commun. Mag. 57 (4) (2019) 88–93.
[96] O.R. Bingol, A. Krishnamurthy, Nurbs-python: an open-source object-oriented nurbs modeling framework in python, SoftwareX 9 (2019) 85–94.
[97] F.A. Nielsen, Data mining with python (working draft) (2017).
[98] I.K. Na, G.Y. Lee, K.W. Song, D.Y. Hawng, Smart home system using portable device, US Patent 9,948,685 (April 17, 2018).
[99] M. Yesilbudak, I. Colak, Main barriers and solution proposals for communication networks and in- formation security in smart grids, in: 2018 International Conference on Smart Grid (icSmartGrid), IEEE, 2018, pp. 58–63.

CHAPTER

ADAPTIVE COMPLEX SYSTEMS: DIGITAL TWINS

11

Tony Clark[1] and Vinay Kulkarni[2]

[1]*School of Engineering and Applied Science, University of Aston, Birmingham, United Kingdom*

[2]*Software Systems & Services Research, TCS Research, Pune, India*

11.1 COMPLEX SYSTEMS

Systems, like production plants, logistics networks, IT service companies, and international financial companies, are characterized as being large complex networks of components with many-to-many communication channels with sophisticated information processing that makes prediction of system states difficult [1]. Such systems operate in highly dynamic environments that need to respond quickly to a variety of change drivers leading to behavior that is inherently uncertain. Typically, our understanding of how such systems operate is limited to localized areas since the entire system is just too big to understand in its entirety [2]. Knowing how to analyze, design, control, and adapt such systems is a difficult problem that lacks suitable mainstream engineering methodologies and technologies.

Conventional techniques for system-wide analysis either lack rigor or are defeated by the scale of the problem, which explains why the current practice often exclusively relies on humans for monitoring and adaptation. The behavior of a complex system is often impossible to express using standard modeling techniques like system dynamics because a complete top-down representation of the system behavior is not available. In most cases there is no single locus of control producing a system with *emergent behavior*.

The scale of enterprise software is leading to large, networked, semiautonomous interdependent systems. As a result, it is increasingly difficult to consider software systems in isolation, instead, they form an ecosystem characterized by a variety of interactions between them. Any new system is thus deployed into a connected world. Execution of the new system involves interacting with previously unknown systems requiring the new system to somehow be future-proofed in terms of maintaining its own goals and taking advantage of new situations being presented by the ecosystem [3]. The deployed system must interact in ways that achieve its required behavior while adapting to new situations. It is no longer possible to consider a single system as having a fixed behavior: it must learn to adapt in order to achieve its intended function and even change its function over time.

Given that such systems often do not have a well-understood single locus of control, desirable outcomes must be achieved through incremental adaptation through learned behavior. A model can be quickly produced that is *good enough* to provide decision-support and even adaptive control.

Artificial Intelligence to Solve Pervasive Internet of Things Issues. DOI: https://doi.org/10.1016/B978-0-12-818576-6.00011-3

A combination of machine learning and human experts can then be used to apply a quality improvement process to the models that leads to a nonoptimal, but satisfactory, *digital twin*.

11.2 APPROACHES TO COMPLEX SYSTEM ENGINEERING

Current industry practice relies principally on human expertise to achieve adaptation; however, this is time consuming, resource intensive, difficult to verify, and relies on tacit knowledge that is transient. Techniques like *system dynamics* [4] can be applied in the form of equations that are either solved or used as the basis of simulation runs. This approach works where a system is *linear* [5]; however, modeling of complex enterprise-scale systems does not lend itself to linearity since behavioral knowledge is uncertain and incomplete and control is highly distributed, as noted in [5]: "A comprehensive analytical framework is needed to address the control of *adaptive* networks."

Agents have been proposed as an alternative to system dynamics for emergent behavioral modeling [6]. Agent-based modeling and simulation is characterized by constructing systems in terms of autonomous, goal-directed individuals where control is distributed amongst a population as opposed to discrete event simulation, which tends to focus on functions and processes [7]. Agent simulations have been applied in a wide range of application domains including: biological systems, social science simulation, healthcare, planning and scheduling, traffic control and planning, geographic simulation, manufacturing, supply chains, defense applications, stock analysis, and super computer performance.

This challenge is addressed in control theory using approaches like model reference adaptive control (MRAC) [8] as shown in Fig. 11.1. MRAC controls a complex system by analyzing the trace of its outputs and providing controls as inputs. A model is used to specify the desirable behavior of the system. A monitoring and sense-making component compares the required model

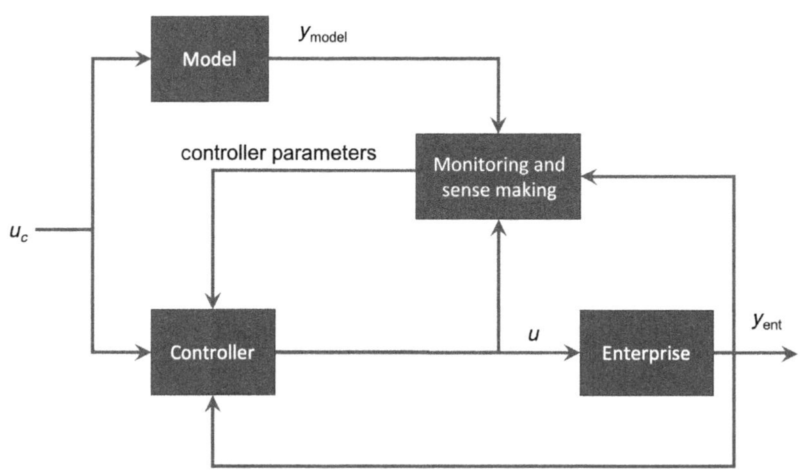

FIGURE 11.1

Model reference adaptive control.

behavior against the actual system behavior and informs a controller that modulates the system through a control interface.

MRAC is often applied to the control of physical systems and signal processing whose complexity arises due to interactions between elements. The properties of each individual component are reasonably well understood, but the combination of their behaviors in order to achieve a collective goal is not. A simple example is the control of a motor engine, which is achieved using a collection of sensors and feedback loops. The overall desired behavior of the system might be difficult to specify completely, however, bounds on the behavior can be precisely defined, for example, in terms of tolerances: the motor engine is maintained within tolerances through actuators that are triggered as a result of sensor information.

Modern software systems have become as complex and unpredictable as the systems that have traditionally been controlled using MRAC, which can therefore provide the basis of a mechanism to design and control large-scale complex software. MRAC techniques have been developed in the traditional domains of application and our challenge is to develop appropriate technologies and methods to support MRAC-style adaptive control for IT solutions.

11.3 DIGITAL TWINS

Digital twins were first proposed at a presentation in 2003 and subsequently published as a white paper [9]. The original concept involved three main parts: a physical product; a virtual product; connections between the two. The original aim of such a concept was to reduce the mental burden on human designers who, up to this point, had performed much of the design of complex systems in terms of raw data without a domain specific conceptual framework. Such a twin also supports comparison since the fidelity of the simulation model and the physical product is high. Finally, the original conception of a digital twin aimed to support collaboration between team members by providing a convenient and practical virtual product as a means for sharing information and decision-making.

Since the original conception, digital twins have been used to predict behavior and used at run time to understand existing behavior, particularly in the context of failure, leading to large cost reduction [10]. Application areas include the Aerospace Industry [11], manufacturing production systems, physical products, Industry 4.0.

Digital twins, as originally conceived, were intended to be applied to complex physical products, and thereby provide a convenient and cost-effective basis for design and verification. Once the product has been engineered, the digital twin is no longer used, or is used only when reengineering is required. The intended use-cases for digital twins involved physical products; our approach aims to generalize the use of digital twins so that they can be used to both engineer complex systems including IT systems and to control and adapt such systems once they have been constructed.

11.3.1 DIGITAL TWIN USE-CASES

A *complex system* is any system whose holistic behavior is difficult or impossible to capture precisely. It is characterized by being made up of many subsystems whose individual behaviors are both autonomous and stochastic, and by existing in an environment, which itself is a complex system. In general, we know *what* the components in a complex system, including the entire system itself, must achieve, but we do not necessarily know *how* it is doing what we observe or how to go about controlling it. A generalized definition for *digital twin* is motived by the following use-cases with respect to complex systems:

C1: Analysis A key requirement is to verify that a complex system is operating as required. A digital twin can provide a cost-effective solution within a simulation environment producing an execution trace that can be examined for desired (and undesired) behavioral patterns. A digital twin can also support what-if and if-what scenario playing to explore the state space.

C2: Design A new complex system can start life as a digital twin that is used as a blueprint. The twin provides a specification of the behavior for the real system and can be integrated with existing systems in the target ecosystem by observing their outputs. The design can then use adaptation to tailor its behavior with respect to real ecosystem data.

C3: Control A digital twin can run alongside the real system and be used to produce control commands based on a comparison of the observed and desired behavior. This leads to the idea of a digital twin being used for continuous improvement or *adaptation* of complex system behavior.

C4: Transformation Maintenance—status quo preserving or evolutionary—is the single most expensive activity in a system lifecycle and can be responsible for over 60% of the overall costs. This is largely due to the present inability to explore the solution space effectively and efficiently. A digital twin can overcome this hurdle through what-if and if-what scenario playing to help arrive at a feasible transformation path from *as is* to the desired *to be* state in silico. Once the transformation path is vindicated, the necessary changes can be introduced into the real system in the right order thus providing assurances of correctness.

The use-cases **C1** and **C2** correspond to those associated with the original definition of a digital twin. However, the generalization of the domain of application to any complex system, especially in the context of complex IT systems, and the introduction of the idea of using a digital twin to transform and adapt an existing system is new. Our vision is based on a conceptual framework based on using digital twins to support the use cases listed earlier. When designing a new complex system that is to be deployed into an existing digital ecosystem, knowledge about as many system components as possible is acquired and used to construct behavioral models. Such models are likely to be incomplete and behavioral responses to given situations will be underspecified; however, they will be a starting point.

11.3.2 DIGITAL TWIN CONCEPTS

The starting point for a digital twin is a *goal behavior*, which takes the form of a description of desirable and undesirable observable behavior. If we are aiming to produce a twin that conforms as-is to an existing system, then the goal may take the form of real-world execution traces. If we are building a twin as a system design then the goal may be a predicate over simulation traces.

Given a clear definition of what the overall system should do, it will be possible to use the simulated components to produce a system history. The history can be compared to what we expect the system to do, and what the system must not do. Such comparison may occur during the simulation or after it has completed; in either case it is likely that the observed behavior is not optimal in which case it can be used as the basis of *behavioral adaptation*. If it occurs during run time it is *dynamic* adaptation and otherwise it is *static* adaptation. In both cases the adaptation incrementally modifies the digital twin toward exhibiting the expected behavior.

An adaptable digital twin can then be run many times to produce a simulation that conforms to the goal. Alternatively, the goal may be a steady state and the twin uses adaptation to reach the steady state and then maintain it. Given the complexity of the systems of interest, it is unlikely that the goal will ever be reached; however, the approach is based on incremental improvement and provides an iterative approach to addressing problems for which we have no other technology. We propose that the approach will lead to a digital twin that is *good enough* to support the desired use cases.

A *self-adaptive system* is able to automatically modify it self in response to changes in its operating environment [12] and relies on features like self-awareness and context-awareness. Research on self-adaptation has led to a taxonomy of features that must be considered like time, goals, control, communication, and context [13] leading to questions where, when and how to adapt within a system. A key problem when developing digital twins for complex adaptive systems is that the behavioral descriptions of the system are not available which makes *engineering* adaptation mechanisms, like the Rainbow Framework [14] difficult. As noted earlier, scale and complexity force us to take an emergent behavioral view of a complex system, which means that it is difficult to know a priori where to place reflection and what is being reflected upon.

Conceptually a digital twin must represent behavior *bottom-up* because we have no way of knowing the behavior *top-down* due to the scale, complexity, and uncertainty of the real system [15]. Our proposal is that it is normally the case that the atomic components of a complex system are reasonably well known and their behavior can be modeled. Further, we propose that it will normally be the case that the *behavioral potential* for each atomic component can be precisely represented, even if the choices between responses to events received by a component are governed by probabilities, or are even completely nondeterministic.

In most cases, the domain knowledge required to construct a digital twin will reside across an organization in policy documents, human experts, existing systems, and so on. This information must be gathered together to form executable models, behavioral specifications. Therefore our proposed digital twin *method* will include appropriate knowledge acquisition techniques, possibly aided by IT systems that use natural language processing to extract conceptual models to be used as the basis of atomic component behavioral models, operating policies, and system goals.

In many cases, the atomic components of a digital twin will need to be *autonomous* because we lack a clear understanding of the locus of control for the complete system. System execution emerges from the independent behavior of the components and their interactions. This implies that each atomic component will have their own goals, behavior, and state, all of which is *encapsulated*. A digital twin can therefore be constructed as a multi agent system (MAS) [16]. If we have no knowledge of the internal structure of the real system then the MAS consists of a collection of independent atomic agents. More realistically, we will have some idea of the internal structure, which can be reflected in a *hierarchical decomposition* of agents. Composite agents may also have goals state and behavior; an analogy is a sales department viewed as an agent containing sales executive agents.

11.3.3 TECHNOLOGY FOR DIGITAL TWINS

The conceptual basis for digital twins outlined earlier may be realized in many different ways. This chapter proposes an approach that uses the actor model of computation [17] to encode agents.

Computation is performed in the actor model by sending asynchronous messages as shown in Fig. 11.2. Each actor encapsulates state and behavior and includes a mailbox that is used to queue messages as they are received. Actors run in their own separate threads of execution that selects the next message that is used to select the most appropriate action.

Once the action has completed, an actor's thread of control inspects the mailbox again, if a message is waiting then the process repeats, otherwise the thread of control waits until a message is received.

The actor model is appropriate as a basis for encoding digital twins because the behavior of an actor system is intrinsically *bottom-up*. Unlike traditional programming architectures, there is no single locus of control: each actor is autonomous and independent. Of course, there is no reason why a more traditional architecture cannot be constructed using actors (and vice versa); however, the starting point for building a digital twin matches the concepts outlined in the previous section.

When selecting an actor-based technology for digital twins we believe there are a number of features that should be considered. Firstly, although a *degenerate* twin will have no composite system structure because we do not have sufficient information about how a complex system is organized, it is realistic to assume that we have some information relating to the internal decomposition of a system. Decomposition of a system with state leads to a consideration of *shared state* where access to the internal state of a parent is shared between the children. The actor model of computation does not support shared state; however, we propose that this is an important feature of the

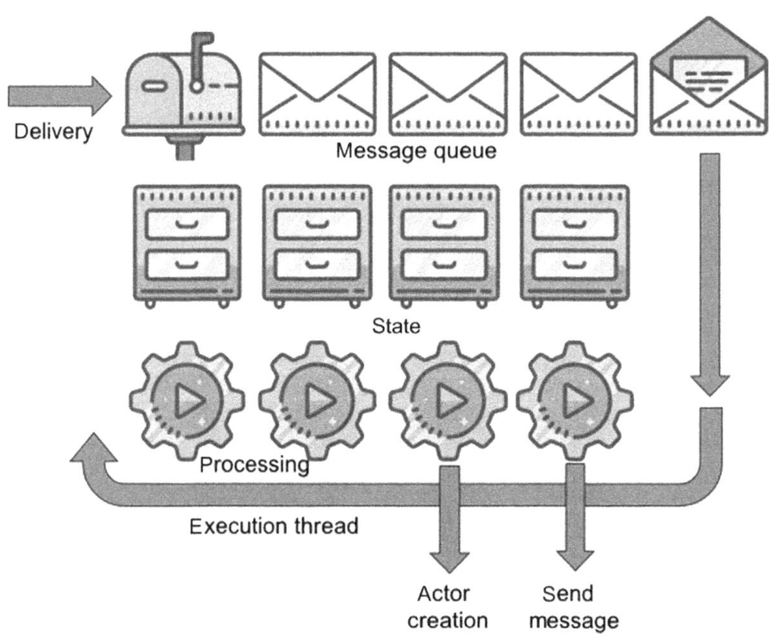

FIGURE 11.2

The actor model [18].

technology base for digital twins as we shall see in the proposed architecture defined in the following text. Providing shared state implies language support for granting exclusive access, for example via locks.

Many use-cases for digital twins will apply the twin model as a simulation of a real system. A simulation is often driven by events that are received from a simulation environment. Events correspond to messages sent to actors and an actor system defines *receptionists* at the interface of the digital twin to receive messages from the environment [19]. Although it is not part of the basic actor model of computation, it is useful for actor-based simulations to receive time events at regular intervals in order to ensure that actors can coordinate and perform future actions.

A *pure* actor-based system performs computation exclusively in terms of asynchronous messages. This provides guarantees about data access, deadlock, and race conditions. Synchronous communication can be implemented in a pure actor model using *continuations* where instead of waiting for a result, a message is sent with a continuation actor to whom the result should be sent. This architecture provides a form of call-and-return while preserving the characteristics of pure actors. While this is attractive, we feel that it can become unwieldy and, assuming shared state is provided as noted above, language features must be provided to address race conditions. Therefore we suggest that it is convenient to provide features for synchronous message passing.

Actors alone cannot realize adaptive MASs [20], there must be some mechanism for representing goals and performing the adaptation that incrementally modifies the system toward a global behavior that has not been directly encoded in any of the individual computational units. MAS is characterized by the following features: communication, goals, beliefs (state), and inference [21]. The last feature, *inference*, aims to allow the autonomous agents to behave intelligently and may take the form of components that perform logical reasoning or planning.

A digital twin model differs from a traditional MAS in that when the twin is constructed, we do not know the intelligence that is often a part of a MAS model. Furthermore, we are not particularly concerned with exhibiting intelligence at the agent level, we are aiming to adapt the overall system, bottom-up, in order to conform to a specified desirable behavior. Therefore while technology for defining agent intelligence may be used when initially constructing atomic components in a digital twin, it is not an intrinsic part of the model.

Instead of inference, agents, individually or collectively, must *adapt*. Therefore we seek a technology for constructing agents that consists of communication, goals, beliefs (state), and adaptation. Actors are sufficient to provide communication and state. We propose that goals can be encoded as predicates over state histories, both of which are supported by actors. Therefore we seek to extended the actor model with features that support adaptation.

11.3.4 ADAPTATION THROUGH MACHINE LEARNING

In order to adapt, agents may use *machine learning*, which is a computational process whereby a data model (often a function) is constructed from data. A digital twin can use machine learning to refine (or *adapt*) the nondeterministic (or underspecified) behavior of the atomic agents in order to achieve the goal. Machine learning may be *supervised*, in which case existing data is used to drive the adaptation. This is equivalent to comparing the behavior of the digital twin against the system goal and the behavior of the real system, provided as an execution trace. Alternatively, machine learning may be unsupervised, in which case there is no existing data; the desired behavior is guided by the goal alone. The third main machine learning approach is *reward* based, which uses the goal and rewards provided by an environment to guide the construction of the data model [22].

As noted in Ref. [22], supervised learning is often challenging due to the lack of historic data. Rewards are an attractive approach in the context of MAS since rewards can be associated with individual actions, which are the key behavioral features of agents that we wish to refine. Such an approach does not *depend* on historic data, although it can benefit from it. A suitable form of reward-based machine learning is reinforcement learning (RL), which exhibits good convergence properties and can be expressed in terms of agent actions [22].

Reinforcement learning involving agents (MARL) proceeds by trial-and-error interactions with an environment [23] as shown in Fig. 11.3. At each step, an agent performs an action and receives a reward. The reward is an indication of how well the action has moved the agent toward a goal, and has prevented the agent from performing an undesirable action. The reward is indicative, in the sense that the absolutely correct action need not be known (as it would be in supervised learning), but it is more informative than completely unsupervised learning.

RL involves a set of probabilistic agent actions that map states to states and a reward function that maps action transitions to rewards. When it is selected to perform an action, an agent uses its policy that maps its current state to probabilistic actions. Performing the action produces a new state and an associated reward. The aim of RL is to use the policy to select actions that maximize the discounted return over time. A discount factor is used to reduce the weighting of future rewards, since typically a high reward received now is better than receiving the same reward later when maximizing the reward over the long run.

Many MARL approaches are derived from Q-Learning [24] that uses a Q-function mapping states and actions to the expected return of performing the action in the given state. If we know all the possible actions in a given state then the policy can choose the action with the highest expected return at any given time. As the system runs (either supervised or unsupervised) the Q-function is updated incrementally by selecting an action using the current policy to move from a current state to a target state, receiving a reward in the process. The value of performing the selected action in the current state in the Q-function is then updated my adding a weighting to the current value calculated in terms of the reward and the discounted estimate of the best value for the target state. A discount is used to ensure that taking the action now is better than deferring until later and thereby we optimize the overall search path.

If the Q-function is updated by exercising all the paths in the search space then it will iterate to an optimal solution. However, in general this is difficult to achieve because the state space is too large. In this case, it may be that Q-learning is *good enough* to provide guidance and human control will select between alternatives. Another option is to use deep learning [25] to make the calculation more efficient through the use of neural networks.

In conclusion, digital twins for complex systems require a bottom up approach that is naturally supported by agent behavior. In particular actors can be used as a technology basis for agent systems. During execution, digital twins will need to adapt. Traditional agent systems use inference to perform reasoning and planning; however, adaptation for digital twins is more appropriately supported by machine learning (perhaps augmented by reasoning). Digital twins will not necessarily have historic system traces that can be used for supervised machine learning. At least one digital twin use-case involves the twin being used in an ongoing situation either as a replacement for the real system or alongside it, therefore learning is likely to be based on unsupervised techniques. Due

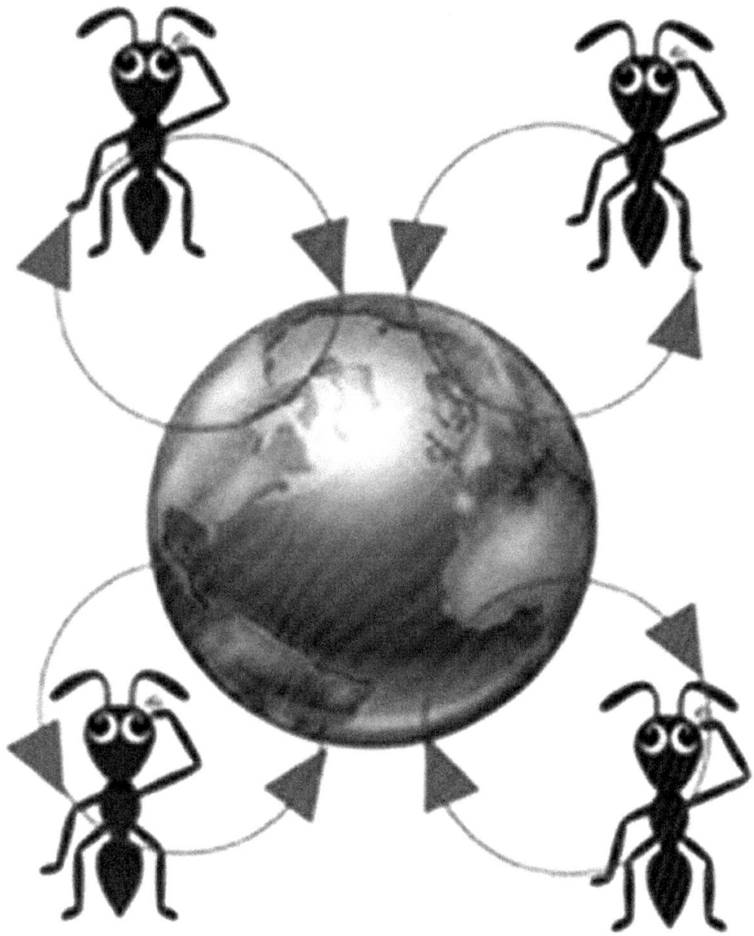

FIGURE 11.3

Agents interacting with a world.

to the nature of a complex system, the agents in a digital twin will have underspecified behavior that are defined in terms of states, events, and actions. This naturally to adaptation through RL, which incrementally develops agent behavior as a policy mapping states to actions that is guided by a system-wide goal and local rewards.

11.3.5 A GENERAL ARCHITECTURE FOR DIGITAL TWINS

It is possible to characterize the proposed architecture for digital twins while leaving open the details of how the adaptation takes place. The architecture is specified in terms of a function that shows the types of the different categories of data and how they are linked together. The function uses the following types:

World	A representation of the shared system state that can be updated by agents and provides a reward when an agent performs an action in a given state that is suitable for an RL algorithm.
Goal	A predicate over system traces. The aim of any digital twin is to achieve the goal over the entire system trace, which consists of a combination of all the agent traces.
History	A system trace that contains the sequence of states generated by each agent. The system trace is a composition of all the individual agent traces.
Action	Something that an agent can perform. An action maps a state to a new state and some output events.
Behavior	A nondeterministic function mapping events and states to actions. This is equivalent to the policy in a reinforcement learning system and provides the individual behavior for each agent.
State	The local state of an agent. This is updated when actions are performed by an agent. Sequences of states form agent traces and merged to form system traces that are referred to as history.
Events	A collection of events received by agents and is produced by agents when they perform actions.

A single agent is of type Agent 0 and is defined by the function agent as shown below:

```
Goal         = (History)                              -> Bool
Agent        = (World, Goal, Behaviour, State) -> Agent0
World        = (Action, State)                     -> Reward
Action       = (State)                                -> (State, Events)
Behaviour = (Events, State)                      -> Action
Agent0       = (History, Events)                   -> (Agent, State, Events)
```

An agent is created in two steps. Firstly it is supplied with a world, a goal, a behavior and a starting state. At each system step, the agent is supplied with a history and a set of events. The agent returns a new version of itself that is created by updating the behavior and state after selecting an action to perform. The action is selected using the current behavior, which is modified using a reinforcement learning function RL:

```
RL::(State, State, Action, Reward, Events, History, Goal, Behavior)-> Behavior
```

The RL function uses the history of the agent together with the state transition it has just performed to update the behavior. Initially, the behavior is a nondeterministic function. By applying the reinforcement function to the behavior in the context of the new transition, the nondeterminism can be reduced. Over time, the RL function will incrementally refine the behavior function so that it becomes deterministic and selects the best action for any given state.

```
agent::Agent
agent(w, g, b, s)(h, e) = (agent(w, g, b', s'), s', e')
   where
      a           = b(e, s)
      (s', e') = a(s)
      r             = w(a, s')
      b'            = RL(s, s', a, r, e, h, g, b)
```

The key idea with an agent is that it receives events and a representation of the current system history. It uses the current version of its behavior in order to produce an action, which updates the state of the agent producing output events. The world is informed of the state change and produces a reward. The agent can then adapt using a RL system, which is provided with the state change, the action, and the history.

Agents can be combined to produce a digital twin as shown in the following text:

```
combine::(Agent O, Agent O)-> Agent O
combine(a1, a2)(h, e)=(combine(a1', a2'), s1+s2, e1+e2)
   where
      (a1', s1, e1)=a1(h, e)
      (a2', s2, e2)=a2(h, e)
```

A run of the system is performed in terms of a combined collection of agents, a system history and a collection of events. A single step of the run produces a new agent, an updated history and new events:

```
run::(Agent O, History, Events) -> (Agent O, History, Events)
run(a, h, e)=(a', h+[s], e')
   where
      (a', s, e')=a(h, e)
```

The digital twin is *saturated* when the goal is satisfied by the history that is produced *in the long run*: `goal(run*(a, [], e0))`. This characterization of the digital twin architecture omits many important implementation details, like how events and histories are managed, and how the RL is implemented. In addition, there is no guarantee, that saturation is possible.

11.4 TECHNOLOGY FOR DIGITAL TWINS

The enterprise simulation language (ESL) and associated EDB development platform has been developed to support the creation of digital twins. ESL is a hybrid of actors and functional programming, with static typing. EDB is a development environment for ESL that performs verification including type checking and supports libraries, including graphics and links to Excel, that are required for certain types of ESL application. ESL is open source and implemented in Java where the ESL compiler translates directly to Java source code that can easily be integrated with other applications. ESL and associated documentation are available at http://www.esl-lang.org.

Fig. 11.4 shows an example ESL program that contains many of the key features. ESL contains algebraic data types like `Direction` that can be used for pattern matching in case-expressions. The behavior type `QSort` defines an interface of messages that are implemented by the behaviors `qwaiter` and `qsorter`. The list `nums` is initially unsorted and is sorted by the Quicksort algorithm.

The Quicksort algorithm selects a pivot element in the list and creates two separate lists: in the first all elements are greater than the pivot and in the second they are less than the pivot. Each of the two lists are then recursively sorted and then joined either wide of the pivot. If this is done in a traditional sequential language then the two lists are sorted one after the other.

```
data Direction = Left | Right | Final;

Act QSort { Sorted([Int],Direction); }

nums::[Int] = [ random(50000) | n::Int <- 0..50000 ];

act qmain(l::[Int])::QSort {                              // The starting point for sorting.
  -> new qsorter(self,l,Final);                           // Think of this as the initial
  Sorted(l::[Int],Final) -> {                             // call frame. When it is returned
    print[Str](now + ' => ');
    print[[Int]](l);                                      // to, it prints the sorted list
    stopAll();                                            // and stops.
  }
}

act qwaiter(parent::QSort,n::Int,dir::Direction)::QSort {  // This is waiting for two separate
  left::[Int] = null[[Int]];                              // lists to be sorted. When both
  right::[Int] = null[[Int]];                             // have been received from the left
  check()::Void =                                         // and right (in any order) then it
    if left <> null[[Int]] and right <> null[[Int]]       // returns to the parent.
    then parent <- Sorted(left+[n]+right,dir);
    else {}
  Sorted(l::[Int],Left)  -> { left := l;  check(); }
  Sorted(l::[Int],Right) -> { right := l; check(); }
}

act qsorter(parent::QSort,l::[Int],dir::Direction)::QSort {  // A request to sort the list l and
  -> case l {                                               // to return to the parent from the
       [] -> parent <- Sorted(l,dir);                       // supplied direction. If the list is
       x::Int:l::[Int] -> {                                 // non-empty then nested sorters are
         new qsorter(self,[ n | n::Int <- l, ?(n<x) ],Left);   // created with this as the parent
         new qsorter(self,[ n | n::Int <- l, ?(n>x) ],Right);  // and this becomes a waiting actor.
         become qwaiter(parent,x,dir);                      // Becoming a waiter is equivalent to
         {}                                                 // a nested stack frame. No-one should
       }                                                    // ever return to this behaviour so
     }                                                      // its Sorted handler is an error.
  Sorted(l::[Int],d::Direction) -> throw[Void] 'error!';
}

act main::Act{} {
  -> new qmain(nums);
}
```

FIGURE 11.4

ESL Quicksort.

In an actor language like ESL, the two lists can be sorted concurrently. This is done in `qsorter`, by creating two new actors with behavior `qsorter`; each actor sorts the two lists independently. Once sorted, the two independent lists must be rejoined to the pivot. This is achieved by supplying a continuation actor to the two new actors; the continuation will receive the independently sorted results.

A feature of actor languages allows an actor to replace its own behavior. This is achieved in ESL using a `become` command as shown in the `qsorter` behavior. By supplying the continuations to newly created actors via the parent argument, a tree of actors is created. Each leaf of this tree is an actor that is supplied with an empty list, which immediately returns the sorted list to its parent. Each parent will have the behavior `qwaiter`, that is, waiting for assorted list.

Each actor with the `qwaiter` behavior is the parent of two `qsorter` actors. It must receive sorted lists from both children before it can return to its own parent. This is managed using the data type `Direction`: each newly created `qsorter` has a direction that it supplies to its parent. When the parent has received results from both directions it can return to its own parent.

This arrangement allows Quicksort to proceed concurrently returning the sorted list to the `qmain` actor.

11.4.1 ESL: EMERGENT BEHAVIOR

An important feature of complex system digital twins is that they exhibit emergent behavior. This can be demonstrated easily by ESL by implementing termites [26]. Termites live in a forest whose floor is littered with twigs.

The behavior of termites is shown in Fig. 11.5. A termite lives to move twigs around and is initially searching for a twig. Termites are quite simple-minded and just move at random until they

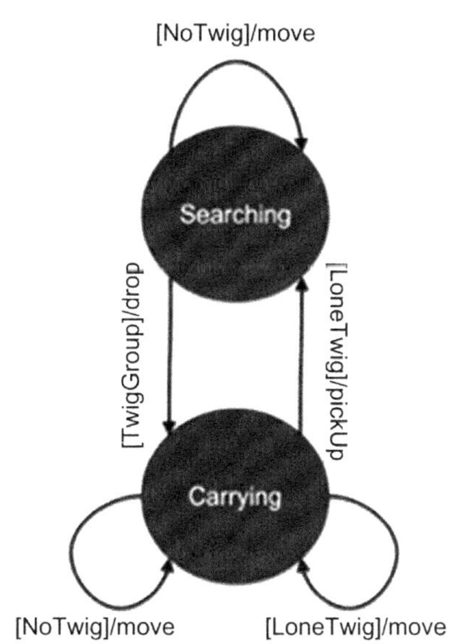

FIGURE 11.5

Termite behavior.

find a twig. In order for the termite to be interested in a twig that it finds, the twig must be by itself on the forest floor.

Once a termite has found a twig it will pick the twig up and start carrying it around. Termite movement is still random; however, if the termite encounters another twig while in carrying mode, it will drop the twig it is carrying. However, it will only drop the twig next to a twig that is not alone on the floor: it must be next to at least one other twig.

Like many simple agent examples, termites live in a simple Grid World where locations are given by coordinates and *next-to* means "1 grid location away."

Notice that the behavior of individual termites is very simple and has no knowledge relating to collaboration and no goal. Even so, when the termites are implemented as separate agents with respect to a world representing the forest floor, the resulting emergent behavior appears to show collaboration. This is shown in Fig. 11.6 where the starting state of the world consists of a random distribution of twigs and termites. The termites all independently interact with the world according to the behavior shown in Fig. 11.5 leading to the state shown on the right in Fig. 11.6. Without having any knowledge of twig-piles or any ability to collaborate with other termites, they have produced an emergent behavior that appears to have been globally coordinated. Although this is a simple example, it is intended to clearly show how ESL can easily encode individual behavior that lead to emergence.

The code shown in Fig. 11.7 provides an overview of how ESL implements actors—in this case the termite Grid World.

An actor definition is like a Java class in that it contains fields and operation definitions, and may be instantiated multiple times. In this case we only need one world whose locations are represented using a two-dimensional array of strings representing twigs and backgrounds. The locations are initialized to contain a random collection of twigs.

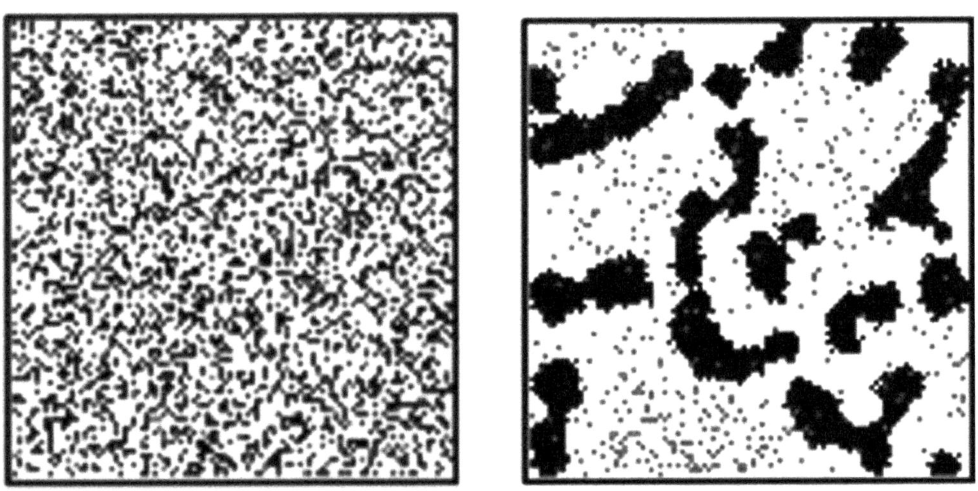

FIGURE 11.6

Termites in a Grid World.

```
act world::World {
  locations::Array[Array[Str]] =
    let a::Array[Array[Str]] = new Array[Array[Str]](width);
    in {
      for x::Int in 0..width do {
        a[x] := new Array[Str](height);
        for y::Int in 0..height do {
          a[x][y] := if random(100) < twigLikelihood then twig else background;
          grid <- SetColour(x,y,a[x][y]);
        }
      }
      edb.display[Grid]('Termites',grid);
      a;
    }
  foundSingleton(x::Int,y::Int)::Bool =  locations[x][y] = twig and twigCount(x,y) < 4;
  foundPile(x::Int,y::Int)::Bool      =  locations[x][y] = twig and twigCount(x,y) > 5;
  isTwig(x::Int,y::Int)::Bool         =  if isLegal(x,y) then locations[x][y] = twig; else false;
  countTwig(x::Int,y::Int)::Int       =  if isTwig(x,y) then 1; else 0;
  twigCount(x::Int,y::Int)::Int       =  sum([ countTwig(x+dx,y+dy) | dx <- -1..2, dy <- -1..2, ?not(dx=0 and dy=0) ]);

  -> for n in 0..numOfTermites do new termite(n,self);

  TryPickup(x::Int,y::Int,t::Termite) ->
    if foundSingleton(x,y)
    then {
      locations[x][y] := background;
      grid <- SetColour(x,y,background);
      t    <- Drop;
    } else t <- Search;

  FindPile(x::Int,y::Int,t::Termite) -> if foundPile(x,y) then t <- FindSpace; else t <- Drop;

  TryDrop(x::Int,y::Int,t::Termite) ->
    if locations[x][y] = background
    then {
      locations[x][y] := twig;
      grid <- SetColour(x,y,twig);
      t    <- GetAway(20);
    } else t <- FindSpace;
}
```

FIGURE 11.7

A world actor.

The number of twigs surrounding a singleton or a pile determines how "neat" the resulting world state is when the termites have completed their task. In this case a singleton in defined as any twig surrounded by at most three other twigs, and a pile is defined by six or more surrounding twigs.

When an actor definition is instantiated, its initialization behavior is performed. This is the command following "->" and in this case creates a number of termite actors that immediately start roaming the world (See the command 'self <- Search' shown below).

Each termite interacts with the world by sending messages including the termite's position and the termite itself. The world is singly threaded, so handling messages from multiple actors guarantees exclusive access to the world state.

The implementation of the termite actor behavior is shown in Fig. 11.8. Unlike the world discussed earlier, many different actors are created with this behavior and will all execute concurrently by selecting the next message from the queue, handling it, and then repeating.

The integrity of the shared world is maintained while allowing termite actors to behave concurrently by sending messages from each termite to the world, which then responds with a subsequent message. This communication protocol continues until the simulation is terminated.

```
act termite(id::Int,w::World)::Termite {

  x::Int  = random(width);
  y::Int  = random(height);

  move()::Void = {
    x := min(max(x + random(3)-1,0),width-1);
    y := min(max(y + random(3)-1,0),height-1);
    grid <- TermiteAt(id,x,y);
  }

  -> self <- Search;

  Search -> {
    move();
    w <- TryPickup(x,y,self);
  }
  Drop -> {
    move();
    w <- FindPile(x,y,self);
  }
  FindSpace -> {
    move();
    w <- TryDrop(x,y,self);
  }
  GetAway(0) -> {
    self <- Search;
  }
  GetAway(n::Int) -> {
    move();
    self <- GetAway(n-1);
  }
}
```

FIGURE 11.8

A termite actor.

A termite actor is initialized with its identity, and a reference to the world. Its initialization behavior sends itself a search message to initialize the search for a twig.

The messages use the local operation move random to select a legal delta to the current termite's position. The new position is sent to the grid actor, which implements the GUI interface. The termite then interacts with the world depending on its current state based on the new position.

11.4.2 ESL: SIMPLE REINFORCEMENT LEARNING

The general architecture for digital twins described earlier requires adaptation in order to refine the agent behavior that is initially underspecified due to incomplete information and uncertainty about atomic system components and their interactions. This section shows a simple example of this in action by defining a generic RL library in ESL and then using it to train a collection of independent agents navigate Grid World. A simple RL algorithm can be defined independently of the types of the states and actions that are to be performed.

The type policy, defined in Fig. 11.9, is supplied with the types of the states and the actions that will be used in the RL. The resulting type is a record of three functions that allows an agent to interact with its policy. The first function allows a new transition to be recorded in terms of a new state, the reward and whether the goal is satisfied or not. The second function requests the next action to be performed, selecting at random if necessary, from the current state. The RL algorithm has a maximum number of learning iterations after which the current episode has completed and a new training episode must be started. The third function is a predicate that returns whether the current episode has completed.

A RL policy record is created using the function `mkPolicy`. There are several subdefinitions to this function, which are described in the rest of this section. Firstly, the arguments that must be supplied for policy creation in Fig. 11.10.

Each new policy has a state that consists of a Q-table, the current state, the most recently performed action, the current time and number of executed episodes, and whether the policy has finished the training (because it has completed the maximum number of episodes). The state of the policy is created using local variables as shown in Fig. 11.11.

The body of the `mkPolicy` function defines several local functions that perform the tasks involved in learning the policy.

The function `maxQ` defined in Fig. 11.12 finds the maximum current value in the Q-table for a supplied state. This is used to calculate the new value for an action when an update to the table takes place.

```
type Policy[State,Action] = {
  transition  ::  (State,Float,Bool) -> Void;
  nextAction  ::                  () -> Action;
  finished    ::                  () -> Bool
};
```

FIGURE 11.9

A polymorphic reinforcement learning policy.

```
mkPolicy[State,Action](states::[State],         // The list of all possible states.
                       actions::[Action],       // The list of all possible actions.
                       startingState::State,    // The starting state.
                       initState::() -> State,  // Creates the initial state.
                       maxEpisodes::Int,        // An episode is a single run.
                       maxSteps::Int,           // A step is an interaction with the world.
                       epsilon::Int,            // A number [0-100] that affects opportunistic selection of actions.
                       learningRate::Float,     // The weighting that is applied to each update that affects its impact.
                       gamma::Float,            // The decay on the learning over transitions.
                       stableEpsilon::Int       // The value of epsilon after learning has stopped.
                       )::Policy[State,Action] =
```

FIGURE 11.10

Policy creation: arguments.

```
let q::Hash[State,Hash[Action,Float]] = new Hash[State,Hash[Action,Float]]; // The policy maps states and actions to the reward
                                                                              // for performing the action in the state
    state::State          = startingState;                                    // The current state.
    action::Action        = null[Action];                                     // The most recently performed action.
    time::Int             = 0;                                                // Number of time steps in the current episode.
    episodes::Int         = 0;                                                // Number of episodes.
    finished::Bool        = false;                                            // Set once maxEpisodes have been performed.
```

FIGURE 11.11

Policy creation: local variables.

```
maxQ(s::State)::Float =
  // Get the maximum value for an action leading from the state s.
  let max::Float = 0.0;
  in {
    for a in actions do {
      if q.get(s).get(a) > max then max := q.get(s).get(a); else {}
    }
    max;
  }
```

FIGURE 11.12

Policy creation: maximum action value.

```
maxAction(s::State)::Void =
  // Find the action from state s with the highest value.
  // Sets the value of the variable 'action'.
  let max::Float = -99999.0;
  in {
   for a in actions do {
     if q.get(s).get(a) >= max
     then {
       action := a;
       max := q.get(s).get(a);
     } else {}
   }
  }
```

FIGURE 11.13

Policy creation: finding the max action.

Given the state s, the function `maxAction`, defined in Fig. 11.13, sets the `variable action` to the action with the highest Q-value. This is used by the policy when choosing an action.

Learning is performed by the function learn that is supplied with two states, an action and the reward that was provided when the action causes the system to perform a state transition. The function is defined in Fig. 11.14.

A transition is supplied to the policy by an actor by calling the function transition defined in Fig. 11.15.

```
learn(state1::State,state2::State,reward::Float,action::Action)::Void = {
// Called to update the Q-table with the transition (action,reward):state1 -> state2
let predict::Float = q.get(state).get(action);
    target::Float = reward + (gamma * maxQ(state2));
in q.get(state1).put(action,predict + (learningRate * (target - predict)));
```

FIGURE 11.14

Policy creation: learning.

```
transition(state2::State,reward::Float,done::Bool)::Void = {
  // Called to record a transition. The starting state and action
  // have been recorded. The target state and the reward are provided.
  if not(finished)
  then {
    learn(state,state2,reward,action);
    state := state2;
    time := time + 1;
    if done or time >= maxSteps
    then {
      episodes := episodes + 1;
      time := 0;
      print[Str]('episodes = ' + episodes);
      if episodes < maxEpisodes
      then state := initState();
      else {
        finished := true;
        //printPolicy();
        state;
      }
    } else {}
  } else {
    if done
    then state := initState();
    else state := state2;
  }
}
```

FIGURE 11.15

Policy creation: record a transition.

Finally, the actor uses the next action function of the policy to request the best next action given the current state of the Q-table. The function is defined in Fig. 11.16. Notice that the epsilon values are used to control exploration by selecting actions at random.

The result of the function `mkPolicy` is a record containing the functions for recording a transition, getting the next action, and checking whether the learning has completed given the number of episodes allowed. The next section shows how `mkPolicy` can be used to create simple learning agents in Grid World.

```
nextAction()::Action = {
  // Returns the next action to perform given the value of 'state'.
  // The values 'epsilon' and 'stableEpsilon' are values [0..100]
  // and control exploratory actions...
  if random(100) < (if finished then stableEpsilon else epsilon)
  then {
    action := nth[Action](actions,random(length[Action](actions)));
    action;
  } else {
    // Otherwise find the action with the maximum Q-value...
    maxAction(state);
    action;
  }
}
```

FIGURE 11.16

Policy creation: requesting the next action.

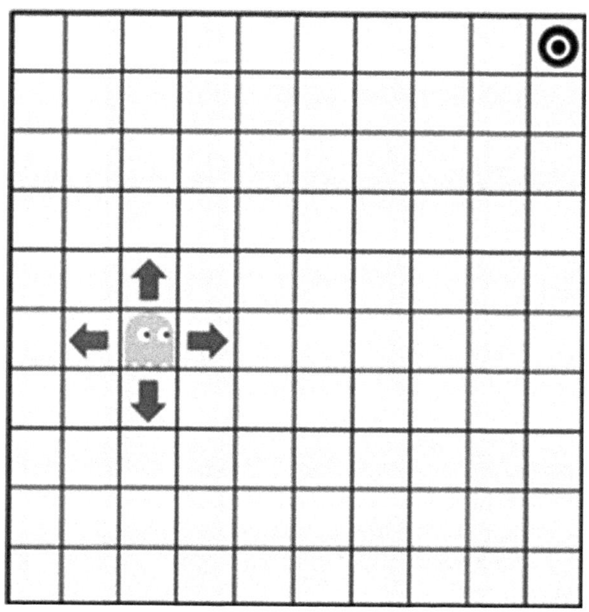

FIGURE 11.17

Grid World with a single agent.

11.4.3 LEARNING IN GRID WORLD

Suppose that we have simple agents living in Grid World shown in Fig. 11.17 that can move horizontally and vertically. Their aim is to get into a particular corner of the world, although they do not have a policy in order to achieve it. If they had a policy, then it would tell them based on their current position, where to move next in order to get to their target location in the least number of moves.

RL can help, because it will gradually learn a policy that gets the agent to the target. Initially the policy will provide an action at random for each of the agent's states. Eventually, the agent must find its goal and the reward for the final step will be larger than the rest. Propagation within the RL algorithm will reinforce the penultimate step that led to the goal and so on. If this process is repeated for many episodes, and if the state space is explored completely then the resulting policy will guide the agent to its target. The worker agent that is supplied with Grid World, the name of an image to show in the world and a goal, is shown in Fig. 11.18.

The worker sends Perform messages to the world that contains the new position for the worker. The message contains the worker and an action to be performed. The actions move the worker north, south, east, or west. The world responds with a State message containing the grid-point for the worker and whether the worker moved or not (some directions may not be legal).

Each worker creates a policy that is initialized with all the worker-states in the grid, and all of the possible actions. Initially, each time the worker requests an action from the policy, it will receive a random direction. Each time the worker performs a legal move, it updates the policy, which learns in terms of the goal and the reward allocated to the current grid-point.

We omit the details of the implementation of the grid-world actor since it deals with displaying the actor on the screen and determining whether the move is legal or not. Multiple actors can be created, each with their own goal and policy as shown in Fig. 11.19.

After the learning has terminated, each of the four workers cycle by randomly positioning themselves on the grid, using the policy to follow a path to the target corner, and then restarting the cycle as shown in Fig. 11.20.

11.5 CASE STUDY

A typical retail supply chain comprises of stores, distribution centers, transporters, manufacturers, and suppliers as shown in Fig. 11.21. Although optimality of the entire support chain if the principle objective, it is seldom achieved due to the scale of the state space, nonlinear dependencies, and inherent uncertainty. Typical practice is to achieve optimality is localized contexts like stock replenishment in stores, picking and packaging in distribution centers, scheduling of trucks, and so on.

Consider a retail chain that sells "P" product types through "S" stores whose stock is replenished every "h" hours in a day. The objective is to compute the replenishment quantity for each product type in each store. This computation is to be done holistically, that is, keeping in mind the entire supply chain and not just the stores. With typically "P" in tens of thousands and "S" in thousands result in tens of million decisions every "h" hours. Given multiple capacity sharing points like warehouse labor, truck volume, and so on, all these decisions are interdependent.

The state of the art is to either (1) take the decisions manually, assuming all products are independent, or (2) encode the business rules in the form of heuristics. In both cases, even if the stores are kept well stocked every other node in the supply chain takes a hit. Moreover, both approaches

```
act worker(w::World,image::Str,goal::(Point)->Bool)::Worker {
  export theImage,stable;                                           // Used by the world to draw the agent when stable.
  stable::Bool = false;                                             // Set when the policy completes learning phase.
  theImage()::Str = image;                                          // Used to draw a picture of the agent.
  startPoint::Point = getPoint(random(width),random(height));
  state0()::Point =                                                 // Required by the policy to reset the state.
    let p::Point = getPoint(random(width),random(height));
    in {
      w <- Start(self,p);                                           // Resets the world, and the screen.
      p;
    }

  reward(p::Point)::Float =                                         // The reward differentiates between achieving
    if goal(p)                                                      // the goal, moving to a legal non-goal position
    then 1.0;                                                       // and moving to an illegal position which is off
    else if onGrid(p)                                               // the grid. These rewards will cause the policy
    then 0.0;                                                       // to learn to make legal moves towards the goal.
    else -1.0;

  policy::Policy[Point,Direction] = mkPolicy[Point,Direction](      // Create a new policy for this worker.
    [ Point(x,y) | x <- 0..width, y <- 0..height ],                 // The states.
    [North,South,East,West],                                        // The actions.
    startPoint,
    state0,                                                         // Called at the start of each episode.
    1000,                                                           // Number of episodes.
    100,                                                            // Max steps per episode.
    90,                                                             // Epsilon (exploration factor).
    0.9,                                                            // The learning rate.
    0.9,                                                            // The discount factor.
    1);                                                             // Epsilon after policy becomes stable.

  -> {
    w <- Start(self,startPoint);
    w <- Perform(self,policy.nextAction());
  }

  State(p::Point,moved::Bool) -> {                                  // State(p,moves) is sent by the world in response
    if policy.finished() then stable:= true; else {}                // to a Perform(w,action) message. The result may be
    policy.transition(p,if moved then reward(p) else -1.0,goal(p)); // a no-move, if the move was illegal.
    let action::Direction = policy.nextAction();                    // Get the next action as defined by the current state
    in w <- Perform(self,action);                                   // of the policy and perform it.
  }
}
```

FIGURE 11.18

A worker actor.

```
act main::Main {
  theWorld::World = new world;
  goal1::Point = Point(0,0);
  goal2::Point = Point(width-1,height-1);
  goal3::Point = Point(width-1,0);
  goal4::Point = Point(0,height-1);
  -> {
    new worker(theWorld,'images/pacman/clyde.png', fun(p::Point)::Bool p = goal1);
    new worker(theWorld,'images/pacman/blinky.png',fun(p::Point)::Bool p = goal2);
    new worker(theWorld,'images/pacman/inky.png',  fun(p::Point)::Bool p = goal3);
    new worker(theWorld,'images/pacman/pinky.png', fun(p::Point)::Bool p = goal4);
  }
}
```

FIGURE 11.19

Starting an adaptive application in ESL.

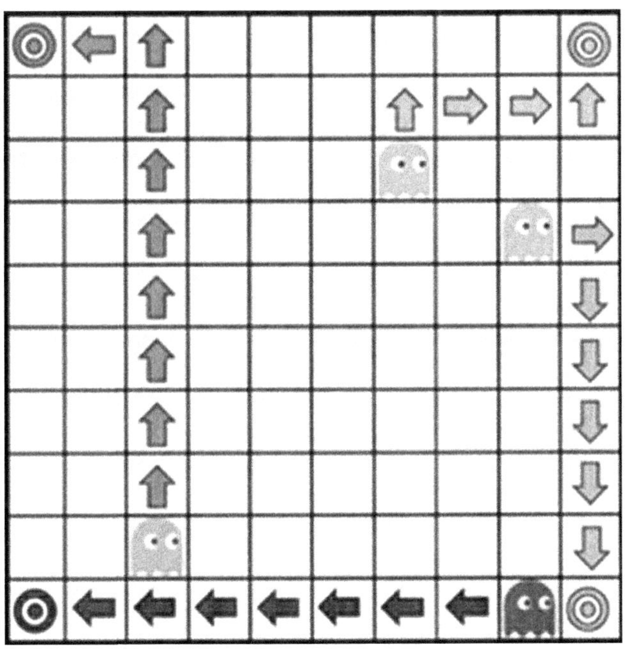

FIGURE 11.20

A stable policy.

rely on past data (i.e., consider only what has happened while ignoring what could have happened) and aggregated heuristics that connote average case behavior (i.e., not the outliers). As a result, both lead to suboptimal performance. In addition, the stock replenishment problem has three

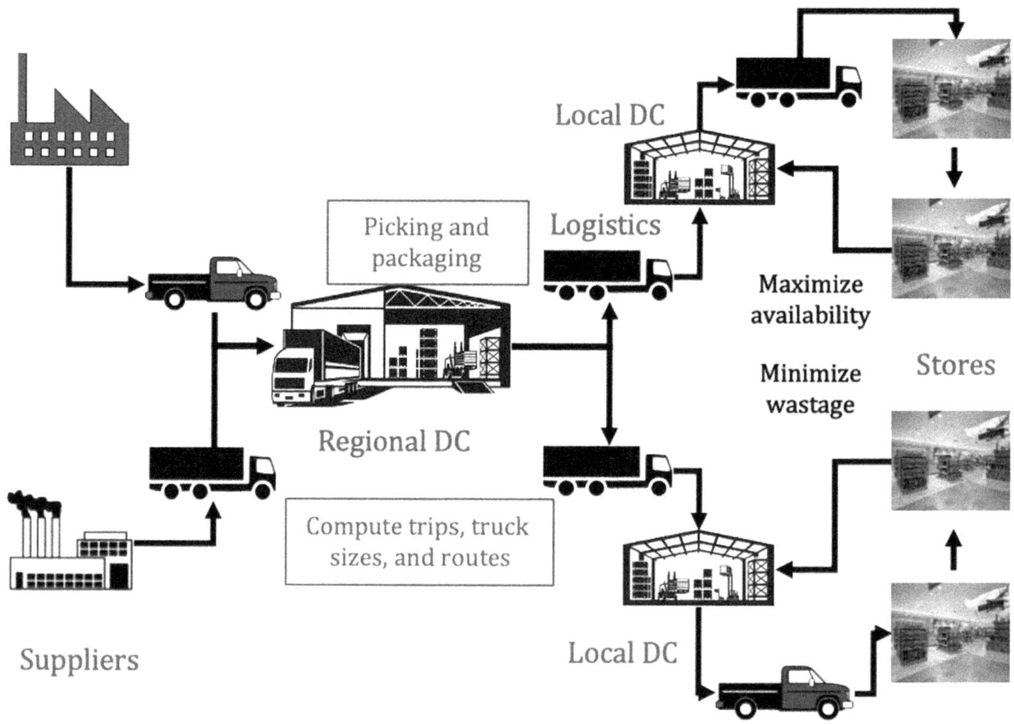

FIGURE 11.21

A typical retail supply chain.

dimensions namely shop, warehouse and logistics thus requiring an integrated solution covering all three dimensions. Current practice relies on independent analysis in local contexts, for instance, shop, warehouse, and logistics. Though it can lead to optimal solutions for the three contexts, there is little or no help available to combine the independently arrived solutions into a holistic solution. In addition, it is not possible to a priori identify the ramifications of decisions taken to achieve local optimality on global robustness.

Our approach is to represent the retail supply chain as a set of autonomous intentional interacting units, that is, a digital twin, as shown in Fig. 11.22.

We used ESL to specify the digital twin in terms of actors like store, distribution center, truck, container, product, shelf, customer, and so on. An actor observes events taking place in its environment and responds to the events of interest by performing an action to achieve its stated goals. A condition may serve as a guard for performing an action.

To account for partial information and uncertainty, we allow the actions to be probabilistic. An actor specification is a set of statements of the form:

```
On event [if condition] do action@probability
```

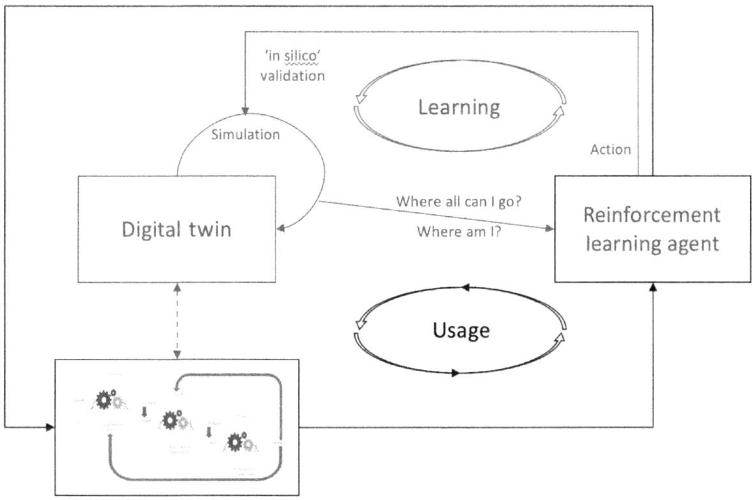

FIGURE 11.22

Digital twin-aided decision making.

Thus digital twin represents the real system as a set of intentional autonomous actors with probabilistic behavior. We simulate the digital twin using past data to generate history consisting of `<state, event, action>` elements for every actor. We distill the behavior of the real system in terms of patterns to be looked for over these streams. We specify the patterns using a pattern language and ascertain their existence in the simulation trace using a pattern matching engine. Thus we establish that the digital twin is a faithful representation of the real system that can be used for decision making.

We use a RL-based recommender system for decision making. We chose RL as it is an experience-based learning technique that does not require a large training corpus annotated with an encoding of "answer key." Our solution has two phases— learning phase and usage phase. The learning phase comprises of several simulation runs wherein the digital twin is subjected to various scenarios of interest, that is, scenarios that could have happened but did not. Data generated from these simulations together with past data, that is, scenarios that did take place, is used to train a RL-based algorithm to compute stock to be replenished for every `<Store, Product>` combination. The usage phase makes use of the trained algorithm that continually learns from the live system.

We used data spanning over one year for validating the digital twin as well as training the RL agent. A total of 220 products were chosen from the data set, and their meta-data (volume, weight, shelf-life which was not available in the original version) was input manually. A single store and a single truck were used for this case study, with the time between successive delivery moments set to 6 hours (leading to four replenishments per day) with lead time as 3 hours. The digital twin was subjected to a set of relevant what-if scenarios thus generating more comprehensive training input for RL-based controller. It resulted in reduced training time as well as improved stock replenishment for all stores.

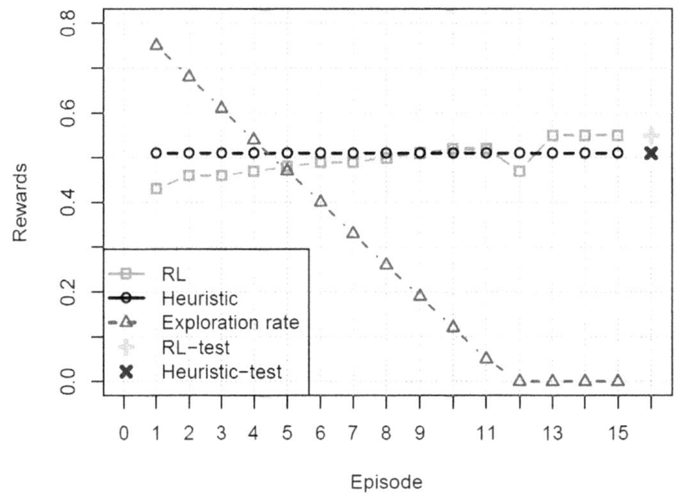

FIGURE 11.23

Digital twin-enabled reinforcement learning-based controller outperforms aggregated heuristics while balancing explore vs exploit.

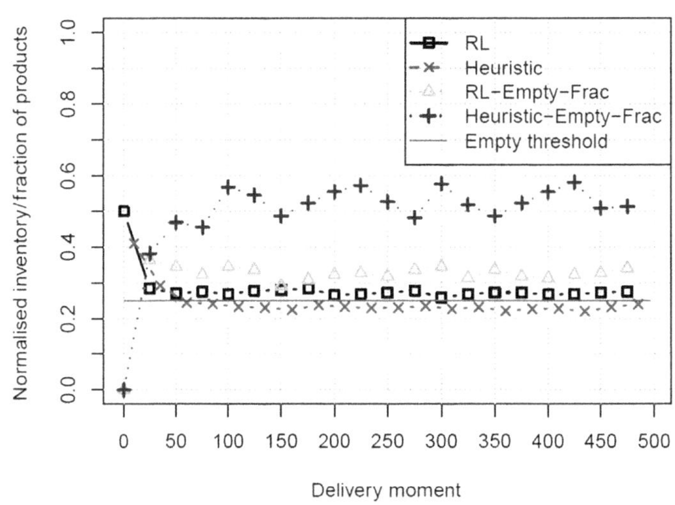

FIGURE 11.24

Digital twin-enabled reinforcement learning-based controller outperforms heuristics as regards inventory.

Fig. 11.23 shows that the reward at the end of the RL-agent training exercise exceeds the heuristic performance, and this advantage is retained on the test data set as well (plotted using separate markers at the ends of the curves). It also depicts the "exploration rate" of RL, which is the probability with which the RL algorithm takes randomized actions (to explore the feature space). This rate is brought to zero toward the end of training, and is also zero for testing. In addition, though both algorithms begin with an initial (normalized) inventory level of 0.5 for all products, RL is able to maintain a higher average inventory level than the heuristic as can be seen from Fig. 11.24.

RL combined with a digital twin performed better than an aggregated heuristic. Use of a digital twin-enabled exploration of the decision space by simulating scenarios that can occur but have not yet occurred. This leads to more comprehensive data to be used for training the RL-based controller and hence better learning. Moreover, it allows the recommendations to be validated in quantitative terms using digital twin itself before being implemented in the real system. In addition, the simulation runs can be examined for desirable and undesirable patterns thus validating if the actor behaviors are leading to the desired system behavior. For instance, we can check a-priori if optimal solutions for local contexts, that is, stock replenishment for stores, loading and packaging for distribution centers, and rostering and scheduling of trucks can make the overall supply chain unreliable.

11.6 CONCLUSION AND RESEARCH ROADMAP

This chapter has proposed an approach to designing controlling and maintaining complex systems using digital twins that consist of agents equipped with RL in order to perform the adaptation that is necessary because of the uncertainty and incompleteness of system behavioral information. We have developed some technology to support our approach and used it on a number of applications including a nationwide financial initiative [27], a business process outsourcing company [28], and the supply chain described in this chapter. Many challenges remain if the digital twin vision is to be achieved. Although it is reasonably straightforward to express agents with underspecified behavior that is controlled via machine learning, it is not clear how to achieve collaboration between the agents where this is necessary to achieve the goals. Furthermore, real-world systems are so large that even if Deep Learning is used, the state space is very large. Digital twin state analysis and transformation is required in order to reduce the amount of nondeterminism that occurs at the outset.

REFERENCES

[1] M. Mitchell, Complexity: A Guided Tour, Oxford University Press, 2009.

[2] Y. Bar-Yam, Dynamics of Complex Systems, 213, Addison-Wesley, Reading, MA, 1997. Vol.

[3] M.D. Norman, Complex systems engineering in a federal IT environment: lessons learned from traditional enterprise-scale system design and change, in: 2015 Annual IEEE Systems Conference (SysCon) Proceedings, IEEE, 2015, pp. 33–36.

[4] B. Andres, R. Poler, L. Camarinha-Matos, H. Afsarmanesh, A simulation approach to assess partners selected for a collaborative network, Int. J. Simul. Model. 16 (3) (2017) 399–411.

[5] Y.Y. Liu, A.L. Barabási, Control principles of complex systems, Rev. Mod. Phys. 88 (3) (2016).

[6] S. Abar, G.K. Theodoropoulos, P. Lemarinier, G.M. O'Hare, Agent based modelling and simulation tools: a review of the state-of-art software, Comput. Sci. Rev. 24 (2017) 13–33.

[7] C.M. Macal, Everything you need to know about agent-based modelling and simulation, J. Simul. 10 (2) (2016) 144–156.
[8] H. Butler, Model reference adaptive control: bridging the gap between theory and practice (Ph.D. thesis), 1990.
[9] M. Grieves, Digital twin: manufacturing excellence through virtual factory replication: a white paper, 2014. Available from: <https://research.fit.edu/media/site-specific/researchfitedu/camid/documents/1411.0_Digital_Twin_White_Paper_Dr_Grieves.pdf>.
[10] M. Grieves, J. Vickers, Digital twin: mitigating unpredictable, undesirable emergent behavior in complex systems, Transdisciplinary Perspectives on Complex Systems, Springer, Cham, 2017, pp. 85–113.
[11] J. Hochhalter, W.P. Leser, J.A. Newman, V.K. Gupta, V. Yamakov, S.R. Cornell, et al., Coupling damage-sensing particles to the digital twin concept, NASA Report NASA/TM-2014-218257, 2014.
[12] P. Oreizy, M.M. Gorlick, R.N. Taylor, D. Heimhigner, G. Johnson, N. Medvidovic, et al., An architecture-based approach to self-adaptive software, IEEE Intell. Syst. Their Appl. 14 (3) (1999) 54–62.
[13] C. Krupitzer, F.M. Roth, S. VanSyckel, G. Schiele, C. Becker, A survey on engineering approaches for self-adaptive systems, Pervasive Mob. Comput. 17 (2015) 184–206.
[14] D. Garlan, S.W. Cheng, A.C. Huang, B. Schmerl, P. Steenkiste, Rainbow: architecture-based self-adaptation with reusable infrastructure, Computer 37 (10) (2004) 46–54.
[15] K. Franziska, A.L.C. Bazzan, Agent-based modeling and simulation, AI Mag. 33 (3) (2012) 29.
[16] M. Wooldridge, An Introduction to Multiagent Systems, John Wiley & Sons, 2009.
[17] C. Hewitt, P. Bishop, R. Steiger, A universal modular ACTOR formalism for artificial intelligence, in: Proceedings of the Third International Joint Conference on Artificial Intelligence, Morgan Kaufmann Publishers Inc., 1973, pp. 235–245.
[18] R.K. Karmani, A. Shali, G. Agha, Actor frameworks for the JVM platform: a comparative analysis, in: Proceedings of the Seventh International Conference on Principles and Practice of Programming in Java, ACM, 2009, pp. 11–20.
[19] G. Agha, I.A. Mason, S. Smith, C. Talcott, Towards a theory of actor computation, in: International Conference on Concurrency Theory, Springer, Berlin, Heidelberg, 1992, pp. 565–579.
[20] C. Bernon, V. Camps, M.P. Gleizes, G. Picard, Engineering adaptive multi-agent systems: the Adelfe methodology, Agent-Oriented Methodologies, IGI Global, 2005, pp. 172–202.
[21] P.G. Balaji, D. Srinivasan, An introduction to multi-agent systems, Innovations in Multi-agent Systems and Applications-1, Springer, Berlin, Heidelberg, 2010, pp. 1–27.
[22] L. Panait, S. Luke, Cooperative multi-agent learning: the state of the art, Autonom. Agents Multi. Syst. 11 (3) (2005) 387–434.
[23] L. Bu, R. Babu, B. De Schutter, A comprehensive survey of multiagent reinforcement learning, IEEE Trans. Syst. Man. Cybernet. Part C (Appl. Rev.) 38 (2) (2008) 156–172.
[24] C.J. Watkins, P. Dayan, Q-learning, Mach. Learn. 8 (3–4) (1992) 279–292.
[25] H. Van Hasselt, A. Guez, D. Silver, Deep reinforcement learning with double q-learning, in: 30th AAAI Conference on Artificial Intelligence, 2016.
[26] D. Feltell, L. Bai, H.J. Jensen, An individual approach to modelling emergent structure in termite swarm systems, Int. J. Model. Identif. Control. 3 (1) (2008) 29–40.
[27] S. Barat, V. Kulkarni, T. Clark, B. Barn, An actor-model based bottom-up simulation: an experiment on Indian demonetisation initiative, in: Proceedings of the 2017 Winter Simulation Conference, IEEE Press, 2017, p. 61.
[28] S. Barat, V. Kulkarni, T. Clark, B. Barn, A model based approach for complex dynamic decision-making, in: International Conference on Model-Driven Engineering and Software Development, Springer, Cham, 2017, pp. 94–118.

CHAPTER

ARTIFICIAL INTELLIGENCE POWERED HEALTHCARE INTERNET OF THINGS DEVICES AND THEIR ROLE

12

Pranjit Deka, Garima Singh and Gurjit Kaur

Department of Electronics and Communication Engineering, Delhi Technological University, Delhi, India

12.1 INTRODUCTION

Human life has become so hectic and autonomous that they are drastically shifting toward a life where time is the only luxury. In the race of fulfilling the responsibilities and priorities in making one's life, not having enough time to consider about health is common. Undoubtedly, health is crucial for the well functioning of one, and hence the autonomous healthcare management is believed to be the next big thing just like any other health and life insurance sector had been in the past.

One can see that artificial intelligence (AI) (e.g., virtual reality) is the next big thing that is going to shape the world like the innovation of television, mobile phone, and the Internet did in past decades, which always has and still making drastic changes in the human world today. Having said that, AI is truly the next big thing to watch out for having enormous capabilities and no wonder why many leading tech giants are desperately working for it.

And therefore everyone would like to grab this opportunity to work for the future by researching the aligned field and ultimately help mankind with newly equipped technologies.

The dream of smart devices taking care of human beings has already been seen by the world and people are working for creating better healthcare with the latest technologies being incorporated.

In the world where driverless cars and home assistance is a reality, it comes with no shock that the health management can be made technically sound too, enriching the capability to the fullest.

12.2 ROLE OF INTERNET OF THINGS IN HEALTHCARE

The Internet of Things (IoT) with the combined capabilities of AI and other emerging technologies like data science, virtual and augmented reality, and blockchain are expected to create the future of medical sciences like "telehealth or smart hospitals," according to recent studies and reports.

Artificial Intelligence to Solve Pervasive Internet of Things Issues. DOI: https://doi.org/10.1016/B978-0-12-818576-6.00012-5

This latest technology is set to swipe the whole healthcare sector and the way of making healthcare better for everyone and anyone overcoming the obstacles like distance, unavailability of a good practitioner, lack of early-stage diagnosis, and the postrecovery tracking of the patient.

The IoT is commonly referred to as the Internet of Medical Things (IoMT) in the healthcare industry domain, which consists of all the medical devices, wearable options, patient sensors, and monitoring tools that can send signals to other devices and systems via the Internet. IoMT is a branch or a submarket of the IoT from which several subsets of the technology have evolved. IoT refers to all web-enabled devices, Internet-enabled appliances for the kitchen and home, whereas IoMT includes only medical devices via Internet connectivity.

Why it is having an edge over is the fact that IoMT technology enables any medical device to collect, analyze, and send data across the web. Simple as well as complex digital devices like heart monitors be connected to the Internet, but so can nondigital items like hospital beds and pills, opening wider variants for the user to have access to multiple simulations and options through a single window.

All of these tools can generate a massive amount of data that needs to be integrated and analyzed to generate fruitful insights for chronic disease management and acute patient care needs and store the data for future acknowledgment.

For instance, Ohio healthcare, as it works with speech recognition of the user, asks you questions and according to one's response, it suggests one with doctor's appointment seeing all schedule and availability of the user that is already being registered on the phone. If the health issue is too serious, it asks for booking urgent care at a nearby hospital, assisting one with traveling options like Uber too. This is where such a module beautifies the real essence of it in day-to-day life.

The IoMT with the help of advancement in connectivity the growing demand for at-home and out-patient care is being carried out properly with the usage of devices that can remotely monitor a patient. Few healthcare insurers are already using the two-way smartphone application to connect clients and users with medical facilities for initial diagnosis and consultation, making it available for anyone and everyone at a lower cost option to avoid any unnecessary in-person consultation for a minor ailment, sustaining time, and curing the ailment at the earliest.

If analysis signifies anything vital, the healthcare practitioner gets the data in real time, enabling multifaceted telehealth services using affordable devices to interface users enabling transmission of data from connected devices like monitors, ECG machines, sensors, pacemakers, wearables, and other devices.

Hence the availability of such vital health data in real time about a client helps the insurers with more calculated risk assessments and in determining and estimating the premiums of the customers to be healthier. An excellent solution for the patient suffering from mobility issues.

12.2.1 HERE THE SIGNIFICANCE OF INTERNET OF THINGS CONNECTIVITY PLAYS A GREAT ROLE

Sensing a patient's health characteristics by the designed sensor is only a part of it. The crucial part, however, is the collected data to be useful; hence it must be accessible by computers and people to be analyzed.

IoMT manufacturer uses a variety of communications protocols to get these IoMT data from point of sensing to the point of analysis. However, the main objective of all is getting the collected data onto the Internet. Once there, any authorized person or computer system can access the data and use it to assist care for the patient accordingly.

Whether an in-home health tracker or an emergency room heart monitor, IoMT devices transmit their data to a nearby available network like LAN (local area network). The initial point of contact for an IoMT device might be a home WiFi connection, a cellular phone network, or an organization's IT network. Finally, IoMT data are stored into a database, which can further be accessed from the Internet and can be worked upon.

12.2.2 WHAT NEXT WITH COLLECTED DATA

With data being collected from sensors, monitors, and all the devices, the data can be just humongous if nothing less. With so much of data comes the problem of classifying and arranging it according to our needs, which is where AI plays an interesting role in it. With the number of IoMT devices propelled to the top 20–30 billion by 2023 [1], the ability to process all that data is crucial to the success of the technology.

AI can use sophisticated algorithms to learn features from huge chunks of data and mimic the behavior of human intelligence. In addition, this very feature can be put to use in the clinical field. AI system can assist a practitioner with learning new and advance research and technologies, and thereby correcting itself as it keeps on learning from its designed algorithms. It is prone to lower the therapeutic and treatment errors like diagnosis, which is apparent to happen with the human clinical process.

The technologies like AI, machine learning, deep learning, data mining, neural networks, and software interfaces can intelligently sort through huge chunks of data from IoMT devices, and provide us with the simplest of understandings and enabling medical practitioners only with data that needs their attention. As the market growth increases, AI will be the trusted and most valued entity that doctors will come to rely on to keep them informed, but not overwhelmed.

The designed algorithm and technique functions the massive collection of data into little segments of data, which will be processed and made desirably useful and simpler. The data hence obtain can be understood by any commoner after compiling it with the software interface.

12.3 EASE OF TREATMENT

12.3.1 EARLY DIAGNOSIS

One such machine learning algorithm is being developed by Google where the retina of a human eye is being studied by taking millions of images of the retina. It works in a way that when a scanned retina image of a person seems to be having few early symptoms, it analyses the same and the algorithm could predict how and when it is going to cause diabetic eye blindness to a patient. If we consider the same case with the diagnosis by a doctor, the doctor might just consider few hundreds of images of diabetic eye blindness according to a doctor's knowledge and experience with such an illness and, might not accurately predict the early symptoms. Hence, AI can help doctors to predict and assist them in curing the patient at the earliest.

The early diagnosis will help cure the illness with less pain and complexity, thereby saving time and money for everyone.

Just imagine how vital a role it can play in the early diagnosis of dangerous illness like cancer. People do not get to know any of the symptoms till a patient reaches a critical stage and the cure gets extremely difficult and complex. Here, the role of AI and IoMT can be a blessing and not less.

12.3.2 REAL-TIME DIAGNOSIS

One of the greatest benefits that the IoT along with medical devices can provide is the ability for the medical practitioner to access and analyze patient health data in real time, just at a fingertip away. Rather than having the patient visits the doctor or vice versa. A cardiologist busy with a tight schedule can get a patient's health stats like heart monitor readings right on his or her smartphone. Doctors can even view medical images like CT scan, MRI scan images as soon as they are taken from across the hospital and even anywhere around the world.

12.3.3 POSTDIAGNOSIS

One of the most important stages of recovery is the posttreatment stage or postoperation stage, where the patient is examined whether the recovery is properly taking place or not, whether there are any chances of retrieving the disease again, whether it is forming another new health hazard or not. Such cases were only limitedly possible to be known only when a patient is called upon for postcheck-up, which might get late for diagnosis. However, with the help of the latest IoMT modules, a patient is being tracked down with a minute to the largest of variation of his or her health. This not only helps in identifying any behavior change in the normal functioning of the human body but also provides one with recovery assistance to the doctor in keeping a track of the patient's postoperation or treatment. Any wrong treatment or malfunction can easily be alarmed at the earliest.

12.4 USES FOR HEALTHCARE ESTABLISHMENTS

Millions of dollars and raw material are being used by the healthcare establishments to assist the patients with types of equipment like crutches, wheelchairs, beds, and oxygen cylinder for critical patients. All of which are lost or kept unused once the recovery is done, which is a loss to both revenue and raw material. We can sustain this by engaging IoMT into the devices. This will ensure that the equipment is well kept and can be reused as the equipment will be under the constant track providing the necessary details of its longevity.

With the smart wearable devices on, knowing the health characteristics like pulse rate, the emergency condition can be dealt with. Calling the ambulance by itself and informing about the emergency of the patient with the location is one of the finest property of IoMT devices. Imagine saving someone's life even when he or she is unconscious.

For instance, "EZOfficeInventory" is one of the leading hospital tracking system utilized by healthcare establishments around the globe. Enabling optimized medical inventory and exercise more control over the healthcare budgetary.

12.5 ENCOUNTERING POSSIBLE CHALLENGES AND VULNERABILITIES

With so much crucial data of billions of people that are being added every second comes with a cost of privacy and security. With the large infuse of connected devices and AI systems to inform decision making, there are obvious concerns for the users and the organizations. A large number of hospital and medical establishments are provided with cloud storage and databases where patient's information are kept, also health insurer and life insurance companies are engaging themselves with the added information of their clients at each minute.

Hence, the vulnerability of data and privacy of every individual and the organization is pretty obvious with so much of devices fetching important information of the user.

Not just the data but also the devices where people working in a sophisticated environment or doing day-to-day life come with a risk. What if a person's device is being hacked onboard a flight or the data provided to the practitioner is malfunctioned. The person might get a stroke by just seeing the faulty data. All of these cannot be taken for granted in today's world of everything digital.

To counter this issue, there are special global standards or norms, which had been established to be followed and maintained by the manufacturer and providers.

This ensures that the devices and services are encrypted and follow minimal safety measures.

AI can also help with a solution to counterattack this vulnerability. AI will also be able to help secure the IoT world in a better way by anticipating and fighting intruders more quickly than human beings can, enabling only the authorized users and devices. Machine learning has already started to be used to predict the behavior of safe IoT options and help sustain cybersecurity. This is likely to increase the trend further. Of course, there will be people trying to penetrate IoT systems using AI to analyze weaknesses and exploit them in many areas; AI is a double-edged sword [2].

As it appears at present, the blockchain technology offers the only framework robust enough to match up to IoMT security challenges. No doubt that the innovation in such new technologies can safeguard the interest of IoMT and efforts have been made for the same.

Companies and providers are coming up with security options that can deal with any of the situations and provide one with faultless and flawless data to the users. The implementation of the latest technologies is the key to safety.

Global healthcare regulations need to be followed which is yet to be approved for IoMT. The international bodies are already providing guidelines that are strict to be followed by the government medical establishment. This may take some time, cause anything related to health has always been a crucial decision, and it might keep many innovations at bay just because of some formalities.

As this technology at its early stage, the infrastructure is yet to build and the cost building is enormous, starting from app development, blockchain, training personals, and supply chain to the devices. However, slowly and steadily the future is bright enough to make it a reality. dedicated IoMT connectivity and IT networks, blockchains, and platforms like cloud computing are all necessary for the well functioning of the same. In addition, while the long-term ROI (return of investment) is undoubtedly impressive, the initial cost for setting up such systems is significant.

12.6 INNOVATION AND BUSINESS PERSPECTIVE OF INTERNET OF MEDICAL THINGS

The recent growth in wearable healthcare devices, from fitness trackers to portable blood pressure and insulin monitors shows the demand for IoT devices in keeping a check on one's well being and health management. There had been actual cases where wearable devices like Apple watch have saved the life of users while having an accident like cardiac arrest by alarming the family members with a panic call and real-time location. Hence Major mobile technology companies are working passionately for the enhancement and updating of their authentic wearables, adding more, and effective health tracking features.

The aforementioned statistics show how the global market is expected to mold and the trends to be believed in. As we all know that the 2000s has been the generation of mobile phone and followed by the Internet, which took the globe by storm. The next big thing is expected to be of the IoT devices. For instance, past years show a significant growth in wearables and devices which is been incorporated in Fig. 12.1. The graph shows a stiff growth elevation in the sales of the products over the years, whereas Fig. 12.2 shows the growth rate of sales in smartphones over the years. The trend it shows is surprising, the growth rate of smartphones, which is too massive, somewhere in billions but somehow it is getting stagnant [3].

Whereas, wearable devices are showcasing a positive elevation in the sales trend, even though in millions. Hence, big companies are investing more in the advancement of these products as they are seeing it as the opportunity to bloom in the upcoming market and wonder IoT devices is going to take up the whole world in every field possible, most importantly the healthcare sector (Fig. 12.3).

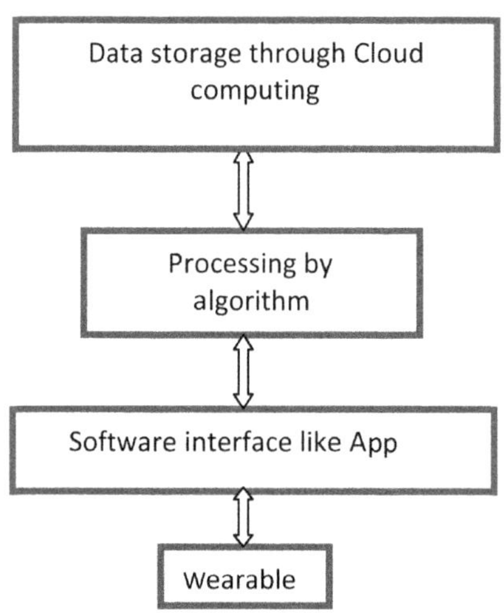

FIGURE 12.1

Artificial intelligence assisting healthcare.

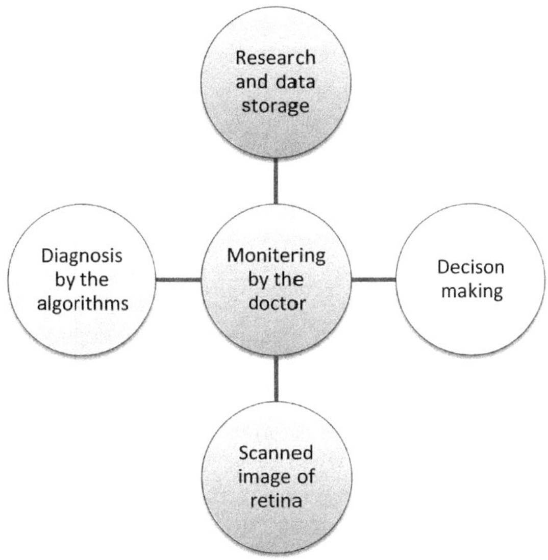

FIGURE 12.2

Retina diagnosis with the help of artificial intelligence.

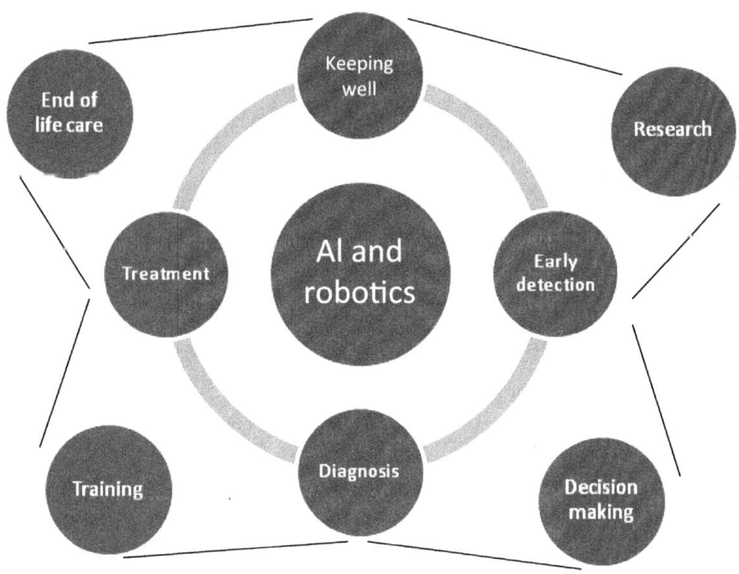

FIGURE 12.3

Benefits of artificial intelligence in healthcare.

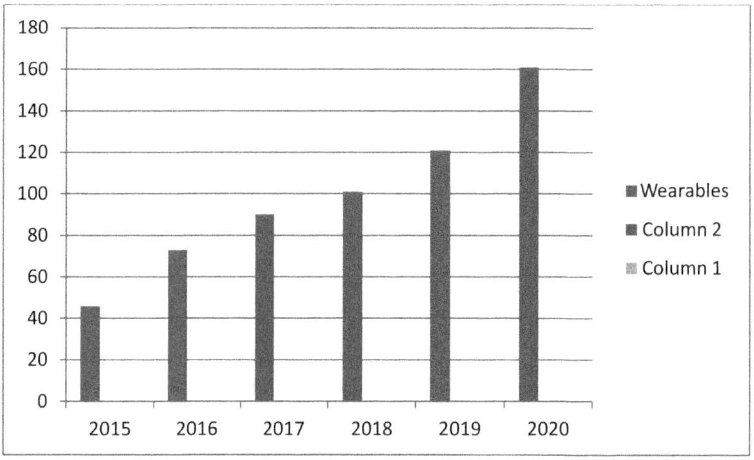

FIGURE 12.4

Projected sales of wearables in millions.

The leading smartphone companies show the same trends of a saturated market, unlike IoT devices (Fig. 12.4).

12.7 ROBOTICS AND NANOTECHNOLOGY AMALGAMATION

Now with the advancement in robotics, surgical robotics became a common reality. Knowing that robotics has already made a remarkable impact in the production of the automobile industry, by cutting down on cost production, and increased batch processing. It is well known for how accurate, error-free, and least time consuming the application of robotics can be in any industry (Fig. 12.5).

In addition, with the interface of IoT and AI, a doctor can perform treatment from far places breaking the time and space bridge. AI-powered robotic surgical means showed to be more precise than real doctors on more than one occasion. Even though there are still limitations and risks involved, but the technology is promising and is looking to become more widespread in the nearest future.

A deeper study into the remote domain enhances the IoT network to be able to connect and track practically any sensor inserted into a human body for medical purposes. This will help prevent serious health issues like cardiac arrests by providing the doctor with real-time motion and ECG data, all sorts of characteristics and provide medical help for critical patients just in time.

In addition, even with the unavailability of doctor, proper treatment can be held by the robotic operative hands saving precious lives of many.

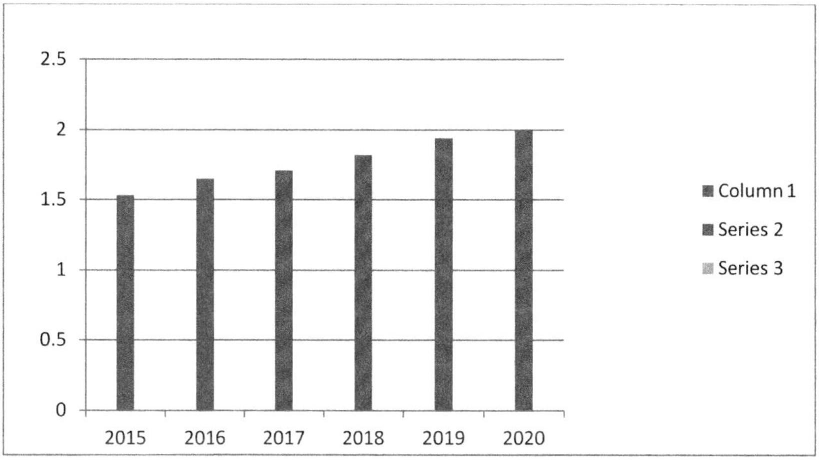

FIGURE 12.5

Projected smartphone sells over the year in billions.

12.8 THE IMPLICATION OF NANOTECHNOLOGY

The amalgamation of nanotechnology and IoMT will lead the medical sciences to a whole new age of advancement. Nanotechnology is already a promising field providing minute devices as a reality. It not only deals with manufacturing a tiny device as small as rice grain but equivalently increases the strength too. And hence, merging these capabilities with IoMT can give a boost to medical sciences.

For instance, the developed smart pill which can be popped into the human body like an ordinary pill, having sensors, and camera inside rather than medicine is where this technology is directed to. This enables a practitioner to observe any differences that the patient is having which can be cured.

Just like sci-fi movies, the reality of biosensors tagged into one's skin is possible now. With the amalgamation of nanotechnology, the sensors can be made as tiny as possible which can get tagged into the skin without any agitation, proving the user with vital information about health characteristics.

In addition, the World Health Organization conducted a study in 2003 to find out that about 50% of prescribed medicines are not taken the right way or completely ignored [4]. A prominent example of resolving this issue is the ingestible sensors solution developed by Proteus.

12.8.1 MEDICAL SENSORS

The affordable cost of sensor technology has evolved IoMT devices and manufacturers to build more economically connected healthcare products.

Biosensors comprise of a healthy size of the IoMT product in the marketplace, with market revenue expected to exceed by US$28 billion by 2025 [5]. Advancement in such devices relies on biological material and sensor to detect characteristics of blood, respiratory function of lungs, tissues, and other parts of the body. In addition, nonbiological medical sensors can measure the electrical activity of the nerves, heart and muscles, body temperature, and other characteristics of the user.

12.8.2 SOLUTION TO DIABETIC

The application of IoMT can be the most influential one in the case of diabetic patients, where a diabetic patient needs to be examined every day of their blood sugar and glucose level. As fluctuation in sugar level is critical and high or low sugar levels could rapture the arteries and valves of the heart. Hence, constant observation is required for the illness. In addition, IoMT can act as a boon for a diabetic patient keeping a continuous check on the blood sugar level with devices like biosensor tagged into the body. The real-time fetched data would alert the user of possible hazard and in any mishap, the family members can be alarmed with real-time location and ambulance facility. The advancement had been made and soon to hit the market for commoners. This could make examination easier for the users as everyday needle injection and checking is time consuming and a complex job affecting the skin with each test, it can be sustained with just the use of a biosensor incorporated inside [6].

12.8.3 SOLUTION TO OLD AGE CARE

IoMT sees a huge potential in old age healthcare management. The critical times of human being are one that of old age where are required to be continuously taken care of with utmost sincerity and this comes with tremendous efforts and hard times. One simple forgotten step would cost the life of someone and this is where IoMT makes things easier.

The caretakers can have a continuous track of the early person in real time and any abnormal fluctuations can be alerted by the device to the family members. The assistance that it can provide like reminding about daily medicine dozes to be taken.

Illness like Alzheimer's, which is one of the most common illnesses over the age of 65 can be treated with AI and innovation has already been made. One in ten people above the age of 65 is suffering from Alzheimer's. Due to the progressive memory loss, the biggest threat to one is wandering here and there, often in their locality or neighborhood [7]. The patient is prone to accidents and can be taken advantage of those less fortunate. Ultimately if Alzheimer's is not found within 24 hours, his or her life is truly in danger.

To deal with this, IoT can provide devices that will keep a track of the person and its whereabouts. The GPS track can help find the person in the time of unavailability. In addition, helping the less fortunate to manage schedule, alerting him or her with what to do next and remind daily motive.

The time-to-time track would also help the caretaker to have detailed health stats [8].

12.9 COMPLEMENTING GOVERNMENT SCHEMES

One domain where IoMT is making drastic changes is that of the healthcare Insurance sector. The business now would be made examining the user on every possible health factors necessary for ensuring and calculating the precise insurance module for the client. In addition, later assist the user with devices and predictive module of health expenses to be taken. One fine example is the UHG Healthcare Organization, namely, United Health Group based in the United States, initially founded as the healthcare insurance company providing insurance but later become one of the largest healthcare services around the globe with the adoption of AI and IOT, where all of the data

FIGURE 12.6

Logo for government's new health scheme.

about user's health are being monitored securely and held to comfort the user with medical attention at prices which are too minimal and could justify the concept of healthcare for everyone and anyone.

The United Nations organization and government in many countries like India and the United States have shown interest in such a concept of healthcare for everyone. For example, Ayushman Bharat of India (Fig. 12.6 shows Ayushman Bharat logo) and the Affordable Care Act (ACA), nicknamed as Obama care of the USA, which has the single objective of better and affordable healthcare for everyone and anyone. And hence, the insurance and healthcare sector is truly having a huge potential to edge the market with trillions of dollars [9,10].

12.10 CONCLUSION

Emphasize of IoT with medical science can be nothing but an important revolution in the field on the global scale as healthcare. Even though there are few technological obstacles and implementation difficulties or peculiarities to be taken care of. Moreover, the technology is at the early stage of its development are the sort to face any of the discussed challenges but, with so many advantages that it has to offer to seem to go very well with the future of healthcare management.

The technology is so promising that even the medical professional community believes that the integration of IoMT and its adaption is only going to make life easier for everyone, providing one with better healthcare.

The obstacles that the practitioner faces in the treatment can be overcome with a huge margin. IoMT, as opinionated by medical science representatives, is simply outstanding and believed to revolutionize the field and named to be the next stage of healthcare.

In addition, not just the medical viewpoint is strong in favor of IoMT, but the reduced time and cost of healthcare is also the vital key point where IoMT plays a huge role. This is going to fulfill the statement of healthcare for everyone and anyone, anywhere!

REFERENCE

[1] <Available from: https://healthitanalytics.com/news/internet-of-things-ai-to-play-key-role-in-future-smart-hospitals.

P. Gope, T. Hwang, BSN-Care: a secure IoT-based modern healthcare system using a body sensor network, IEEE Sensor. J. 16 (5) (2015) 1368–1376.

G.J. Joyia, R.M. Liaqat, A. Farooq, S. Rehman, Internet of Medical Things (IOMT): applications, benefits and future challenges in the healthcare domain, J. Commun. 12 (4) (2017). 240-7.

L. Haoyu, L. Jianxing, et al., Algorithms for the automated detection of diabetic retinopathy using digital fundus images: a review, J. Med. Syst. 36 (1) (2012) 145–157.

[2] S. Goel, P. Tomar, G. Kaur, A fuzzy based approach for denoising of ECG signal using wavelet transform, Int. J. Bio. Sci. Bio Tech. 8 (2) (2016) 143–156.

[3] <Available from: https://gs.statcounter.com/vendor-market-share/mobile/worldwide/2018.

<Available from: https://www.statista.com/statistics/274658/forecast-of-mobile-phone-users-in-india/.

<Available from: https://gadgets.ndtv.com/mobiles/news/smartphone-growth-in-emerging-markets-to-continue-in-2019-counterpoint-2053410.

[4] Y. Xin, L. Kong, Z. Liu, C. Wang, H. Zhu, M. Gao, et al., Multimodal feature-level fusion for biometrics identification system on IoMT platform, IEEE Access 6 (2018) 21418–21426.

<Available from: https://existek.com/blog/iot-in-healthcare/.

[5] <Available from: https://healthitanalytics.com/news/internet-of-things-ai-to-play-key-role-in-future-smart-hospitals.

S. Patel, R. Nanda, S. Sahoo, Nanotechnology in healthcare: applications and challenges, Med. Chem. 5 (21) (2015) 528–533.

V. Raffa, O. Vittorio, C. Riggio, A. Cuschieri, Progress in nanotechnology for healthcare, Min. Invasive Ther. Allied Technol. 19 (3) (2010) 127–135.

[6] D. Tomar, J.P. Bhati, P. Tomar, G. Kaur, Migration of healthcare relational database to NoSQL cloud database for healthcare analytics and management, Healthcare Data Analytics and Management, Academic Press, 2019, pp. 59–87.

[7] Helmy, J., & Helmy, A., The Alzimio app for dementia, autism & alzheimer's: using novel activity recognition algorithms and geofencing, in: 2016 IEEE International Conference on Smart Computing (SMARTCOMP), IEEE, 2016, pp. 1–6;

D.S. Abd-Elminaam, Smart life saver system for Alzheimer patients, down syndromes, and child missing using IoT, Asian J. Appl. Sci. 6 (01) (2018).

<Available from: https://blog.aeris.com/neo/how-iot-solutions-help-alzheimers-patients-stay-independent.

<Available from: https://kidshealth.org/en/kids/alzheimers.html.

[8] G. Kaur, N. Gupta, E-health: a new perspective on global health, J. Evolution Technol. 15 (1) (2006) 23–35.

[9] <https://en.wikipedia.org/wiki/Patient_Protection_and_Affordable_Care_Act>.

[10] <https://www.pmjay.gov.in/>.

CHAPTER

IOIT: INTEGRATING ARTIFICIAL INTELLIGENCE WITH IOT TO SOLVE PERVASIVE IOT ISSUES

13

Lakshita Aggarwal[1], Prateek Singh[1], Rashbir Singh[2] and Latika Kharb[1]

[1]*Jagan Institute of Management Studies, Rohini, New Delhi, India* [2]*RMIT- Royal Melbourne Institute of Technology, Melbourne, Australia*

13.1 INTRODUCTION

Artificial intelligent makes the device Internet of Intelligent Things (IoIT) rather than Internet of Things (IoT) with significant intelligence added to things [1]. Today, AI and IoT have together embedded to each and every place as these sensor networks are bringing a new changing phase from the areas of virtual reality to the reality especially in areas such as health, medicine, military, observing seismic activity observation in volcanoes, smart cities, remote monitoring, cloud services [2], and many more. There were the ages when people use to communicate with each other via letters with the ongoing technological developments Internet has changed the way we communicate. Artificial intelligence (AI) brings not only new challenges but also comes with the developments in the nation with the global labor, economic growth, capital competition, and also solves moral, supervisory, and legal issues. As the technologies advanced, era passed by Internet raised with IoT, which shows how different objects are connected to each other with the Internet. For instance, refrigerators are connected to Internet and respond about the people with the information of the existing fresh food stock available or the food is getting rotten, clothes we wear might able to tell all the useful biomedical information required about the people if any necessary action is required to be taken. The concept behind IoT is "the extensive use of the variety of things or objects that are around us such as sensors and other smart devices that are able to interact and produce intelligent results adjusting with the environment" [3]. This is the widespread in the society as it changes life of one and all with the changing demands of the society. Integration of AI with IoT brings smart and intelligent devices acting logistic and automated help to the human.

In this chapter, we have introduced the concept of "Smart Stores" which uses the concepts of IoT as well as machine learning models inculcating AI in it. With the help of IoT we can set temperature of particular section in store according to number of consumers in particular section and train our machine learning models to maintain the temperature in that section so that consumer can shop in more comfortable environment, thus it will also help store to reduce electricity cost. With

Artificial Intelligence to Solve Pervasive Internet of Things Issues. DOI: https://doi.org/10.1016/B978-0-12-818576-6.00013-7

the help of machine learning we can also predict amount being spend by particular consumer so that store can attract them by introducing several offers. AI could help in adjusting light of different sections present in store according to number of consumers present.

In this chapter, author will discuss about expansion of AI and IoT and then propose IoIT and then evaluate the importance of integrated concept of IoIT in our daily life through concept of random forest regression model of machine learning and evaluation of proposed IoIT model is done with variance score.

13.1.1 EXPANSION OF ARTIFICIAL INTELLIGENCE

The expansion of AI with the changing environment is a major concern for one and all as it poses a wide range of issues from the government proposals to the individual usage. Artificial intelligence just not only makes our life easier but has a wide range of application such as:

- *Wider use of AI*: To expand the use of AI further, it should have wider support from the new technological fields especially by small-scale companies to improve economic growth and productive development to have a competitive and a healthy market. AI just not only helps companies to expand further but also the employees working in the company by giving the ease at workplace.
- *Solving problems*: The AI revolution will deeply affect the work and wages of people everywhere. As the algorithms used by AI to perform the task would be much more efficient than the same task done by the employees. It would give much more ubiquitous results in measuring the efficiency of algorithms.
- *Ethical, legal, and regulatory issues:* AI works on the transparency and accountability of the algorithms. For instance, we need to protect the highly skilled data the data of the medical report of the patient without harming the sensitivity of the data. AI will be able to perform such jobs with more transparency showing the accurate data of a person with the effective actions need to be taken.
- *Training data:* First the large amount of data is needed to train an artificial intelligent device. So that when the device is needed to work for the large petabytes of data it could prove its efficiency and a training for the labors [4] is also required at a large scale for further implementations and their development. The data is transformed into information.

To summarize, we can say that AI has given us solutions like:

- Resolving IoT hurdles and emerging companies enable interoperability through open source development.
- Imposing standards and regulations for the industry to ensure that data does not gets misused or pirated by government and other industrial bodies. Data should be secure, reliable imposed on regulatory standards.
- Providing strong authentication methods, that is, it acts like a platform that can track shortcomings or irregularities of the network which IoT lacks.

So, we can say that AI just not only makes our life easier but also supports the application of IoT with it.

13.1.2 EXPANSION OF IOT

IoT is increasing the power of the Internet beyond computers even to short range of things, devices, processes, and environment. Based on Ref. [5], IoT is a system of physical things embedded with sensors, software, electronics, and connectivity to provide an infrastructure to perform better by exchanging information with other connected devices [6]. IoT consists of interconnected computing devices, mechanical/electrical devices, and objects/animals/people that have information exchanging mechanism and the ability of devices to interact and transfer data over a network without requiring any human effort, that is, human-to-human or human-to-computer interactions. The connected devices are used to gather information, send information back or forth. IoT allows people to communicate and produce desired results in an efficient manner. In IoT things connected to the Internet can be put into three broad classifications:

- Things that collect information and then send the data.
- Things that receive information and then act on it.
- Things that communicate with both the data or collector and information or receiver.

The all three are used to feed each other input and output.

13.1.3 ALONG WITH THE EFFICIENCY OF IOT, RISKS ARE ALSO ASSOCIATED, NAMELY

- *Data privacy*—As all the things are open to the social media so the data we stream on the internet always gets stored which leads to piracy of data or other misuse of personal information which leads to malicious attacks. Steps must be taken in a direction so that the personal information could be kept separate so that it is safe from identity losses and other attacks.
- *Physical attacks*—As nowadays, cybercrimes have increase its way the amount of cybercrimes are more vulnerably increasing. So, it might happen that someone controls our car it could be hacked leading to some physical attack.
- *Grid collapse*—These devices just need power to operate when the power falls steps must be taken so that the running of these devices continue. Networks must be as robust as possible so that the devices continue to navigate as per their own functioning of the system.

13.1.4 DUE TO PREVALENT RISKS, IOT FACES VARIOUS CHALLENGES LIKE

- *Scalability*—IoT faces the challenge of time and size of data growing immensely. The ability of data to grow in size and scale the performance immensely.
- *Security*—Security is a major challenge with the devices related smart enough to interact and produce reliable and efficient results.
- *Technical requirements*—Devices require proper techniques and ways to manage those technical requirements should be ubiquitously analyzed.
- *Technological standardization*—Standards for devices to meet IEEE technology standardization to interact and make technological results.
- *Software complexity*—IoT faces challenge of time and size complex enough to handle the growing demand of customers.

But these challenges can be challenged by including AI and IoT together in our proposed IoIT. IoIT is capable to overcome above problems through the algorithmic structures.

13.1.5 PROPOSED ARCHITECTURE OF IOIT

There is no single architecture that is universally followed in IoT architecture. Different architectures have been proposed by different researchers. The available IoT architecture is not sufficient for idea of IoIT as it does not include all the basic idea from initial stage to the development stage so it has been further expanded IoT architecture to proposed five-layer IoIT architecture. We have proposed the five-layered architecture for IoIT.

1. *Perception layer*: Perception layer acts as the physical layer which is responsible for sensing and then it gathers information from the environment. It examines and identifies smart objects present in the environment from their physical parameters.
2. *Network layer*: It acts as a connection for connecting the devices with the servers and network devices. It is used for transmitting and processing server's data. It acts as a communication link between client and the server making the devices interact via network in this layer. It is also used for transmitting and processing the server's data.
3. *Processing layer*: It acts as the middle ware layer by storing, analyzing, and processing huge amount of data that comes from network layer. It manages all lower layer services.
4. *Application layer*: It is concerned with all the responsibilities for delivering applications specific services to the users. It includes various applications in which different IoT devices are deployed for the help to human kind such as smart cities, smart homes providing smart health services.
5. *Business layer*: It includes the logic of IoT system, that is, applications, business, profit models, and other users privacy. It enables sensors to act smart and analogous results from the data centers. It allows various technologies to interact with databases including the logic of the system inculcating profit models so that devices could interact together in a better and smart way (Fig. 13.1).

The five-layer model works as first perception layer identifies smart objects then it is connected with the network layer for further processing and transport of data from client's end to the server

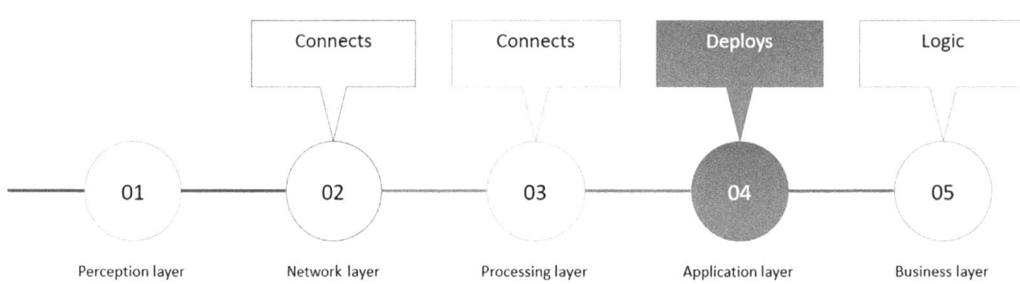

FIGURE 13.1

Architecture of IoIT.

and further transmission and processing of the clients data take place in this layer. Client and the server then on the application layer devices are deployed in the environment (Fig. 13.2).

As the data is gathered at the perception layer from this it is connected with the network layer here the processing is done on the processing layer managing all lower layer services, modules, and connecting with other databases. Then it goes for presentation and interaction to the application layer from here it is added with the business logics and intelligence to the devices for further transformation to enhance the devices to the intelligent and smart devices.

The integration of all five layers in IoT and AI makes the devices integrated together has allowed devices to enter into the world of smart gadgets but the integration of intelligence with IoT and AI has universally accepted the welfare of all, bringing into actuation by making it a reality. By the intelligence AI and IoT will spread their wings more high making the devices IoIT. It will let humans enable those devices to act smart and transforming raw data sets into information. All five layers completely transform the data sets into relevant information. These all five layers helps us to study about finer and deeper aspects of the IoIT.

13.1.6 EXPANSION OF IOIT

IoT is concerned with making connection among machines and using the data generated after some processing from those machines. The proposed architecture of IoIT is made of integrating the concepts of IoT and AI together to get intelligent results. AI and IoT together as IoIT makes intelligent behavior in machines of all kinds. IoT in whole generates vast amount of data so to manage the enormous amount of data AI is functionally essential to deal with the massive volume of records to produce sensitive information from the data.

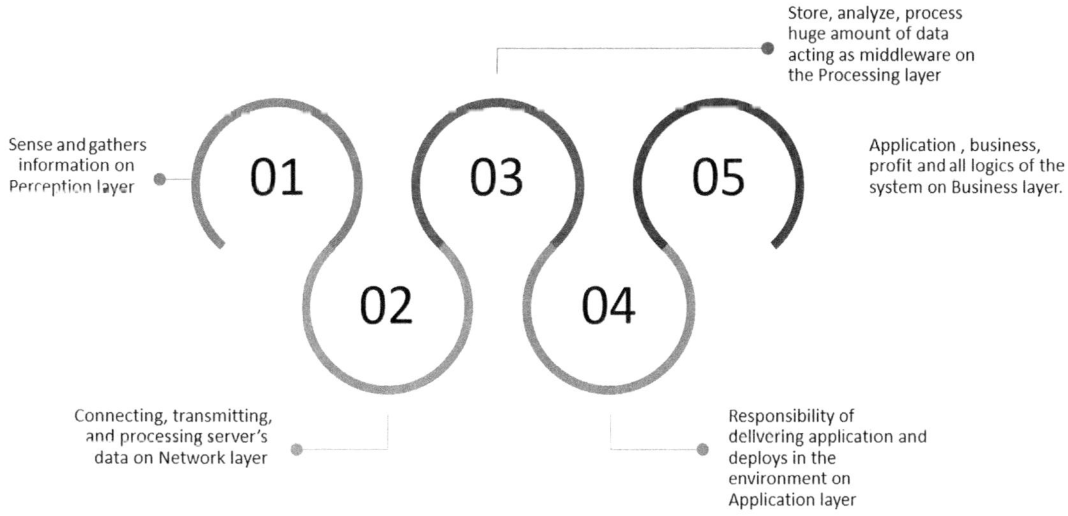

FIGURE 13.2

Understanding flow in IoIT architecture.

Here, we will highlight the key points lacking in IoT that led to need of proposed IoIT:

- *Connectivity*: IoT alone is not able to connect all the devices to the Internet uniquely. It has been proven to be insufficient in addressing of ipv4 to provide unique addresses to the things connected [7] over the Internet but the addressing through ipv6 have not met with the majority of the users. The range of addressing through ipv6 is able to connect each device uniquely over the Internet providing different address to all the devices over the Internet so it improves the features of security as well which is a major concern nowadays.
- *Energy management:* Energy management is also a big concern in the era of technological development using the devices alone is not sufficient but using the devices ubiquitously is necessary. For example: recharging the cells without the need of power source or to produce power through sensing the human gestures and behaviors to act accordingly. This also improves the aspect of appearance of technologies in a smart environment.
- IoT is evolving so that the various networks and sensors integrated with each other and become standardized.

So, these challenges are overcome only when we integrate IoT with the AI to make the devices IoIT to act smart, that is (Fig. 13.3).

In Table 13.1, we have taken a case study of how an alarm operates with IoT, that is, without intelligence and then with IoIT, that is, with intelligence. IoT without intelligence does not care about when I got up rather all things were done according to preset timings but in the case when IoT was added with intelligence then it is proven to be ubiquitous choice of every making devices to act smart and intelligent. The tabular data clearly shows:

- Tasks that are done without intelligence create a mismatch between actual requirement and real-time outputs.
- Tasks that are done with intelligence create a perfect output as per real-time requirement.

FIGURE 13.3

IoIT = AI + IoT.

Table 13.1 A case study for IoT and IoIT.

Time Slots	IoT-Without Intelligence	IoIT-With Intelligence
Time T1	Alarm rings on mobile, I wake up	Alarm rings on mobile, I wake up
Time T2	Take a bath, make food in microwave with time set.	Smart bathroom is providing me lukewarm water for bath, my smart kitchen gets food ready just on time after I had a bath.
Time T3	Work hard to recall remaining tasks.	After breakfast, my smart mirror, shows me list of today's tasks.

Table 13.2 Comparative analysis of IoT, AI and IoIT.

IoT (Internet of Things)	AI (Artificial Intelligence)	IoIT (Internet of Intelligent Devices)
Works on interrelated devices	Adds intelligence to devices	Provides business logic and intelligence to devices.
No business logic	Not a complete solution to device	Intelligence + IoT + AI
Do not process data	**Processing of large amount of data**	**Intelligent processing of large amount of data**
Unified way of handling the data	**Self learning of the data**	**Intelligent way to extract information from data.**
Do not require any kind of human interaction	**Provides human interaction**	**Well-defined interaction with humans**
Data is not repaired; it just transforms the data over a network	**It does not adds complete business logic or complete repairing of data does not occur.**	**Self healing or repairing of data making it as information.**
Simple decision-making progress	**Complex decision-making progress.**	**Reliable, efficient but complex decision making progress.**
Nonautonomous process	**Autonomous process**	**Autonomous process**
Not much encrypted	**Encrypted process**	**Highly encrypted**

Tasks [8] first those were sequential now got adjusted according to other parallel activities. All processes are now taking place concurrent to each other. This is all the power of intelligence with IoT, that is, proposed IoIT, it is going to create a world which is totally free of human intervention in order to save our valuable time from erroneous mistakes.

In Table 13.2, we have given a comparative analysis of IoT, AI, and IoIT to clearly show the need of proposed IoIT.

Only integration of IoT and AI, that is, IoIT makes the devices to enter into the world of gadgets for the welfare of all and bringing dreams into actuation by making it a reality.

IoIT will let humans enable those devices to act smart and transforming raw data sets into information.

We have worked in various real-time problems and analyzed following applications of IoIT:

- *Education*—In today's technological era learning is not only limited to text and images from the textbook but it is an infinite combination from more of the web-based sites which incorporates additional videos, materials, animations, assessments, and other materials that aids the learning process from the IoIT devices. IoIT in education can be regarded as a powerful creative tool to transform the ways of teaching and learning which enables teachers and students to create 3D textbooks and create graphical notes which increases efficiency.
- *Management*—IoT device management is to increase the device capabilities and reduce the cost and effort of devices. Effective device management is necessary for maintaining the health, connectivity, and security of IoT devices. IoIT is the asset to the device management.
- *Logistics*—IoIT offers easy location and monitor key assets to optimize logistics, prevent quality issues, and monitor inventory levels in the logistics department. Embedded sensors, connected devices, and analytics technologies intelligently exploit rich sets to optimize operations and make new possibilities.
- *Food*—It offers connected consumers and supply chain of food items in the open market such as baking cakes requires proper techniques to cook the cake with proper interactive sensors and devices to make the devices intelligent enough.
- *Pharmaceuticals*—IoT has the potential to revolutionize the manufacturing of medicines by enabling the patients sensing and preventive maintenance of the patient from the IoIT devices.

We have worked over three case studies for implementation of IoIT. These case studies tell how different areas would change its shape after integrating intelligence with IoT and AI. We have summarized the results in Table 13.3.

Table 13.3 Implementation of IoIT.

Medical and Healthcare	Education	Entertainment
AI helps in better diagnosis through its planned schemes.	AI has improved ways of teaching education and it is now entertainment.	AI has added advancements in games, music, videos, etc.
One can track fitness using smart devices.	Real like virtual environment makes education easy.	Digital technology has made movies a thrilling experience.
By using data from smart devices, health management is easy.	One can interact with virtual world in realistic education platforms.	Smart and digital techniques save time and energy of film producers.

13.1.7 DATA ANALYSIS FOR IOIT

Data is useful only when it becomes useful information. To make data useful we need to convert it into the information first for action-oriented response, it needs to be added with context and creativity to make it connected intelligence and not only just connected devices. The devices using IoT need lots of data which is then transported to the information, converted to action-oriented response producing context, and creative results transforming into AI. IoT devices generate lots of data which might not be that useful to mankind. The data needs to be turned into actionable format which is known as an information. The huge volume of data should be handled carefully that the data may include (Fig. 13.4):

- analytics
- statistics operation
- metric calculations
- event correlation

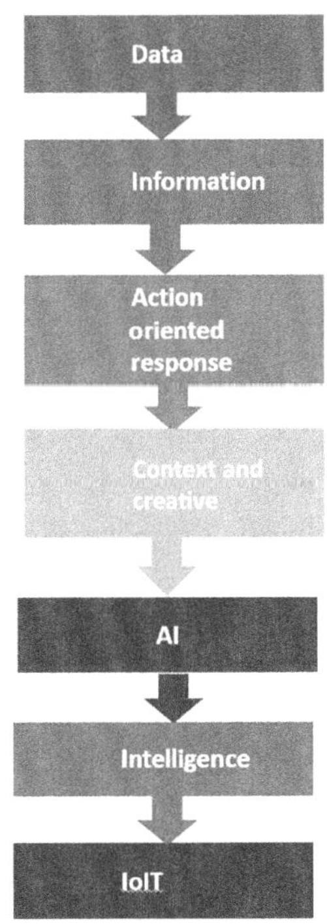

FIGURE 13.4

Conversion from Data to IoIT.

Data analysis improves the traditional methods of analyzing only the optimal value from the data sets. AI is ubiquitous for real-time processes and postevent processes [4].

1. Postevent processes—It identifies patterns found in data sets and making predictive analysis going to occur in recent future. For example, correlation between traffic on roads, polluted air level going to cause harm to people, and more recently new algorithm developed on seeing the same process to control the number of people entering for particular kumbh-mela and all so that number of people required to control could be predicted and where to post those people for safe journey throughout.
2. Real-time processes—It responds quickly to all analyzed conditions and builds-up knowledge of decisions drawn from those events. For example, information collected from remotes, video cameras they all predict the real-time reports if the reports drawn are delayed they are not more useful.

With the use of various tools and processes such as machine learning and many other provides the ability to detect and decode patterns in the form data should be presented. So, machine learning provides ways to analyze those actions and produce the correct form of data. Analytics takes place in the following manner:

- *Predictive analytics*—It tells what will happen.
- *Prescriptive analytics*—It tells what should we do.
- *Adaptive/continuous analytics*—What are the actions to be performed and what decisions should be taken.

IoT and AI combined together would act as trigger to the devices involving pervasive business houses to grow much as it would make intelligent IoT devices all together.

13.1.8 EXPLORATORY DATA ANALYSIS OF IOIT

In this chapter, we have used machine learning techniques to show the relation between AI and IoT. We have used sample data sets from Kaggle to have a better insight into integrated IoIT. We have chosen a dataset of transactions in a retail store named as "Smart Store."

13.1.9 SENSORS USED FOR SMART STORE

AI and IoT together will constitute machine learning models to be trained according to purchasing behavior and age category of consumer so that air-conditioning of particular section in store can be adjusted. Model uses LM35 sensor to adjust the air-conditioning system in the store. Lights of store can also be adjusted according to trained machine learning model with the help of LDR sensor and with the help of motion sensor lights of particular section will be turned off, if no one is present in the store to use and spend the light in a ubiquitous manner.

IoIT will help the devices to act smart and produce smart and reliable results so that sensors act smart enough.

- *Light-dependent resistors (LDR)—Also called as* photoresistors are employed in circuits where it is important to recognize the nearness and the dimension of light. LDRs are an especially advantageous hardware segment to utilize in the picture aspect.
- *MotionSensor*—The fundamental reason for movement recognition is to detect a gate crasher and send a caution to your control board, which alarms your checking focus. Sensors work when you are not home, or when you tell the framework you are not there. Some security frameworks can be customized to record occasions by means of a surveillance camera when movement is recognized. Movement sensors stand watch, prepared to respond to different circumstances, for example, development in your lounge room, windows, or entryways being opened or shut, or a broken window.
- *LM35*-LM35 gadget has preference over straight temperature sensors aligned in Kelvin scale, as the client is not required to subtract an expensive steady voltage from the output to get helpful Centigrade scaling. The LM35 gadget does not require any outer adjustment or cutting to give normal exactness's of $\pm$ ¼°C at room temperature and $\pm$ ¾°C over a full -55°C to 150°C temperature go.

We thought about various cases that could happen in a store like:

- Customer purchase behavior against different products
- Prediction of amount of purchase
- Prediction of age of consumer
- Prediction of category of goods bought
  - Who will spend more: Male/Female?
  - Which products will be sold more?
  - Which products are common among male and female consumers?

Then, we will come up with four suggestions about Smart Store:

- Based on age
- Based on location
- Based on gender
- Based on products

Fig. 13.5 given here shows the attributes of a consumer viz. user ID, product ID, gender, age, city, marital status, etc. To summarize main attributes from our dataset we used several python libraries in Fig. 13.6.

Fig. 13.7 shows that there are 3623 products in the store with unique product ID.

Fig. 13.8 shows the top 10 highest selling products in the Smart Store. The Product ID P00265242 is highest selling product in store we have created a data frame but product ID does not belong to product category 3 so we cannot replace NaN values with any statistical values (i.e., mean, median, and mode) (Fig. 13.9).

Since product with Product ID P00265242 only belong to product categories 1 and 2, we have chosen second highest occurring product which also belongs in all three product categories (Figs. 13.10 and 13.11).

```
In [129]: import pandas as pd
          import numpy as np
          import seaborn as sns
          import matplotlib.pyplot as plt
          import math
```

FIGURE 13.5

Libraries used for EDA.

```
In [3]: df.loc[1, :]

Out[3]: User_ID                              1000001
        Product_ID                         P00248942
        Gender                                     F
        Age                                     0-17
        Occupation                                10
        City_Category                              A
        Stay_In_Current_City_Years                 2
        Marital_Status                             0
        Product_Category_1                         1
        Product_Category_2                         6
        Product_Category_3                        14
        Purchase                               15200
        Name: 1, dtype: object
```

FIGURE 13.6

Dataset attributes information.

```
In [5]: print(len(np.unique(df.loc[:, 'Product_ID'])))
        df.shape

        3623

Out[5]: (537577, 12)
```

FIGURE 13.7

3623 Products with unique product ID.

```
In [10]: pd.set_option('display.max_rows', 20)
         Sorted_by_ProductID = df.groupby(by='Product_ID').size().sort_values(ascending=False)
         print(Sorted_by_ProductID)

         Product_ID
         P00265242    1858
         P00110742    1591
         P00025442    1586
         P00112142    1539
         P00057642    1430
         P00184942    1424
         P00046742    1417
         P00058042    1396
         P00145042    1384
         P00059442    1384
```

FIGURE 13.8

Top 10 highest selling products.

```
In [11]: count = 0
         df_P00265242  = pd.DataFrame()
         for i in df['Product_ID']:
             if i == 'P00265242':
                 df_P00265242 = df_P00265242.append(df.loc[count, :])
             count = count + 1

In [12]: df_P00265242
```

FIGURE 13.9

(1) Code to check occurrence of Product ID P00265242 in three categories. (2) Occurrence of Product ID P00265242 in all three categories.

Making different data frame columns based on age 100 is returned in case there is no upper age, that is. In case of 55 + 55 is the lower age and 100 is positive infinity, assuming that 55 + is 100 (Figs. 13.12 and 13.13).

13.1.10 MACHINE LEARNING MODELS USED

- Prediction

  In this chapter, we have applied random forest regression model and evaluation of our model is done with variance score.
- Random forest regression

It uses several machine learning models (i.e., decision tree regression) to predict and land on single conclusion. It learns from decision of several models and conclude its results in the form of predictions (Fig. 13.14).

Index	User_ID	Product_ID	Gender	Age	Occupation	City_Category	_In_Current_City_'	Marital_Status	roduct_Category_	roduct_Category_	roduct_Category_	Purchase
399	1000066	P00265242	M	26-35	18	C	2	0	5	8.000000000	nan	8652
1191	1000196	P00265242	F	36-45	9	C	4+	0	5	8.000000000	nan	8767
1372	1000222	P00265242	M	26-35	1	A	1	0	5	8.000000000	nan	6944
1845	1000301	P00265242	M	18-25	4	B	4+	0	5	8.000000000	nan	8628
2209	1000345	P00265242	M	26-35	12	A	2	1	5	8.000000000	nan	8593
2404	1000383	P00265242	F	26-35	7	A	4+	1	5	8.000000000	nan	6998
2466	1000398	P00265242	M	26-35	17	B	3	1	5	8.000000000	nan	8646
2682	1000426	P00265242	M	18-25	4	A	1	1	5	8.000000000	nan	8890
2851	1000459	P00265242	F	18-25	4	A	3	0	5	8.000000000	nan	8770
3148	1000518	P00265242	F	36-45	12	C	4+	0	5	8.000000000	nan	8896
3158	1000520	P00265242	F	36-45	20	A	2	0	5	8.000000000	nan	8694
3574	1000582	P00265242	M	18-25	4	C	4+	0	5	8.000000000	nan	8761
3748	1000617	P00265242	F	26-35	7	C	1	0	5	8.000000000	nan	8568
4085	1000675	P00265242	M	26-35	20	C	1	1	5	8.000000000	nan	8689
4087	1000676	P00265242	M	18-25	16	B	1	1	5	8.000000000	nan	8714
5920	1000960	P00265242	M	0-17	10	C	4+	0	5	8.000000000	nan	8864
5958	1000970	P00265242	M	26-35	0	B	4+	1	5	8.000000000	nan	7127
6355	1001023	P00265242	M	55+	13	C	2	1	5	8.000000000	nan	3752
7238	1001139	P00265242	M	26-35	1	C	2	0	5	8.000000000	nan	8788
7465	1001165	P00265242	M	26-35	12	B	1	1	5	8.000000000	nan	8882
7526	1001174	P00265242	F	18-25	16	C	0	0	5	8.000000000	nan	7180

FIGURE 13.10

Occurrence of Product ID P00110742 in all three categories.

```
In [14]: count = 0
         df_P00110742  = pd.DataFrame()
         for i in df['Product_ID']:
             if i == 'P00110742':
                 df_P00110742 = df_P00110742.append(df.loc[count, :])
             count = count + 1

In [15]: df_P00110742
Out[15]:
```

	Age	City_Category	Gender	Marital_Status	Occupation	Product_Category_1	Product_Category_2	Product_Category_3	Product_ID	Purchase	Stay_I
342	26-35	B	M	0.0	2.0	1.0	2.0	8.0	P00110742	15824.0	
784	18-25	B	M	0.0	4.0	1.0	2.0	8.0	P00110742	15308.0	
1415	18-25	C	M	0.0	10.0	1.0	2.0	8.0	P00110742	19273.0	
2750	36-45	B	M	1.0	14.0	1.0	2.0	8.0	P00110742	19147.0	
2794	55+	C	M	1.0	13.0	1.0	2.0	8.0	P00110742	19506.0	
3221	36-45	C	M	0.0	12.0	1.0	2.0	8.0	P00110742	11754.0	
3594	36-45	C	M	1.0	14.0	1.0	2.0	8.0	P00110742	19240.0	

FIGURE 13.11

Removing − signs from age columns for better model prediction.

13.1.11 METRICS EVALUATION FOR IOIT

For evaluation of our model, we have used variance score to check accuracy of our regression model. We have tested several machine learning model but random forest shows us the best accuracy prediction with 67.4%. Thus this model can help store for predicting the purchase amount spent by the

```
In [16]: Age_df_P00265242= df_P00265242.loc[:, 'Age'].str.split('[-+]', n = 1, expand = True)
```

FIGURE 13.12

Dividing age attribute in two categories: lower age and upper age.

```
In [23]: List_of_thousand_sales = [a for a, b in Sorted_by_ProductID.items() if b >=1000]

In [24]: Product_dict = {}
         Age_Product_df = {}
         for Product in List_of_thousand_sales:
             count = 0
             Product_dict[Product] = pd.DataFrame()
             for i in df['Product_ID']:
                 if i == Product:
                     Product_dict[Product] = Product_dict[Product].append(df.loc[count, :])
                 count = count + 1
             Age_Product_df[Product]= Product_dict[Product].loc[:, 'Age'].str.split('[-+]', n = 1, expand = True)
             Age_Product_df[Product]['Lower_age'] =  [int(i) for i in Age_Product_df[Product][0]]
             Age_Product_df[Product]['Upper_age'] =  [100 if i == '' else int(i) for i in Age_Product_df[Product][1]]
             plt.subplot(2,1,1)
             sns.distplot(Age_Product_df[Product]['Lower_age'])
             plt.subplot(2,1,2)
             sns.distplot(Age_Product_df[Product]['Upper_age'])
             plt.subplots_adjust(hspace=0.5)
             plt.legend(List_of_thousand_sales, loc='center left', bbox_to_anchor=(1, 0.5))
         KeyboardInterrupt:
```

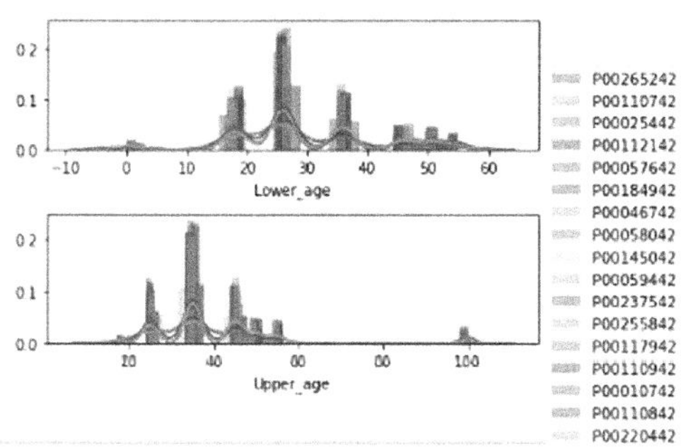

FIGURE 13.13

Demand for top 35 products in different age groups.

consumer according to different attributes (i.e., age, gender occupation, and stay in current city years) so that stores can attract more consumers using different marketing techniques (Fig. 13.15).

Further with IoT and AI, machine learning models can be trained so that they can help store to predict the number of consumers that can present in particular section and adjust the air-conditioning in various sections of stores to adjust the light according to number and age of consumer that could shop in particular section. AI will provide intelligent sensing to the devices so that each section consumes the amount of light and air-conditioning required according to the number of consumers present and automatically gets off when no consumer is present. This will provide pervasive and ubiquitous use of devices making smart enough to predict the results and act accordingly, which is known as IoIT.

```
In [23]:  List_of_thousand_sales = [a for a, b in Sorted_by_ProductID.items() if b >=1000]

In [24]: Product_dict = {}
         Age_Product_df = {}
         for Product in List_of_thousand_sales:
             count = 0
             Product_dict[Product] = pd.DataFrame()
             for i in df['Product_ID']:
                 if i == Product:
                     Product_dict[Product] = Product_dict[Product].append(df.loc[count, :])
                 count = count + 1
             Age_Product_df[Product]= Product_dict[Product].loc[:, 'Age'].str.split('[-+]', n = 1, expand = True)
             Age_Product_df[Product]['Lower_age'] =  [int(i) for i in Age_Product_df[Product][0]]
             Age_Product_df[Product]['Upper_age'] =  [100 if i == '' else int(i) for i in Age_Product_df[Product][1]]
             plt.subplot(2,1,1)
             sns.distplot(Age_Product_df[Product]['Lower_age'])
             plt.subplot(2,1,2)
             sns.distplot(Age_Product_df[Product]['Upper_age'])
             plt.subplots_adjust(hspace=0.5)
             plt.legend(List_of_thousand_sales, loc='center left', bbox_to_anchor=(1, 0.5))
         KeyboardInterrupt:
```

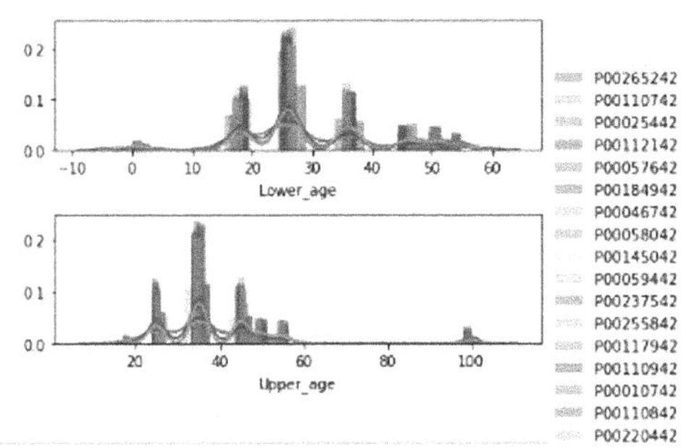

FIGURE 13.14

Random forest regressor.

```
from sklearn.ensemble import RandomForestRegressor
rfr = RandomForestRegressor()
rfr.fit(X_Train, Y_Train)
pred_RFR = rfr.predict(X_Test)
```

FIGURE 13.15

Accuracy prediction results.

13.2 CONCLUSION

We wrap up by concluding the solution that AI is the real application of IoT-enabled devices. It would surely help in increasing the size of business and would make a benchmarking effect on the economic. AI adds intelligence to the IoT devices making it IoIT. IoT with AI making it IoIT brings applications of technology in real life and make human life easy and fast. IoIT is directly related with pervasive technologies and it blends the computation skills with reality of the life. AI is the paradigm to transform everyday real-life objects by bringing intelligence to the devices for easy communication of devices and humans well. Nowadays, IoT has turned its phase from IoT to IoIT as it's a ubiquitous need of an hour. It acts as a common platform that integrates both by transforming the data from sensors into the machines which understands the needs of the changing world and renovates AI into requirements of existent life.

REFERENCES

[1] J. Gubbi, R. Buyya, S. Marusic, M. Palaniswami, Internet of Things (IoT): a vision, architectural elements, and future directions, Future Gener. Computer Syst. 29 (7) (2013) 1645–1660.

[2] I. Lee, K. Lee, The Internet of Things (IoT): applications, investments, and challenges for enterprises, Bus. Horiz. 58 (4) (2015) 431–440.

[3] S.D.T. Kelly, N.K. Suryadevara, S.C. Mukhopadhyay, Towards the implementation of IoT for environmental condition monitoring in homes, IEEE Sens. J. 13 (10) (2013) 3846–3853.

[4] M. Song, K. Zhong, J. Zhang, Y. Hu, D. Liu, W. Zhang, et al., In-situ AI: towards autonomous and incremental deep learning for IoT systems, in: 2018 IEEE International Symposium on High Performance Computer Architecture (HPCA), 2018, February, pp. 92–103.

[5] D. Evans, The Internet of Things: how the next evolution of the Internet is changing everything, CISCO White Paper, April 2011, pp. 1–11.

[6] H.F. Azgomi, M. Jamshidi, A brief survey on smart community and smart transportation, in: 2018 IEEE 30th International Conference on Tools with Artificial Intelligence (ICTAI), Volos, 2018, pp. 932–939.

[7] A. Arsénio, H. Serra, R. Francisco, F. Nabais, J. Andrade, E. Serrano, Internet of Intelligent Things: bringing artificial intelligence into things and communication networks, in: F. Xhafa, N. Bessis (Eds.), Inter-cooperative Collective Intelligence: Techniques and Applications, Studies in Computational Intelligence, vol. 495, Springer, Berlin, Heidelberg, 2014.

[8] A.W. Blackett, M.E. Teachman, B.J. Forth, U.S. Patent No. 6,944,555, Washington, DC, U.S. Patent and Trademark Office, 2005.

CHAPTER

INTELLIGENT ENERGY-ORIENTED HOME 14

Zita Vale, Luís Gomes, Pedro Faria and Carlos Ramos
Polytechnic of Porto, Porto, Portugal

14.1 INTRODUCTION

The main goals of the chapter are to make readers aware of the progress already done in the field of smart homes and to make them aware of all the new possibilities open by the cross-fertilization between smart homes, intelligent energy systems, Internet of Things (IoT), and artificial intelligence.

To accomplish its goals, the chapter starts by providing readers with insights on smart homes, in which current limitations are discussed. The possibilities for evolving to intelligent energy-oriented homes are addressed, by looking at smart homes as building blocks of smart grids and intelligent energy systems and by taking full advantage of ambient intelligence (AmI) in the home environment. Smart homes are largely relevant in the scope of smart grids and intelligent energy systems [1]. AmI is also a very related topic that can help to improve the effectiveness of an intelligent energy-oriented home [2].

The chapter proves that by combining diverse existing concepts and technologies from the referred fields it is possible to provide intelligence to IoT systems to control devices and appliances effectively and efficiently, namely, in terms of the use of the energy, in a smart home dynamic environment.

Smart grids and intelligent energy systems are relatively new concepts from which some aspects have very successfully and quickly been widely adopted by the power and energy sector and put to practice [3]. Smart meters are currently being rollout at high rates and distributed renewable-based energy generation has been increasing and proving its technologic and economic importance [4]. However, effective consumers' engagement for their active participation is just emerging. With nondispatchable, intermittent energy sources as wind and solar radiation, demand flexibility becomes a crucial energy resource to make the intensive use of renewables a success [5].

AmI enables human interaction with day life objects intelligently and unobtrusively [6]. The environment will be able to have an awareness of people's needs and provide the required customization. In order to make the intelligent energy-oriented home a reality, a comprehensive approach is required. The authors consider the seven important roles of ambient intelligent, namely, supporting in the interpretation of the environment contexts; representing the information and the knowledge associated with each environment; modelling, simulating, and representing the entities involved in the environment; planning and supporting the decisions and actions to be taken;

Artificial Intelligence to Solve Pervasive Internet of Things Issues. DOI: https://doi.org/10.1016/B978-0-12-818576-6.00014-9

learning about the environment and respected information; interacting with the human beings; and acting on the environment according to the context [7].

In this chapter, it is demonstrated the principles of the intelligent energy-oriented home employing a six-layered platform, responsible for connectivity, persistence, unification, IoT integration, subsystem integration, and user interface. The model is illustrated with a real-world application in a living demo environment. The intelligent energy-oriented home allows the integration of subsystems that are already built systems that provide integration. In this chapter, seven subsystems will be integrated; they are described in Section 14.2. In the proposed platform, multiple systems and technologies can be used, from simple hardware sensors to complex artificial intelligence systems using machine learning techniques, computer vision, or complex optimization algorithms.

After this introduction, Section 14.2 brings insights into the background related to smart homes, AmI, and smart grids. The architecture for an intelligent energy-oriented home is discussed in Section 14.3. Then, in Section 14.4 is presented several examples of implementation of systems for illustration of intelligent behaviors in an intelligent energy-oriented home. After the test and analysis of the intelligent energy-oriented home in Section 14.5, the conclusions and further developments are presented in Section 14.6.

14.2 SMART HOMES, AMBIENT INTELLIGENCE, AND SMART GRIDS

Smart homes can change the way we interact with the inanimate resources located in our homes. Focusing on specific fields, like energy management systems, assisted living, and health. The current solutions for smart homes usually lack the integration capability to provide a unique system able to control the building combining all the devices, resources, and systems already installed.

Smart homes for energy management and smart grid participation can be found in the literature [8]. A system to manage the users' resources using their preferences is proposed in [8]. A learning methodology, for autonomous learning of the users' preferences, is proposed in [9,10]. The local management of resources inside the building enables the efficient use of energy, a reduction of consumption, a decrease in energy costs, the efficient use of renewable energy sources, and the active participation of buildings/end-used in the smart grid environment. The active participation in the smart grid brings advantages for end-users and grid management, as can be seen in [11].

AmI systems result in complete, and usually complex, systems that provide resources management/control using a seamless approach while producing an impact in the building environment. The combination of smart homes and AmI systems is possible and desired. According to [12], an AmI application for smart home needs: heterogeneity; self-configurable; extensibility; context awareness; usability; security and privacy protection; and intelligence.

AmI system can improve the users' comfort, and this can be used in a smart home system to improve the users' satisfaction and acceptance. A good example of AmI in buildings can be seen in [13], where the management of an office and a meeting room is presented. AmI can be used to improve users' comfort and support them on a daily basis. Three examples can be found in [14] where domain-based modelling for AmI (DoMAIns) is proposed, or in [15] where the virtual, social, multisensorial, and adaptive system for the rehabilitation of people with acquired brain injury (Vi-SMARt) is proposed, or in the smart home proposed in [16]. These are great examples of how AmI can change how users interact with systems and how they can improve our life.

The possibility to remotely control and monitor resources is a key aspect of AmI systems and smart homes. Without remote monitoring and control, it will not be possible to manage anything. There are a few proposed architectures for frameworks, like the ones proposed by [17] and [18] that use middleware solutions that allow the integration of resources. Another major part of AmI system is domain models that enable the world interpretation, enhance the system knowledge. In [19], the CONSERT context model is proposed for modelling the interaction between persons, places, and resources. In [14] is detailed how the combination of ontologies is used in DoMAIns system. A merge of ontologies is proposed in [20] to enhance the world knowledge used for assisted living.

The integration of smart homes and AmI systems is beneficial. However, these two concepts are possible to add artificial intelligence techniques [7] and IoT. This combination enables a complete solution that takes advantage of different scientific fields.

IoT can be seen as a system of physical devices, smart machines, or objects all connected [21]. Devices like sensors and transducers with different characteristics and purposes focus mainly on collect and transfer data. These devices can and must be used in smart homes and AmI systems; they are natural enablers of such concepts. The use of the already deployed IoT devices enables the deployment of smart home and AmI systems without the need for adding additional hardware. IoT devices can adopt several types; for instance, our mobile phones are good examples of IoT devices that embed several sensors [22]. All the referred hardware, including sensors, processors, and communication means, handle different amounts of data according to the specific application. The collected data can be processed locally or using services provided through the Internet to produce knowledge about that data [23]. Several approaches have been applied to IoT data, namely by applying artificial intelligence [24–29].

In [27] it is made a survey on the machine learning methods, which include deep learning, support vector machines, and Naïve Bayes, as the ones categorized as supervised learning. Unsupervised learning methods include clustering and vector quantification. The authors point out the main objectives of such methods applied to IoT to provide pattern recognition, anomaly detection, computer vision, and speech processing. Other data mining methods, most of the family of artificial neural networks, are compared in [24]. Similar studies have been made in Bishop [26], Tsai et al. [28], and Feng et al. [29] where authors agree in several common aspects like:

- Big data is the future trend as mostly like all the devices will be connected to the Internet so enhancing the available data from IoT devices.
- Taxonomy, ontology, and semantic web technologies will help to improve the accuracy and the organization of the features that are respective to each data set, avoiding the disadvantages of "bad data" that results from the inadequate data treatment.
- While environment and ecology aspects are having increased relevance, an efficient study on the IoT data, namely, regarding energy use, can help to identify simple and effective behavior changes that will contribute to addressing those aspects.
- Applications related to multimedia devices and to control the energy at home will be the most relevant in the future, as well as the results of the respective data treatment.
- The locally distributed processing of data in IoT devices will require more adequate optimization methods, namely, heuristic ones in order to take local decisions.

Other very relevant aspects to take into consideration include data privacy and security as IoT devices will be monitoring people all the time. The authors in [23] propose a reflection around the long-lasting ambition and curiosity of humans about the possibility of having a machine that can learn independently as humans themselves do. It is concluded that in the scope of IoT data, machines will be able to collect and process more information about humans' habits and thus be able to learn with the existing sensors.

With more focus on energy aspects, at the home level, several solutions can be listed. An energy management system enabled by IoT devices is proposed in [30]. A centralized approach is also proposed in [31], where an energy management system uses smart plugs for resource integration. A new smart plug is proposed in [32] for on-peak and off-peak grid periods detection and provide resources management according to the period. Another smart plug for distributed energy management methodology, using deep learning algorithms, is proposed in [33].

Energy-related aspects can largely benefit from IoT, namely, in the scope of smart grids where enhanced control and data processing contribute to improve efficiency, reliability, and reduce energy prices for the consumers [34]. Special focus can be given to the demand response concept where consumer plays as a resource to manage the consumption according to renewables-based generation, for example [35]. Demand response has a huge potential in improving electricity market efficiency, bringing benefits to the whole system, including the final consumers, like the domestic ones at home. However, to achieve the full potential of demand response, the learning of consumers' preferences and the automation of decisions are needed. IoT at home will largely improve smart grids and demand response.

14.3 ARCHITECTURE FOR AN INTELLIGENT ENERGY-ORIENTED HOME

In this section, an architecture is presented, enabling the combination of smart homes, AmI solutions, and intelligent energy systems. The proposed architecture was idealized and designed to use third-party solutions/systems, using a modular and dynamic approach. The existence of multiple solutions, namely, in the IoT field, cannot be ignored nor discarded; they should be used, taking advantage of third-party solutions. The Internet of Home (IoH) architecture, Fig. 14.1, uses a layered architecture where five layers are used; one of those being divided into two components [36]. This architecture was inspired by the energy management system proposed in [37], which had a focus on the integration of IoT devices in order to control a building. All the IoH architectural layers are detailed below regarding their motivation and functionalities.

14.3.1 CONNECTIVITY LAYER

The connectivity to others can bring security risks, but it can enable the enhancement of a system. A system that can communicate uses the functionalities, knowledge, and data from others without the need of doing everything in a standalone approach. The connectivity layer is responsible for the system communication with peers and third-party solutions. The communication with peers can enhance the system knowledge about context, promoting the possibility of having an AmI

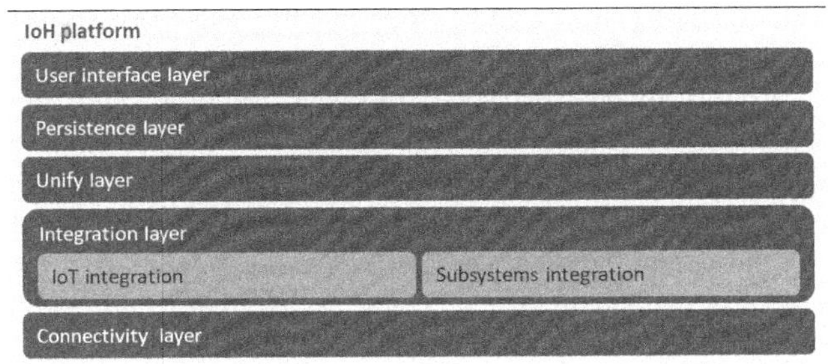

FIGURE 14.1

IoH architecture. *IoH*, Internet of Home; *IoT*, Internet of Things.

environment. The connectivity layer is mandatory and provides the communication ability to IoH platform. Currently, TCP/IP and universal asynchronous receiver-transmitter serial communication are available in IoH.

14.3.2 INTEGRATION LAYER

The integration layer is responsible to integrate all the resources and systems/solutions. The IoH platform does not demand proprietary hardware and does not provide hardware. The platform is a pure software solution that integrates third-party hardware. The platform allows the integration of two types of resources: IoT devices, and complete subsystems. An IoT device is unique hardware that provides reading actions using a cloud-oriented approach. The IoT device usually does not provide intelligence and other management actions without the integration with a gateway and/or a cloud. This type of devices can be integrated into the IoH platform directly. The platform uses such devices to measure the context or control the environment. In the IoH perspective, a subsystem is defined by being a complete system that provides resource management on its own without the need for extra hardware or software.

The IoT market is growing, and in 2022 it is expected 216.9 million homes worldwide with at least one smart device [38]. Projections like this and the growth of this market cannot be ignored or took lightly by academic researchers. The IoT devices are in our homes and this can be used to bring the academic research to our homes, putting it closer to the market and providing vital tests and validations. Concepts, like smart homes, AmI, and intelligent energy systems can be enabled by IoT devices and can be deployed using them. The direct integration of IoT devices can be challenging because not all manufacturers provide appropriated documentation. To simplify this integration and release work from developers, third-party solutions can be found in the open-source community. Home automation systems, like Home Assistant, Domoticz, and openHAB can be easily configured in a building. These solutions are not only able to integrate IoT devices, but also provide RESTful application programming interface (APIs) that can be used by the IoH platform.

The integration of subsystems is also important in the IoH platform. This platform was idealized and designed for research proposes where multiple solutions could be combined to provide a context-aware platform resulting in an AmI environment. Therefore full systems must have the opportunity to be part of IoH platform.

These subsystems are already installed, or tailor-made solutions built for the building, users, or resources. The IoH platform will integrate the subsystems and use the data do improve its database. In addition, IoH will enable the control of subsystems. To enable the integration of subsystems, they must provide an API with a reading method—enabling the reading of all sensors—and a control method—enabling the control of the resources and/or the activation of the subsystem.

The integration layer is dependent on the connectivity layer. Only protocols available in the connectivity layer can then be used by the integration layer. Each instance of the integration layer, IoT and subsystem, are unique threads, meaning that each IoT device and each subsystem will run a thread. This allows the continuous monitoring of such devices and subsystems.

14.3.3 UNIFY LAYER

In order to make sense to all the IoT devices and subsystems integrated, the unify layer is used to organize and catalogue all data. On top of the data, provided by the hardware integration, the unify layer will add metadata that describes the data and how it affects the system. These metadata can be latter used for rules that IoH platform uses to manage the building. The unified layer can be seen as the IoH core, where the novelty and main functionalities are located. The data is processed in this layer according to its metadata and the context.

Internally, IoH works with sensors and actuators; these represent the IoT devices and subsystems. However, a subsystem can result in multiple sensors and/or multiple actuators. Each sensor and actuator are described according to the following parameters: device name, device type (i.e., IoT device or subsystem), building's zone, parameters that triggers the device, which parameters are impacted by the device, measurements (i.e., sensors), and information that the device can provide (e.g., there are persons in the building's zone where the sensor is). Following the above description, a smart plug connected to a fan heater can be described as follows: located at the meeting room, does not have a trigger because it is manually controlled, create an impact in the temperature, measures consumption, and indicate that the user wants a higher temperature. Using the previous example of the autonomous light control subsystem, it can be described as follows: located at the office room, is triggered by movement, creates impact in the room's clarity, measures clarity and movement, and indicates human presence.

14.3.4 PERSISTENCE LAYER

To provide efficient and intelligent resource management within an ambient intelligent environment, the system needs data. These data are extremely important, otherwise, the system cannot provide intelligence because it simply does not understand the system context and the users' preferences. The data can be divided into three temporal types: past data, real-time data, and forecasted data. The past data is all the data that were measured in the past and cannot reflect the current state. In addition, the past data can have subtype divisions where the aggregation of time is built. The real-time data define the current state of the system, providing vital information for context-awareness solutions. The forecasted data are provided by forecast algorithms that demand the knowledge of past data, and sometimes the real-time data. The forecasted data have an error value associated that the system must be considered when dealing with such data sources. The past and real-time data can also have an error association; representing the error of the hardware sensor. All these three data types are needed and vital in IoH. The past allows to understand the users, the real-time allows the understanding of the context, and the forecast allows efficient energy-oriented resource optimization.

Currently, the persistence layer uses a remote PostgreSQL database where the three types of data are stored. The persistence layer has routines that calculate the forecast and forecast errors according to past data. The IoH platform can freely query any of the data temporal types.

14.3.5 USER INTERFACE LAYER

User interfaces are good options to display information to users and receive their feedback. The IoH will manage the building resources and, consequently, impact the users. The idea of IoH is to promote a good impact on the users' daily life, providing an ambient intelligent environment. However, some bad and wrong decisions can be made by the system; this is part of the learning process. Although, this is a normal situation that cannot block the deployment of the system. The IoH provides a user interface that lets the users protest and override the platform if they intend to. The user interface provides the user with a monitor center were all data and metadata are available.

The user interface layer uses a web-based approach where a website is used to communicate with the IoH platform. Therefore this layer uses the connectivity layer to enable the use of the TCP/IP protocol and communication.

14.4 IMPLEMENTATION OF SYSTEMS FOR ILLUSTRATION OF INTELLIGENT BEHAVIORS IN AN INTELLIGENT ENERGY-ORIENTED HOME

In this section, some IoT devices and subsystems integrated into IoH will be described. The IoT devices are energy-oriented and provide vital information for energy management systems. The subsystems are standalone systems that were developed and deployed for resource management in an intelligent energy-oriented approach. The presented IoT devices and subsystems were developed outside IoH context; they were later updated to provide the necessary communication and integration with the IoH platform. A total of seven developments are described.

14.4.1 IOT DEVICE FOR PHOTOVOLTAIC GENERATION MONITORING

The continuous monitor of energy sources is important for energy management systems. Therefore detailed monitoring of the photovoltaic panels' generation benefits the overall system. The shown IoT device was developed to monitor a DC/AC inverter that manages the injection of photovoltaic energy into the building network. Currently, the Fronius inverters provide a cloud connection using an RJ45 connection and data can be remotely visualized using the manufacture website. However, no API was available to read the data autocratically, periodically or in real time. The inverter also provides an RS485 connection using Modbus/RTU protocol to read only the real-time values of generation.

The IoT device will monitor the inverter values using the Modbus/RTU protocol using the RS485 network. The IoT device will also read and publish the clarity data from an outdoor clarity sensor directly connected to the device. The IoT device was developed using the lightweight architecture proposed in [39]. The data is published in an external server using the message queuing telemetry transport (MQTT) protocol. IoH, at its current state, does not allow MQTT protocol. This is simply because of the development time need to add new protocols, not allowing the team to put IoH at the conception level. At this moment the presented IoT device communicates to IoH using the Home Assistant as an intermediary. The home automation system, Home Assistant, enables the integration of MQTT devices and provides the desired RESTful API that IoH can use.

Fig. 14.2 shows the deployment of the presented IoT device in our laboratories. As a power supplier, it is used a simple universal serial bus (USB) charger of a 1 A for 5 V DC. On the right is possible to see the outside clarity sensor installed in the wall pointing south, the same direction as the photovoltaic panels. The MAX485 board that enables the connection with the RS485 network of the invertor is placed inside the inverter. The WiFi board, NodeMCU, is placed outside to maximize the wireless range.

14.4.2 INTERNET OF THINGS DEVICE FOR BATTERY MONITORING

Storage units are a viable resource for energy management systems, providing the ability to sift an energy source or reduce the energy from the grid without producing a decrease of consumption inside the building. Therefore an IoT device for battery monitoring was developed, providing real-time information regarding the battery energy flow. The developed IoT device was developed using the proposed lightweight architecture in [39].

The storage unit used is a Fronius Solar Battery of 12 kWh, with the Fronius Symo Hybrid inverter. The installed system uses an energy analyzer that communicated with the inverter using the RS485 network and Modbus/RTU protocol, the developed device uses this network to monitor all the network information. The system acts in a sniffing mode, only reading data from others. The IoT device is capable to read the RS485 data and understand the parameters. The IoT device uses the MQTT protocol to publish the data. Fig. 14.3 shows the deployment of the IoT device

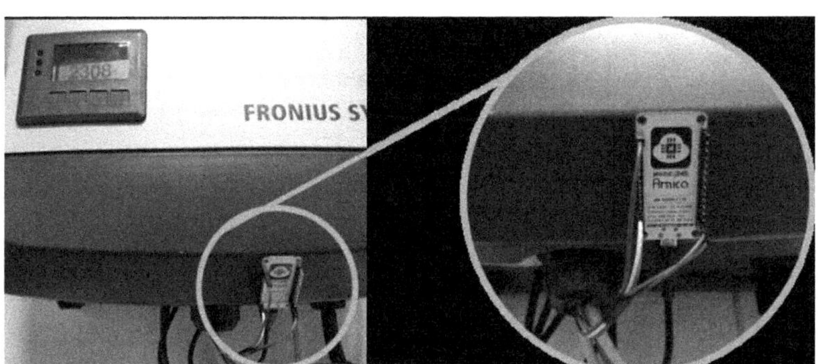

FIGURE 14.2

IoT device deployed in the photovoltaic inverter. *IoT*, Internet of Things.

FIGURE 14.3

IoT device deployed in the battery inverter. *IoT*, Internet of Things.

developed for battery monitoring. The real-time data can identify if the battery is charging or discharging and it can measure the energy flow "going to"/"coming from" the battery.

14.4.3 INTELLIGENT LIGHT CONTROL

The intelligent light control was one of the first developments deployed in our laboratories. This simple solution has a single idea: control the intensity of light in the laboratory according to users' position and room's clarity. The laboratory has analogic lamp ballasts that could dim the lamps individually, controlled by a programmable logic controller. However, the idea was to develop an autonomous system that could in a fast way control the lamps according to several sensors spread in the laboratory. The laboratory was divided into three different zones. In these zones, each has two lamps and one sensor box composed of one motion sensor and one clarity sensor. Lamp 2 is part of zone 1 and zone 2 at the same time. In this system was also integrated the control of three window shutters available in the lab and an outside clarity sensor. Fig. 14.4 shows the architecture used in this development. A total of two Arduinos Mega 2560 R3 are used: one to control the lamps, and another to control the shutters. Both communicate using serial communication. The control over the shutters is done using relays activated with 5 V DC. To enable fine control over the shutters, encoders were added. The used amplifier, top of the light control hardware board of Fig. 14.4, is needed because the used ballast expects a 0–10 V digital signal; that Arduino cannot provide.

The target value of clarity needed inside the laboratory is set in the system. Then the system runs 24/7 providing autonomous control over the light. The system will monitor the present on users inside the laboratory and turn on the lamps accordingly. Before the lamps are turned on, the system checks if outside clarity exists and if so, the shutters are open. If the shutters and lamps are on, and if no user presence is detected over seven minutes, the lamps will decrease their intensity to warn possible users that the system is not seeing anyone. If over the next 2 minutes no movement is detected, then the lamps will be turned off, at this point, if no outside clarity is measured, then the system will not only turn off the lamps, but it also closes the shutters.

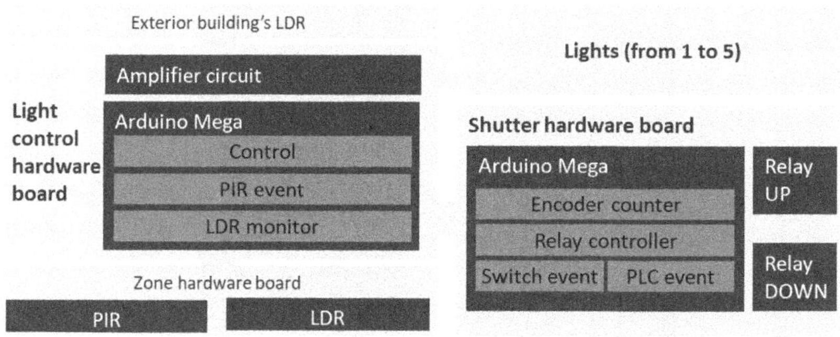

FIGURE 14.4

Intelligent light control.

14.4.4 CONSUMPTION OPTIMIZATION FOR DEMAND RESPONSE PARTICIPATION

This section proposes an optimization algorithm to reduce energy consumption in a building to enable active participation in demand response programs. The optimization will provide the usage targets for lighting and air-conditioner units having into account the maximum consumption values that the building can consume. For this purpose, it is considered there are 10 offices in the building, each of which includes one AC and two lights. Therefore there would be 10 ACs and 20 lights in total. For controlling these loads, it is also considered that there is a supervisory control and data acquisition (SCADA) system responsible to control and monitor the ACs and lights. The ACs are controlled through infrared light-emitting diode (IR LED) connected to a microcontroller, and the lights are controlled through the digital addressable lighting interface. Furthermore, several energy meters monitor the power consumption of the ACs and lights separately.

The optimization algorithm proposed for power consumption minimization of lights and AC is here explained. In the first stage, the objective function of the optimization algorithm can be seen in (14.1), and the constraints subjected to the objective function are shown by (14.2)–(14.6):

$$\text{Minimize Objective Function} = \sum_{t=1}^{T}\left[\sum_{l=1}^{L} W\_L_{(l,t)} \times P\_L_{(l,t)} + \sum_{a=1}^{A} W\_\text{AC}_{(a,t)} \times P\_\text{AC}_{(a,t)}\right] \tag{14.1}$$

$$\sum_{l=1}^{L} P\_L_{(l,t)} + \sum_{a=1}^{A} P\_\text{AC}_{(a,t)} \leq \text{RR}; \forall 1 \leq t \leq T \tag{14.2}$$

$$\sum_{t=1}^{T} P\_L_{(l,t)} \leq \sum_{t=1}^{T} \text{PRR}\_L_{(l,t)} \times P\_L\_\text{Nom}_{(l,t)}; \forall 1 \leq l \leq L \tag{14.3}$$

$$\sum_{t=1}^{T} P\_\text{AC}_{(a,t)} \leq \sum_{t=1}^{T} \text{PRR}\_\text{AC}_{(a,t)} \times P\_\text{AC}\_\text{Nom}_{(a,t)}; \forall 1 \leq a \leq A \tag{14.4}$$

$$P\_L_{(l,t)} + P\_L_{(l,t-1)} \leq \text{MaxRed}\_L; \forall 1 \leq l \leq L; \forall 2 \leq t \leq T \tag{14.5}$$

$$P\_\text{AC}_{(a,t)} + P\_\text{AC}_{(a,t-1)} \leq \text{MaxRed}\_\text{AC}; \forall 1 \leq a \leq A; \forall 2 \leq t \leq T \tag{14.6}$$

In (14.1), *W_L* and *W*_AC show the importance weights of lights and the importance weights of ACs for the users respectively. *W_L* and *W*_AC are the numbers between 0 and 1 that the higher values are correlated to more important devices. *P_L* and *P*_AC are the decision variables of the optimization algorithm that indicate the amount of power that should be decreased from lights and ACs, respectively. It should also be noted that *T* variable indicates the maximum number of periods, *L* illustrates the maximum number of lights, and *A* illustrates the maximum number of ACs.

Regarding the constraints, in (14.2), RR is the abbreviation of required reduction (RR) that presents the total amount of power that should be decreased in each time period from all the lights and ACs. This amount can be defined based on several aspects, like market prices, renewable energy production, demand response programs, or user preferences.

Equations (14.3), (14.4), (14.5), and (14.6) are defined to maintain the user comforts by limiting the power reduction in various aspects. PRR_*L* is the power reduction rate of each light that limits the light during all periods to avoid exorbitance power reduction. PRR_AC is the power reduction rate of each AC that limits each individual AC to avoid exorbitance power reduction during all periods of optimization. P_L_Nom is the nominal power consumption of the lights, and P_AC_Nom stands for nominal power reduction of ACs. Moreover, MaxRed_L has been defined to avoid consecutive power reduction from certain lights, and MaxRed_AC is considered to prevent power reduction from certain AC in two consecutive periods.

14.4.5 INTELLIGENT PERSONS COUNTER SYSTEM

The detection of persons in a building is not an easy task. It easy to detect movement but is not easy, using nonintrusive technology, to get the information if a person is in a certain room; special is no apparent movement is made. In the other side, the detection of persons inside a building is almost a mandatory parameter as input in any energy management system that manages users' resources that have an impact on users' comfort and daily life. This creates an ambiguous situation that is an issue for intelligent energy-based systems.

The intelligent persons counter system tries to overpass this issue by detecting the entrance and leaving of persons in a building zone. Knowing how many persons are inside, avoids the need for presence sensors; of course, this is not true for everything, the counter does not provide a detail distribution of the persons inside the building.

The system was developed using computer vision to detect persons and the direction that they take in a hallway. The hallway where the system was deployed is the only access point to that size of the building; with six independent offices accessed by the hallway. The Python computer vision library (OpenCV) was used to process the image. Connected to the system is a normal USB webcam placed in the celling of the hallway.

Fig. 14.5 shows the architecture of the system and the GUI where is visible the webcam image being processed by the system. The system uses four pairs of horizontal lines to trigger events in the code, working as perimeters to detect the passage of persons. The vertical lines identify the person. On the right is the same image where changes are represented in the white and static image is represented in black. Using the events triggered by the horizontal pairs, it is possible to obtain the direction of the person passing in the hallway.

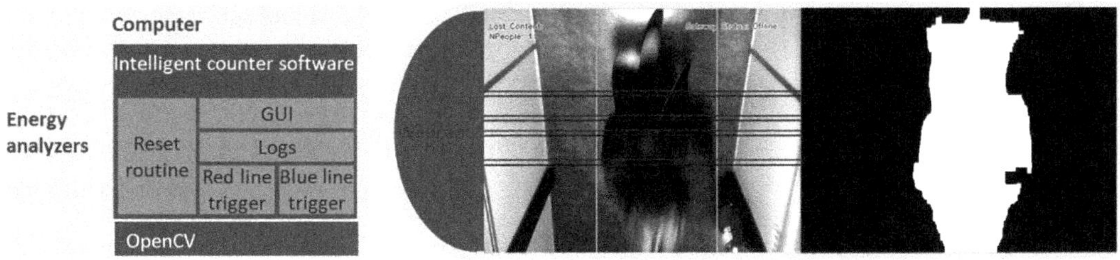

FIGURE 14.5

Intelligent persons counter system.

The system is also able to reset itself. All systems fail, especially those on how to run 24/7 in uncontrollable environments. Therefore the key is to work around this issue and provide ways for autorecovery in the event of a failure. The reset routine uses data from energy analyzers to understand if the lamps inside the building are turned on and if there is outside clarity. If no clarity outside is zero (i.e., it is night) and if no lamp consumption is measured, then the system assumes that no one is present in the building and the counter is set to zero. For the reset, the routine has the assumption that no one is inside the building in the complete dark. The system uses HTTP requests to read the energy analyzers and the outside clarity. The outside clarity uses the sensor of the IoT device developed for photovoltaic generation monitors.

The counting of persons can be a challenge. The system is still not perfect and has some issues. The hallway is too narrow and short, and the used webcam lens is not enough to provide a wider image. This is an issue when more than one person crosses the hallway at the same time, misleading the system to count only one person

14.4.6 INTELLIGENT DESK LIGHT CONTROL

The intelligent desk light control uses a combination of hardware and software that enables its correct operation. An Arduino Nano is used to integrate hardware sensors. A total of four sensors are integrated: a temperature sensor, a light-dependent resistor (i.e., clarity sensor), a passive infrared sensor (PIR) (i.e., movement sensor), and a button. The Arduino Nano reads sensor data and sends it to the computer where a Java software reads the data and processes it. There, movement sensor data is combined with the keyboard and mouse usage data; this data is monitored by the Java software. Fig. 14.6 shows the architecture and deployment of the system.

The user specifies the clarity level that it wants to have in his/her working desk using the system interface. This interface can be accessed by pressing the hardware button on the board; the green button in Fig. 14.6. The system will read the clarity sensor and control the lamp to achieve the clarity defined. The movement sensor is used to detect the user arriving at the desk. If the motion sensor is not activated, then the system considers that there is no person in the desk, even if the keyboard and mouse are activated, to prevent the turn-on of the lamp when users use the operating system remote control. If a user is detected arriving at the desk the system turns on the lamp and starts measuring physical movement and keyboard and mouse movements. It is assumed that a user is present while there is still movement at least one of the sensors: PIR, keyboard, and mouse. After 5 minutes without any user detection, the system will show on the monitor a small window displaying the message "move." This provides feedback to the user that is not being detected by the system. This message displays for two minutes and then, without any user detection, the system will turn off the desk lamp. At this moment, the message "out" is displayed in the monitor; this will disappear when a user arrives at the desk.

14.4.7 INTELLIGENT TELEVISION BRIGHTNESS CONTROL

In televisions, brightness can be used to decrease consumption without negatively impact the users. The intelligent television brightness control built according to the architecture proposed in [39] is used to control the brightness of televisions that are used for demonstrations and exhibitions.

The television is controlled by a smart plug that provides remote control and energy monitoring. The smart plug is scheduled to turn on at 08:00 and turn off at 20:00. The idea behind the intelligent television brightness control is to control this television according to the presence of persons in the hall, resulting in a decrease in energy consumptions and, therefore energy costs.

A NodeMCU board is used as a processing unit, providing WiFi communication and MQTT support. Fig. 14.7 shows the architecture and deployment of the system. The system uses data from: a movement sensor (i.e., PIR), the television smart plug (i.e., TP-Link HS110), and a

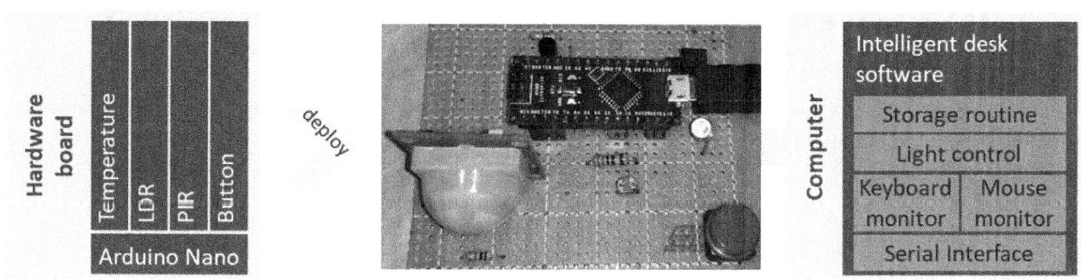

FIGURE 14.6

Intelligent desk light control.

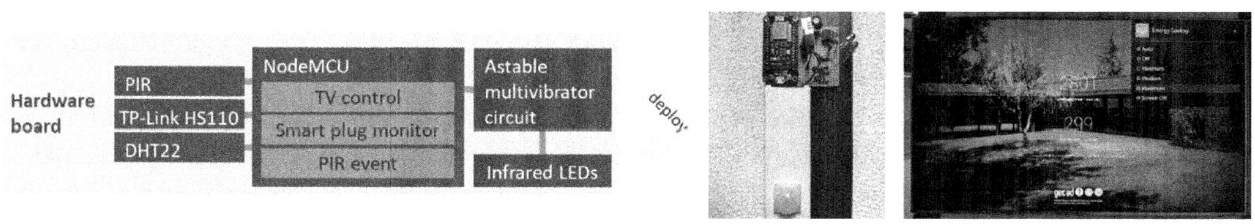

FIGURE 14.7

Intelligent television brightness control.

temperature and humidity sensor (i.e., DHT22). Using this data, the system can identify the movement in the hall and act on the television's brightness.

The control over the television is not straightforward. The NodeMCU did not provide the needed 38 kHz frequency to send the right infrared signals to the television; demanding the implementation of an astable multivibrator circuit to allow 38 kHz signals. The other issue that was faced is the lack of a unique infrared signal to control the television brightness. The control of the brightness is possible but by using the energy-saving menu. This imposes the sending of multiple infrared signals in order to successfully change the television brightness. First, the menu must be open, then the up and down key signals must be sent—to choose the right brightness state—and then the ok key signal must be sent. In this system, the sequence of signals can fail, or users could change the setting on the television. For this reason, a correction mechanism was defined. The system periodically reads the consumption values of the TP-Link HS110 smart plug and identifies the set brightness in the television. With this identification, the system can correct itself.

The savings on the television are significant in its higher brightness level the television has a consumption of around 95 W, while in its lower brightness level the television shows consumption of 35 W. The lowest brightness level used in this system still allows the visualization of the television. Therefore if the reaction of the system is not enough, the persons passing by the hall are still capable of reading the building's energy values.

14.5 TEST AND ANALYSIS OF THE INTELLIGENT ENERGY-ORIENTED HOME

In this section, some examples of intelligent energy-oriented home applications are presented. A total of four applications are illustrated.

14.5.1 INTELLIGENT PERSONS COUNTER

The intelligent persons counter system provides the counting of persons inside the building using computer vision. Fig. 14.8 shows the results of this system for 24 hours. The results of the graph were measured each minute. On this day, a purposeful counter error was made at 18:11, counting only one person and not two persons at that minute. This enables the visualization of the reset routine at 20:44, setting the counter to zero. The reset of the counter occurs when outside light is null and no lamp's consumption is found in the building. The outside light uses the IoT device developed for photovoltaic generation monitoring.

In Fig. 14.8 is possible to see the counter working during the day. Our building receives daily from 14 to 15 researchers and students. In the graph is possible to see the income of people, the lunch hour, the afternoon snack, and the left of the people. Even though, an error was purposefully provoked in the system, at 23:59 the counter corrected itself and shows a total of zero persons inside the building. Analyzing the data, it can be suggested an automatic reset at midnight. However, this can bring errors because researches have unlimited access to the building.

14.5.2 INTELLIGENT TELEVISION BRIGHTNESS CONTROL

Being located near the intelligent persons counter, the intelligent television brightness control has a similar profile. Fig. 14.9 shows the same 24-hour period as the previous results; of Fig. 14.8. The television will increase the brightness every time a person passes by the hall and will decrease the brightness when there are no persons in the hall. The consumptions of Fig. 14.9 start at 08:00, when the smart plug turn on the television. Currently, the television starts at its maximum brightness. A few seconds the system realizes this, using the data from the smart plug, and changes the brightness to its right state.

The low brightness is kept until a person is detected in the hall, at 09:15. After observing movement in the hall, the system maintains the television at its highest brightness for 5 minutes, until it changes to the lowest brightness. Around 10:00, several persons passed by the hall at different times, keeping the high brightness and high consumption for 34 minutes straight. At 20:00, the smart plug turns off the television. The scheduled control programmed in the smart plug only applies to weekdays; during the weekend the television does not turn on.

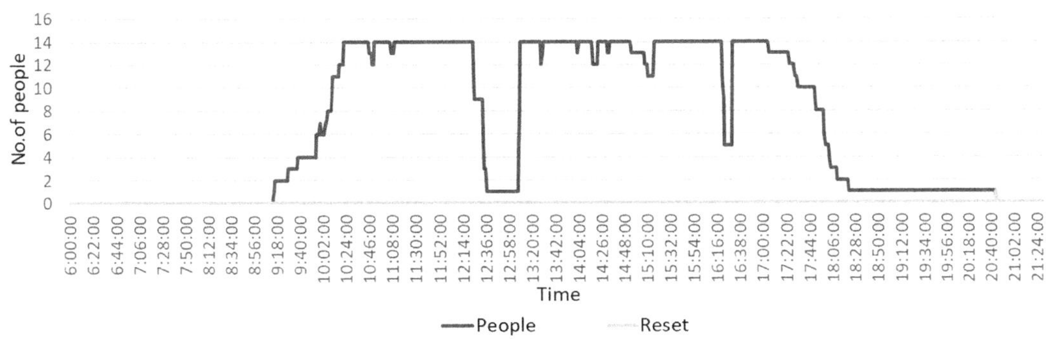

FIGURE 14.8

Intelligent persons counter results.

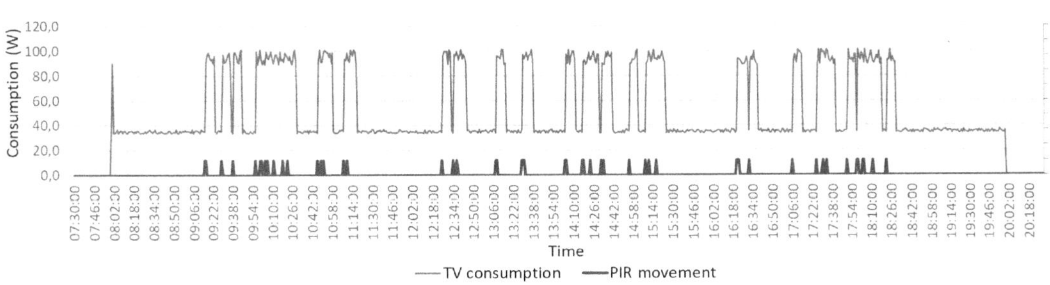

FIGURE 14.9

Intelligent television brightness control.

14.5.3 INTELLIGENT LIGHT CONTROL

The present case study focusses on 48 periods of a day, from 08:00 to 20:00 with 15 minutes interval. It is considered that the power consumption of the ACs and lights contribute to achieve the desired power reduction of the algorithm. The algorithm considers the lighting system and the ACs of the building to accomplish its purposes. Since the required power reduction is considered as a high value, it is intended that the algorithm uses ACs and lights of the building. Fig. 14.10 illustrates the initial consumption of the lights and AC units, the uncontrolled devices consumption, the building's consumption, and the required power reduction in all periods. As it can be seen in Fig. 14.10, the initial power consumption of the ACs is significantly more than the power consumption of the lights. However, the contribution of both systems would be needed to maintain the comfort of users. It should be noted that the comfort parameters (PRR_L and PRR_AC) are considered equal to 30%. In addition, the Maxred_L and MaxRed_AC are assumed as 60 W and 1000 W, respectively. Fig. 14.11 illustrates the total amount of power reduction of each light during all periods.

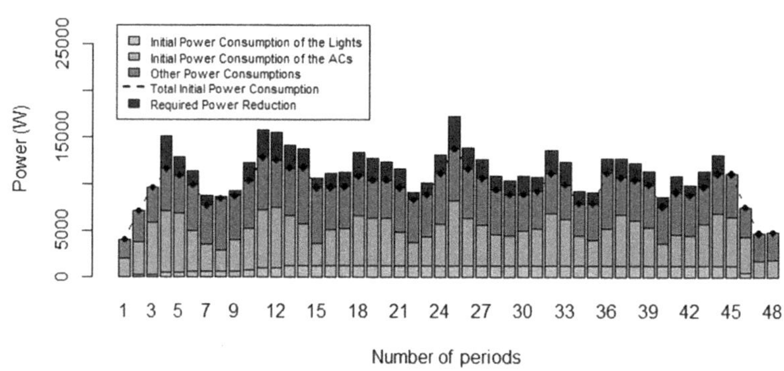

FIGURE 14.10

Initial power consumption of the building and required power reduction.

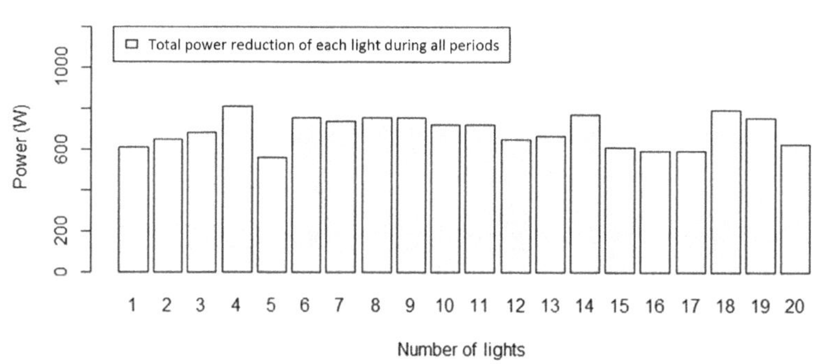

FIGURE 14.11

Total power reduction of each light.

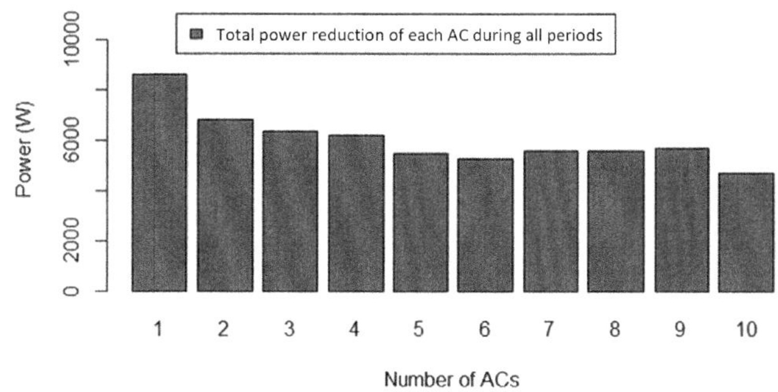

FIGURE 14.12

Total power reduction of ACs.

As it can be seen in Fig. 14.11, all the lights have been participated in the optimization to fulfil the algorithm goals. The balanced reduced power among the lights shows that PRR_L and MaxRed_L have been restricted the lights to excessive power reduction. However, there are slight differences in a power reduction of each light that depends on the priority and power availability of each device. Fig. 14.12 shows the total power reduction of each AC during all periods.

According to Fig. 14.12, the required power reduction has been reduced balanced among all the ACs. However, the AC number 1 has the most power reduction, the power reduction of the other ACs are approximately in the same range. Regarding the purposes of the present algorithm, which are opposing to reduce excessive power from a certain device while the other devices are working in a normal situation, it can be interpreted that the desired purposes are met.

14.5.4 INTERNET OF HOME ALERT SYSTEM

The IoH platform integrates the presented IoT devices and systems. The platform enables the descriptions of devices and systems using metadata that can be used directly in its alert system, as seen in Table 14.1. The system provides two types of alerts: not secure warning, and intrusion alarm. These alarms are triggered by the sensors connected to the IoH platform—directly connected or using a subsystem. In this scenario, the door sensors were added in IoH platform. These sensors indicate if the door is open or closed; they do not know if the door is locked or unlocked. The sensors were already installed in our building and integrated using a SCADA system. IoT will integrate the SCADA system using HTTP requests to get information regarding the door sensors.

The not secure warning will use the intelligent persons counter system, the door sensors, and the intelligent light control. The system will use the subsystem that indicates if the building is empty (i.e., intelligent persons counter system); using the "indicates" metadata tag. After realizing the building is empty, the IoH platform will start to monitor subsystems and devices, which indicate the status of the rooms (i.e., open/closed). The door sensors are used also because they "indicate" that the building is not properly closed. In addition, the intelligent light control is used because it has the knowledge regarding the shutters. The intrusion alarm uses the intelligent persons counter system, the door sensors, the intelligent light control, and the intelligent desk light control.

Table 14.1 Integration Metadata.

Device	Type	Zone	Triggered by	Impacts	Measures	Indicates
Fan heater	IoT device	N 115	Movement	Room's temperature	Consumption	-
Door locker	IoT device	Each door	-	Room closed status	Door status	Room closed status and human presence
Intelligent light control	Subsystem	N 114	Movement	Room's light and shutters status	Movement and shutters status	Room closed status and human presence
Intelligent television brightness	Subsystem	Hall	Movement	Television brightness	Movement	Human presence
Intelligent desk light	Subsystem	N 115	Movement	Room's light	Temperature, clarity and movement	Human presence
Intelligent persons counter	Subsystem	Hallway	-	-	Number of people	-
Intelligent HVAC control	Subsystem	N 111	Occupancy	Room's temperature	Temperature	Room closed status

The intelligent persons counter system is used to detect when the building is empty. At this moment, the system will monitor all devices and subsystems with a movement sensor. Except for the intelligent television brightness control, this is because its movement sensor is placed before the intelligent persons counter; having the building's door at the starting point, meaning that the movement sensor of the intelligent television brightness control is always activated before the counting of the intelligent persons counter. If movement is detected inside the building while the persons counter indicates that no one is inside, then the IoH platform will fire an intrusion alarm. The "not secure" warning is fired by the system every time the persons counter is zero, but the building is not properly closed. The "intrusion" alarm is fired by IoT platform every time the persons counter is zero and there is movement inside the building.

14.6 CONCLUSIONS AND FURTHER DEVELOPMENTS

IoT and smart homes have a very promising future together due to the great support that the first can provide to the second. This chapter addressed the articulation of two fields: intelligent energy systems and smart homes. Several developments and applications upon an architecture for an intelligent energy-oriented home were illustrated and discussed. It covered intelligent energy-oriented home aspects with the support of the AmI concept, considering the efficient use of energy according to the context and to the residents' profile and behavior, and the coordination of renewable energy, demand flexibility, and energy prices.

It can be concluded that the use of sensing, reasoning, and actuation on several devices and appliances, considering IoT systems from diverse manufacturers will allow dealing with questions like context awareness, and user adaptation, namely, in the smart grids context. However, the Integration of systems and devices from multiple manufactures usually brings compatibility issues, which can make the development of complex systems particularly complex. In the presented work, it was difficult to integrate some resources because there were not available solutions in the market that could be used for that purpose, like the intelligent desk light controller mechanism. In such cases, new hardware had to be developed, enabling the development of full compliant devices. Although this has required more effort than initially estimated, avoiding market available devices, made the presented system to go a step forward, by providing a better contextual description of the devices and improving the system metadata. Nonetheless, using tailored solutions does not allow to ensure that the system is fully interoperable.

Future developments can improve the presented system, by enhancing the knowledge base and providing new metadata. Regarding the system use and configuration, the rule system can be expand to include more metadata that can be used for the configuration of alarms and interoperability among devices. Open-source models must also be studied and considered to open the present system to the community, allowing third-party developments.

ACKNOWLEDGMENTS

The present work was done and funded in the scope of the following projects: COLORS Project PTDC/EEI-EEE/28967/2017 and UID/EEA/00760/2019 funded by FEDER Funds through COMPETE program and by National Funds through FCT.

REFERENCES

[1] M. Shakeri, M. Shayestegan, H. Abunima, S.M.S. Reza, M. Akhtaruzzaman, A.R.M. Alamoud, et al., An intelligent system architecture in home energy management systems (HEMS) for efficient demand response in smart grid, Energy Build. 138 (2017) 154–164. Available from: https://doi.org/10.1016/j.enbuild.2016.12.026.

[2] Z. Wu, K. Yang, J. Yang, Y. Cao, Y. Gan, Energy-efficiency-oriented scheduling in smart manufacturing, J. Ambient. Intell. Humanized Comput. 10 (2019) 969–978. Available from: https://doi.org/10.1007/s12652-018-1022-x.

[3] G. Dileep, A survey on smart grid technologies and applications, Renew. Energy 146 (2020) 2589–2625. Available from: https://doi.org/10.1016/j.renene.2019.08.092.

[4] Y. Wang, Q. Chen, T. Hong, C. Kang, Review of smart meter data analytics: applications, methodologies, and challenges, IEEE Trans. Smart Grid 10 (2019) 3125–3148. Available from: https://doi.org/10.1109/TSG.2018.2818167.

[5] S. Gottwalt, J. Gärttner, H. Schmeck, C. Weinhardt, Modeling and valuation of residential demand flexibility for renewable energy integration, IEEE Trans. Smart Grid 8 (2017) 2565–2574. Available from: https://doi.org/10.1109/TSG.2016.2529424.

[6] I.A. Doush, I. Damaj, M.A. Al-Betar, M.A. Awadallah, M. Al-khatib Ra'ed, A.E. Alchalabi, et al., A survey on accessible context-aware systems, EAI/Springer Innov. Commun. Comput. (2019) 29–63. Available from: https://doi.org/10.1007/978-3-030-16450-8_2.

[7] C. Ramos, J.C. Augusto, D. Shapiro, Ambient intelligence-the next step for artificial intelligence, IEEE Intell. Syst. 23 (2) (2008) 15–18. Available from: https://doi.org/10.1109/MIS.2008.19.

[8] F. Fernandes, H. Morais, Z. Vale, C. Ramos, Dynamic load management in a smart home to participate in demand response events, Energy Build. 82 (2014) 592–606. Available from: https://doi.org/10.1016/j.enbuild.2014.07.067.

[9] L. Gomes, P. Faria, H. Morais, Z. Vale, C. Ramos, Distributed, agent-based intelligent system for demand response program simulation in smart grids, IEEE Intell. Syst. 29 (1) (2014) 56–65. Available from: https://doi.org/10.1109/MIS.2013.2.

[10] L. Gomes, F. Fernandes, Z. Vale, P. Faria, C. Ramos, A learning algorithm and system approach to address exceptional events in domestic consumption management, in: Proceedings of the 2014 IEEE Symposium on Computational Intelligence Applications in Smart Grid (CIASG), Orlando, FL, 2014, pp. 1–7. doi: 10.1109/CIASG.2014.7011564.

[11] P. Faria, Z. Vale, Demand response in electrical energy supply: an optimal real time pricing approach, Energy 36 (8) (2011) 5374–5384. Available from: https://doi.org/10.1016/j.energy.2011.06.049.

[12] T.K.L. Hui, R. Simon Sherratt, D.D. Sánchez, Major requirements for building smart homes in smart cities based on Internet of Things technologies, Future Gener. Computer Syst. 76 (2016) 358–369. Available from: https://doi.org/10.1016/j.future.2016.10.026.

[13] C. Ramos, G. Marreiros, R. Santos, C.F. Freitas, Smart offices and intelligent decision rooms, in: H. Nakashima, H. Aghajan, J.C. Augusto (Eds.), Handbook of Ambient Intelligence and Smart Environments, Springer, Boston, MA, 2010. Available from: http://dx.doi.org/10.1007/978-0-387-93808-0_32.

[14] D. Bonino, C. Fulvio, DoMAIns: domain-based modeling for ambient intelligence, Pervasive Mob. Comput. 8 (4) (2012) 614–628. Available from: https://doi.org/10.1016/j.pmcj.2011.10.009.

[15] C. Roda, A.C. Rodríguez, V. López-Jaquero, E. Navarro, P. González, A multi-agent system for acquired brain injury rehabilitation in ambient intelligence environments, Neurocomputing 231 (2017) 11–18. Available from: https://doi.org/10.1016/j.neucom.2016.04.066.

[16] S. Makonin, L. Bartram, F. Popowich, A smarter smart home: case studies of ambient intelligence, IEEE Pervasive Comput. 12 (1) (2013) 58–66. Available from: https://doi.org/10.1109/MPRV.2012.58.

[17] A. Paola, S. Gaglio, G. Lo Re, M. Ortolani, Sensor9k: a testbed for designing and experimenting with WSN-based ambient intelligence applications, Pervasive Mob. Comput. 8 (3) (2012) 448–466. Available from: https://doi.org/10.1016/j.pmcj.2011.02.006.

[18] H. Alex, M. Kumar, B. Shirazi, MidFusion: an adaptive middleware for information fusion in sensor network applications, Inf. Fusion. 9 (3) (2008) 332–343. Available from: https://doi.org/10.1016/j.inffus.2005.05.007.

[19] A. Sorici, G. Picard, O. Boissier, A. Zimmermann, A. Florea, CONSERT: Applying semantic web technologies to context modeling in ambient intelligence, Computers Electr. Eng. 44 (2015) 280–306. Available from: https://doi.org/10.1016/j.compeleceng.2015.03.012.

[20] A. Machado, V. Maran, I. Augustin, L. Krug Wives, J.P. Moreira de Oliveira, Reactive, proactive, and extensible situation-awareness in ambient assisted living, Expert. Syst. Appl. 76 (15) (2017) 21–35. Available from: https://doi.org/10.1016/j.eswa.2017.01.033.

[21] J. Gubbi, et al., Internet of Things (IoT): a vision, architectural elements, and future directions, Future Gener. Computer Syst. 29 (7) (2013) 1645–1660. Available from: https://doi.org/10.1016/j.future.2013.01.010.

[22] A. Zanella, et al., Internet of Things for smart cities, IEEE Internet of Things J. 1 (1) (2014). Available from: https://doi.org/10.1109/JIOT.2014.2306328.

[23] B.M. Tanskanen, M.I. Consultant, Applying Machine Learning to IoT Data, SAS Finland, 2016, pp. 1–7.

[24] F. Alam, et al., Analysis of eight data mining algorithms for smarter Internet of Things (IoT), Procedia Computer Sci. 58 (2016) 437–442. Available from: https://doi.org/10.1016/j.procs.2016.09.068.

[25] M.S. Mahdavinejad, et al., Machine learning for internet of things data analysis: a survey, Digital Commun. Netw. 4 (3) (2018) 161–175. Available from: https://doi.org/10.1016/j.dcan.2017.10.002.

[26] D. Bishop, How to use machine learning for IoT analysis, 2018. <https://jaxenter.com/use-machine-learning-iot-analysis-150765.html>, 2019 (accessed 31.05.19).

[27] U.S. Shanthamallu, et al., A brief survey of machine learning methods and their sensor and IoT applications, in: Proceedings of the 2017 Eighth International Conference on Information, Intelligence, Systems and Applications, IISA 2017, 2018, pp. 1–8. doi: 10.1109/IISA.2017.8316459.

[28] C.W. Tsai, et al., Data mining for Internet of Things: a survey, IEEE Commun. Surv. Tutor. 16 (1) (2014) 77–97. Available from: https://doi.org/10.1109/SURV.2013.103013.00206.

[29] C. Feng, et al., Data mining for the Internet of Things: literature review and challenges, Int. J. Distrib. Sens. Netw. 11 (2015) 431047. Available from: https://doi.org/10.1155/2015/431047.

[30] H. Morsali, S.M. Shekarabi, K. Ardekani, H. Khayamim, A. Fereidunian, M. Ghassemian, et al., Smart plugs for building energy management systems, in: Proceedings of the Iranian Conference on Smart Grids, Tehran, May 24–25, 2012, pp. 1–5.

[31] S. Heo, W. Park, I. Lee, Energy management based on communication of smart plugs and inverter for smart home systems, in: Proceedings of the 2017 International Conference on Information and Communication Technology Convergence (ICTC), Jeju, South Korea, October 18–20, 2017, pp. 810–812. doi: 10.1109/ICTC.2017.8190788.

[32] T. Ganu, J. Hazra, D.P. Seetharam, S.A. Husain, V. Arya, L. Chandratilake De Silva, et al., S. nPlug: a smart plug for alleviating peak loads, in: Proceedings of the 2012 Third International Conference on Future Systems: Where Energy, Computing and Communication Meet (e-Energy), Madrid, May 9–11, 2012, pp. 1–10. https://doi.org/10.1145/2208828.2208858.

[33] L. Gomes, F. Sousa, Z. Vale, An intelligent smart plug with shared knowledge capabilities, Sensors 18 (2018) 3961. Available from: https://doi.org/10.3390/s18113961.

[34] P. Faria, J. Spínola, Z. Vale, Aggregation and remuneration of electricity consumers and producers for the definition of demand-response programs, IEEE Trans. Ind. Inform. 12 (3) (2016) 952–961. Available from: https://doi.org/10.1109/TII.2016.2541542.

[35] M.A.F. Ghazvini, J. Soares, O. Abrishambaf, R. Castro, Z. Vale, Demand response implementation in smart households, Energy Build. 143 (2017) 129–148. Available from: https://doi.org/10.1016/j.enbuild.2017.03.020.

[36] L. Gomes, C. Ramos, A. Jozi, B. Serra, L. Paiva, Z. Vale, IoH: a platform for the intelligence of home with a context awareness and ambient intelligence approach, Future Internet 11 (2019) 58. Available from: https://doi.org/10.3390/fi11030058.

[37] L. Gomes, J. Spínola, Z. Vale, J.M. Corchado, Agent-based architecture for demand side management using real-time resources' priorities and a deterministic optimization algorithm, J. Clean. Prod. 241 (2019) 118154. Available from: https://doi.org/10.1016/j.jclepro.2019.118154.

[38] Statista, Smart home report 2018 – control and connectivity, Statista, Hamburg, 2018.

[39] B. Serra, L. Gomes, Z. Vale, Lightweight architecture for IoT devices with context-aware autonomous control, in: Proceedings of the IEEE Wireless Communications and Networking Conference, April 15–19, 2019. https://doi.org/10.1109/WCNCW.2019.8902882.

FURTHER READING

Ramos and Liu, 2011 C. Ramos, C. Liu, AI in power systems and energy markets, IEEE Intell. Syst. 26 (2) (2011) 5–8. Available from: https://doi.org/10.1109/MIS.2011.26.

CHAPTER

CORPORATE CYBERSECURITY STRATEGY TO ENABLE ARTIFICIAL INTELLIGENCE AND INTERNET OF THINGS 15

Michele Myauo

Microsoft, Reston, VA, United States

15.1 INTRODUCTION

All the data shows that the most effective way to protect ourselves and our companies is by making cybersecurity something that is not a one-off decision. Cybersecurity is not some IT project to be implemented, but something we practice as part of our everyday routine. If we can think differently about cybersecurity, we will act differently. Cybersecurity is a C-Suite decision. To be sure, the solutions are there, and the technology exists to help solve these problems. However, for cybersecurity measures to be most effective, they need to be methodical, thoughtful, consistent, and above all strategic.

Like the field of cybersecurity, artificial intelligence (AI) and Internet of Things (IoT) technologies are rapidly evolving. Integrating cybersecurity into the fabric of corporate strategy will increase the velocity and successful implementation of AI and IoT capabilities. In turn, AI and IoT capabilities will enhance cybersecurity mitigations.

This chapter will begin by providing an overview of foundational cybersecurity principles and concepts. We will examine the cyber-adversarial system to better understand how to infiltrate the system and mitigate cyber-attack to include nonmalicious noncompliance and malicious noncompliance like financially motivated, ideologically and politically motivated cyber-attackers. We will then examine the anatomy of a cyber-attack via a key attack vector seen in numerous attacks on multinational corporations and governments, the Pass-the-Hash attack.

Understanding the underlying motivational barriers, will help executives and cybersecurity professionals alike to customize their approach to landing cybersecurity implementation with stakeholders. The chapter will address how we can begin to overcome the inertia we face around cybersecurity by embracing the Three I's, of cybersecurity strategy, innovation, intelligence, and investment. Furthermore, this chapter will provide an overview of key cybersecurity laws and provide practical real-world examples on how corporate cybersecurity strategy can be accelerated through leveraging cybersecurity systems engineering, architectural frameworks, and cybersecurity IT portfolio management. To get started on the corporate cybersecurity journey or to assist organizations on their current journey we will walk through tactical considerations for assessing the current

Artificial Intelligence to Solve Pervasive Internet of Things Issues. DOI: https://doi.org/10.1016/B978-0-12-818576-6.00015-0

state against Cybersecurity-Protect-Detect-Respond-Recover Framework; identifying pain points, and developing that future state to mitigate those pain points [1].

By approaching cybersecurity not as a technology issue but a change management issue, we can begin to knock down the barriers around cybersecurity adoption. To stop thinking of cybersecurity tactically and start thinking of it strategically. In the process, we can also create opportunity for our businesses: We can feel confident our IP is safe. Our customers can feel confident their data is safe. In addition, our employees and prospective employees can know their employer values their personal information as well. Most important of all, by approaching cybersecurity as a long-term corporate strategy, our brand is safe.

Because stakeholders across the spectrum, from customers to regulators to the media, will know that if there is an attack that our company has the protocols and protections in place to respond quickly and appropriately. In that sense cybersecurity is not simply a threat mitigation, but an opportunity to create a true competitive advantage, one a lot of our competitors may not yet fully embrace.

We do not have to snuff out every possible cybersecurity threat to our business, we cannot. We may still get attacked anyway on that point the data is very clear. However, if we are smart and if we are strategic about how we approach these threats. If we motivate those around us to collectively take action, we can implement an effective corporate cybersecurity strategy necessary to enable disruptive AI and IoT technology advancements.

15.2 CYBERSECURITY

Let us start by level setting on definitions of some key cybersecurity terms. Cyberspace is "a global domain within the information environment consisting of the interdependent network of information systems infrastructures including the Internet, telecommunications networks, computer systems, and embedded processors and controllers" [2]. A cyber-attack is "an attack, via cyberspace, targeting an enterprise's use of cyberspace for the purpose of disrupting, disabling, destroying, or maliciously controlling a computing environment/infrastructure; or destroying the integrity of the data or stealing controlled information" [2].

Cyber-attacks are increasingly prevalent across government, academia, and private sector organizations. Cyber-attacks threaten organizations across industries from banking, retail, casinos, IT companies, media outlets, oil and gas companies, to retailers, universities, and governments. Cyber-attackers are exploiting any and all weak links in the IT supply chain from networks, software and hardware, to physical security and personnel.

Cybersecurity is the ability to protect or defend an organization's information technology (IT) environment against cyber-attacks [1]. IT security standards and secure development practices play an important role in cybersecurity. System security must be considered when a new system is created or functionality is added to an existing system [3]. The need for IT security standards and secure system development processes is recognized across industry by academia, government, and private industry organizations.

IT standards organizations including the International Organization for Standards (ISO), International Electrotechnical Commission (IEC), and National Institute of Standards and Technology identify the importance of secure systems development lifecycle processes [4]. The US Federal Government has reinforced the need to implement security standards through federal law The Federal Information Security Management Act (FISMA) 2002 requires US Federal Government agencies to assess their risks and implement appropriate security requirements and controls tailored to meet organizational mission needs [2].

Although important, scholars have discovered that implementing IT security standards alone is not effective in mitigating cyber-attacks; therefore standards are only one part of the solution [3,5–10]. Cyber-attacks exploit vulnerabilities in people, processes, and technologies; therefore it is necessary for systems engineers to integrate cybersecurity into systems engineering processes, like enterprise architecture, to more effectively mitigate attacks [3,5,8].

In 2013, US President Barack Obama signed an Executive Order into law to strengthen US critical infrastructure against cyber-attacks [11]. The National Institute of Standards and Technology Cybersecurity Framework focuses on framing and adopting a cybersecurity program for US government organizations with critical infrastructure. The framework spans five key cybersecurity areas: knowing cyber risks; preventing cyber-attack; detecting cyber-attack; responding to cyber-attack; and recovering from a cyber-attack [12].

Cybersecurity experts create and subsequently model the path cyber-attack scenarios promulgate throughout an enterprise system to more clearly define the attacker's likely path to compromising a system. Researchers employ various methods to document cyber-attack scenarios, from graphical to textual. A threat model is a graphical technique used to document potential cyber-attacks on an IT system [13], whereas, the Common Weakness Enumeration is an open source textual database of software vulnerabilities [14].

Several researchers have demonstrated that considering a cyber-attack scenario during the systems engineering process is not only beneficial but also critical. By way of example, researchers have asserted that cybersecurity intrusion detection systems must be designed and constructed based on architectural considerations in order to better mitigate cyber-attacks [10,15]. Jones and Horowitz [7] demonstrated how considering likely organizational cybersecurity adversaries and tactics in designing the system architecture can enable organizations to mitigate these types of attack. Furthermore, Zafar et al. (2013) identified cybersecurity requirements by exercising a smart grid system-of-systems case study.

15.3 UNDERSTANDING THE CYBER-ADVERSARIAL SYSTEM

To understand how to incorporate cybersecurity into our corporate strategy we must first understand the cyber threat landscape. Cyber adversaries exist both internal and external to organizations. Internal threats to organizations include nonmalicious, unintentional computer misuse and malicious, and intentional corruption [16]. An internal organizational cyber adversary, often called an insider threat, is an internal member of the organization with special access to organizational data that persons outside the organization do not have [16].

Cyber-Attackers exist both internal and external to organizations. Threats internal to organizations include nonmalicious unintentional computer misuse and malicious intentional corruption like the insider threat at Saudi Aramco that resulted in three-quarters of organizational data loss [18].

Financial and political motivations are key motives of both the malicious insider threat and threats from groups external to the organization. Financially motivated cyber-attacks are typically committed by Cybercrime groups, as demonstrated by the attacks on Target [17].

Ideologically and politically motivated cyber-attacks included attacks conducted by hacktivists against The New York Times [19] and the cyber-warhead directed at a Iranian Nuclear Facility [20]. These threat vectors, cyber-attacker groups, and motives are discussed in the following text. An overview of these threat vectors, cyber-attacker groups, and their motives are provided in the following text via examples of these real-world cyber-attacks.

15.4 THREAT VECTORS: INTERNAL AND EXTERNAL THREATS

Cyber adversaries exist both internal and external to organizations. Threats internal to organizations include nonmalicious unintentional computer misuse and malicious intentional corruption [16]. According to Willison and Warkentin [16], motives behind cyber threats internal to organizations are either nonmalicious or malicious noncompliance. An internal organizational cyber adversary often called the insider threat is someone internal to the organization with special access to organizational data that persons outside the organization do not [16].

Financial, ideological, and political motivations are also key motives of both the malicious insider threat and threats from groups external to the organization. Financially motivated attacks are typically committed by cybercrime groups, as demonstrated by the cyber-attack on target [17]. Ideologically and politically motivated attacks included attacks conducted by hacktivists against The New York Times [19] and the cyber warhead directed at a Iranian Nuclear Facility [20]. The following sections provide an overview of these threat vectors, cyber-attacker groups, and their motives via examples of real-world cyber-attacks.

15.5 NONMALICIOUS NONCOMPLIANCE

Nonmalicious noncompliance is unintentional (e.g., accidental and forgetful oversights). Nonmalicious noncompliance has a spectrum of negative consequences to the organization, from minor violations, like entering data incorrectly, to more volatile consequences, and impacts to business, like failing to log off when leaving a computer, resulting in delayed backups, or not changing passwords regularly. Although harmful, nonmalicious noncompliance would not be considered a cybercrime; rather, training on organizational processes and baking security into systems engineering initiatives could address it [16].

15.6 MALICIOUS NONCOMPLIANCE

Willison and Warkentin [16] describe malicious noncompliance as intentional, harmful computer abuse (e.g., sabotage, data theft or corruption, embezzlement, fraud, and deliberate policy). Malicious noncompliance can occur from individuals and groups both internal and external to an organization [16]. Malicious noncompliance is considered a form of cybercrime. Cybercrime refers to criminal attacks using cyberspace [21]. The intent of cybercrime is typically referred to as cyber espionage. Cyber espionage is an intrusion into a network in order to access or tamper with sensitive political or economic information or processes [22].

In the August 2012 attack against the Saudi oil company Aramco, a malicious insider rendered more than 30,000 computers unusable [22]. The insider initiated a virus that erased three quarters of Aramco's corporate data, including documents, spreadsheets, e-mails, and other files. The insider replaced the data with an image of a burning American flag [18].

According to Choo [21], malicious cyber-attackers are typically motivated by financial gain and political reasons. Financial and political motivations are key motives of both malicious insider threats and threats from groups external to the organization. These threat vectors, cyber-attacker groups, and their motives are discussed next.

15.7 FINANCIALLY MOTIVATED CYBER-ATTACKERS

Financially motivated cyber-attacks are typically more targeted, focusing on financial institutions and top executives. Three key groups of financially motivated hackers exist: (1) organized cybercrime groups motivated by financial gain, including traditional and Internet-only crime groups; (2) traditional cybercrime groups, which typically operate outside of the Internet, but augment their methods by exploiting and executing crime via the Internet; and (3) Internet-only cybercrime groups, which exclusively leverage the Internet for crime [21].

In December 2013, cyber criminals stole 110 million customer records from the retail chain Target. Records stolen included 40 million debit and credit card numbers and personal information of 70 million more people. After the attack, Target announced that the intruder stole a vendor's credentials to access Target's system [17].

15.8 IDEOLOGICALLY AND POLITICALLY MOTIVATED CYBER-ATTACKERS

Ideologically and politically motivated hackers comprise four key groups: (1) ideologically and politically motivated cybercrime groups, (2) hacktivists, (3) cyberterrorists and cybermilitias, and (4) nation-states. Choo [21] states that over time, cybercrime groups can develop ideological and political motivations to entice followers and cover actions.

Hacktivists are ideologically and politically motivations hackers. Hacktivist groups tend to be composed of geographically dispersed individuals who share similar beliefs and work together to conduct illegal IT activity to target a single issue, person, or organization [21].

Hacktivists have targeted a wide array of industries and organizations. Anonymous is a well-known hacktivist group. Tactics frequently leveraged by hacktivists include short-term denial-of-service operations, exposure of personally identifiable information, and more radically systematic impacts, like disrupting financial systems [22]. In August 2013, a hacktivist group that supports Syrian President Bashar al-Assad claimed responsibility for a cyber-attack that crashed The New

York Times website for several hours. The cyber-attackers rerouted Internet traffic from The New York Times to other websites [19].

Ideologically and politically motivated groups can become terroristic [21]. Cyberterrorism is the use of cyberspace for terrorist activities [22]. Hacktivists, cyberterrorists, and cybermilitias can be used by nation-states to carry out cyberwarfare in order to shield a nation from attribution and retribution [23]. Cyberwarfare refers to the use of cyberspace for warfare by nation-states [22].

Applegate [23] identifies that nation-states engaging in cyberwarfare include Iran, Turkey, Israel, North and South Korea, People's Republic of China, and the Russian Federation. China and Russia are building cyberwarfare programs and have a large volume of new doctrine and professional publications [23]. Benefits of cyber-attacks versus other traditional military attacks include: the element of surprise; conducting the attack at the time an attacker chooses; easy modification of attack techniques on short notice; and the legal ambiguity and lack of applicable international law covering cyberwarfare, which mitigates retribution [23].

An example of cyberwarfare is Stuxnet, a cyberwarhead that manipulated and dismantled the controller that ran the physical production process at an Iranian nuclear facility. Stuxnet was discovered in June 2010 and is believed to have been created by the United States and Israel to attack Iran's nuclear facilities reaching and disabling the controller. Stuxnet targeted controllers from the manufacturer Siemens [20].

15.9 ANATOMY OF A CYBER-ATTACK

Cyber-attacks leverage a cyber-adversarial system with many players. One must understand the cyber-adversarial system and the interaction of the key players within the system in order to infiltrate and fragment such a system [24]. This section examines the cyber-adversarial system in order to better understand how to infiltrate the system and mitigate the cyber-attack. Key concepts examined include the players, motives, and methods.

Holt [25] examines the social dynamics and business processes of the cybercrime market by analyzing transactions between buyers and sellers in 909 threads from 10 open, publicly accessible online Russian Web forums that facilitate the distribution of malware and attack tools. At a high-level overview of the sales process is provided in Fig. 15.1.

In this underground online cybercrime community, sellers post malware and attack tools for purchase. Buyers and sellers then communicate online and negotiate a price. The exploitation and payment are conducted online, typically via an online currency. Price, trust, and customer service are important in the interaction between the buyers and sellers. Hot items for sale in hacker forums

FIGURE 15.1

Underground online cybercrime sales process.

include cybercrime services, personal information, malware and related services, and stolen personal information [25].

Sood et al. [26] state that the underground cybercrime marketplace is built e-currency, a form of digital currency. The general concept "is to start with an established national currency (known as a fiat currency) like the USD or Euro and convert these currencies to an intermediate digital currency until the transaction is over, at which point the currency is converted back into a fiat currency" [26]. Exploits purchased online are typically leveraged in targeted cyber-attacks.

A targeted cyber-attack is directed toward a specific entity (individual, group, business, and government body) [24]. The more sophisticated cyber-attacks leverage a combination of cyber-adversarial "tools, social engineering tricks and tactics" [26]. Sood and Enbody [24] identify the following three phases of a targeted cyber-attack:

1. Intelligence gathering: The cyber-attacker gathers information on the target of the attack from publicly available sources, also known as open source intelligence (OSINT).
2. Develop attack model: The cyber-attacker analyzes information gathered and reconstructs the target environment to plan the attack. The cyber-attacker identifies the most vulnerable employees and networks and takes the path of least resistance.
3. Launch the attack: The cyber-attacker launches the attack against the target. The attack patterns vary, depending on information gathered and the environment.

Security standards compliance and vulnerability mitigations are not sufficient to protect organizational assets against live, advanced, and persistent attackers. Live attackers respond to changes in tactics and techniques in real time, focusing on exploiting the weakest links in the enterprise architecture, which are often the people in the organization [24]. For example, a Pass-the-Hash attack is identified as one of the top cyber threats by government, academia, and industries. The US Computer Emergency Readiness Team, the SANS Institute, McAfee, and Microsoft all regard Pass-the-Hash as a commonly used technique for credential theft.

In a Pass-the-Hash attack, the attacker targets workstations en masse via various means, including phishing e-mails. A phishing e-mail is an e-mail that tries to trick a user into providing personal information or access to his/her computer network by impersonating a legitimate business. Once the unsuspecting user clicks on the e-mail, he or she, running as the local administrator on his or her computer, is compromised and the attacker harvests his or her credentials. The attacker can now impersonate this user. The attacker uses the credentials for lateral movement and privilege escalation. Once the attacker acquires domain administrator credentials, he or she can exercise full control of data and systems in the environment [27].

For an organization to mitigate a Pass-the-Hash attack, the organization must teach its employees how to identify a phishing e-mail and what to do when a phishing e-mail is received. If an employee receives a phishing e-mail in his or her corporate e-mail account, the attacker knows the employee's business industry and company. The employee should alert his or her company that he or she may be under attack. The employee may also report phishing to the US Federal Trade Commission and to the US Computer Emergency Readiness Team (2013).

Given the likelihood that some employees will eventually click on a malicious e-mail, organizations should architect their infrastructure in a manner that best mitigates domain administrator credentials from being stolen. For example, Microsoft provides publicly available mitigations in the "Mitigating Pass-the-Hash Attacks and Other Credential Theft Techniques" [28]. Mitigations found in the whitepaper include restricting and protecting high-privileged domain accounts and local accounts within administrative privileges and restricting inbound traffic using Windows Firewall. More sophisticated methods are also available for purchase from Microsoft consulting services [28]. Cyber-attack-specific mitigations should be a component of a larger Cybersecurity attack mitigation strategy.

15.10 WHY CYBERSECURITY AND WHY CORPORATE STRATEGY

So why corporate strategy? Well, a few reasons. For one, we take corporate strategy seriously. Think about it. We review our corporate strategy once or twice a year. We bring all the major players to the table and we use it as a forcing mechanism to plan for everything. From our HR needs to our succession planning, to our capacity planning.

We also use corporate strategy to review all our various portfolios. From our financial portfolio, to ensure we have enough resources to fund our ongoing business needs. To our strategic portfolio, to ensure we can acquire the capacity we do not have to meet those needs. In addition, we decide on new business initiatives we need to be successful, the big bets we take. We figure out which initiatives we need to shore up and we wind down the ones that are not turning out the way we had hoped.

By contrast, too often we think of cyber as a one-time expenditure. We think, "Let's get us some of that cyber stuff and be done with it." Well, here is the thing: it is not that simple. If we know anything about the threat environment it is that well, we do not really know anything about the threat environment! Because threats are always evolving, and we do not control the threat environment. What we can control is our defensive posture, how we approach defending ourselves.

15.11 CYBERSECURITY LAWS

Organizations may also want to incorporate cybersecurity into their corporate strategy to more effectively plan to meet appropriate cyber laws. The US federal government has reinforced the need to implement cybersecurity mitigations through federal law. The Federal Information Security Management Act (FISMA) 2002 requires US federal government agencies to assess their risks and implement appropriate security requirements and security controls tailored to meet organizational mission needs. The Federal Information Processing Standards (FIPS) are the US government's minimum set of security controls for use in computer systems by nonmilitary US government agencies and their contractors [2].

In 2009, US President Obama created the "Comprehensive National Cybersecurity Initiative" to help secure US assets in cyberspace [29]. The 2012, US President Barack Obama signed Executive Order 13636 "Improving Critical Infrastructure Cybersecurity" into law to strengthen US critical infrastructure against cyber-attacks. The same year, President Obama directed the Department of Defense (DoD) to defend the nation against cyber-attack. In response, the DoD built the Cyber Mission Force, comprised of over 6200 military, civilian, and contractor personnel across the military and Defense Department to carry out DoD's cyber missions [30].

In the 2013 Worldwide Threat Assessment, US Director of National Intelligence James Clapper identified cyber threats as the number one threat to the United States [22]. Other US government regulations and standards were instituted to facilitate cybersecurity information sharing in 2014 to include the "Cybersecurity Information Sharing Act" [31]. In the National Defense Authorization Act (NDAA) of 2014, Congress required the DoD to designate a Principal Cyber Advisor to the Secretary of Defense to review and govern the development of DoD cyberspace policy and strategy.

The National Institute of Standards and Technology "Framework for Improving Critical Infrastructure Cybersecurity" was also created in 2014 and provides a structure organization can employ to develop a comprehensive cybersecurity program. The framework is now in version 1.1 and addresses the need for enterprise architecture and incorporates technical and nontechnical (managerial) IT security standards from organizations, like the International Organization for Standardization (ISO)/Institute of Electrical and Electronics Engineers (IEEE), the International Electrotechnical Commission (IEC), and the National Institute of Standards and Technology (NIST) [12].

Finally, the Department of Defense Cyber Strategy (2015) is a call to action for partnerships with governments, academia, and industry, and is a key thread weaved throughout the DoD Cyber Strategy. The need for this "partnership" is cited over 50 times throughout the document [30]. "As a matter of first principle, cybersecurity is a team effort within the US Federal government. To succeed in its missions, the Defense Department must operate in partnership with other Departments and Agencies, international allies and partners, state and local governments, and, most importantly, the private sector" [30].

The 27 nations of the European Union (EU) released the EU's Cybersecurity Strategy for "An Open, Safe and Secure Cyberspace" [32]. The Regulation (EU) 2016/679 of the European Parliament and of the Council, the European Union's (EU) new General Data Protection Regulation (GDPR), regulates the processing of personal data relating to individuals in the EU by an individual, a company or an organization (2016).

EU residents are provided control over their personal data through GDPR to include the right to:

- Receive a copy of personal data
- Access personal data held by an organization
- Have incorrect personal data deleted or corrected
- Access information about how personal data is used
- Restrict or object to automated processing of personal data
- Have personal data rectified and erased in certain circumstances [33].

For example, GDPR would apply in the case where a company with an establishment in the EU provides travel services to customers based in the Baltic countries processes personal data of natural persons [33]. GDPR would not apply in the case where an individual uses their own private address book to invite friends via email to a party that they are organizing. Provided there is no connection to a professional or commercial activity, GDPR does not apply to data processed by an individual for purely personal reasons or for activities carried out in one's home. GDPR does have to be followed when an individual uses personal data outside the personal sphere, for socio-cultural or financial activities [33].

The European Commission also provides standard terms that can be used to transfer data outside the European Economic Area in a compliant manner. Additionally, the EU-US Privacy Shield helps customers that want to transfer their data to the United States do so in a manner consistent with their data protection obligations [33].

15.12 CAUSES OF CYBERSECURITY INERTIA

Even with the heightened threat landscape and legal requirements, we often do not act. Or do not do enough. I call it "Cyber Inertia" and most of us start with a pretty bad case of it. Now I believe that is due to a variety of reasons.

First, the complexity of the issue itself—cyber is not generally an issue that is understood at the C-Suite level. I like to give noobs to cyber a copy of *Cybersecurity for Dummies*. It is a good book, but let us be honest. I am not sure most people get through DoS, DDoS, and botnets in Chapter 1, before their eyes start to glaze over. Rootkits, Vectors, Proxies, Anonymizers, Circumventors, SSL, these are not concepts or words most normal people understand. Now, I would argue that decision makers do not need to understand them. However, the point is, every time we try to get them to, I think we make it harder, not easier, to address the threat environment.

The second reason we do not act is cost. According to the IBM, the average cost of a data breach is $3.86 million ($148 per lost or stolen record) with a 28% likelihood of a recurring material breach over the next 2 years. Whereas the average cost savings with an Incident Response team is $14 per record. The cost of proactive mitigations far outweigh the costs [34].

The third reason executives do not act on cybersecurity is productivity, the concern that an end-to-end data protection program will slow how much their employees can accomplish. It is a fair point, right? Anyone who has worked from home on their laptop, with all the apps and services they want on them, knows how much easier it is to get things done when you do not have someone from IT breathing down your neck about what you can and cannot use. To say nothing about all the time and effort spent training people how to avoid malware or various other attacks.

Now, some executives also believe they are immune or might just get lucky when it comes to cyber-attacks. This despite reports that about half of all businesses have been attacked and according to IBM. These are just some of the reasons we do not act. Cybersecurity is like going to the gym, something you have to do every day. And when you think about it, a lot of the reasons we do not go to the gym, it costs too much, does it really help me, it is a pain in the neck, are awfully similar to reasons we do not act on cybersecurity.

15.13 THREE I'S OF CORPORATE CYBERSECURITY STRATEGY

In my experience I find that if we think differently about cybersecurity, we tend to act differently. I call it "The Three I's of Cybersecurity." The first I is "innovation."

Today, almost every company is obsessed with innovation. With understanding what is next, with peeking around corners, with learning how we can evolve our products, our services, even our brands. As companies, we put a lot of our resources behind innovation. In addition, we do it for a simple reason, as we see innovation as a matter of survival. Well, cybersecurity is similar to innovation in a lot of ways. Like our competitors, cyber criminals are constantly reinventing themselves. In that sense, we need to not only keep up with the pace of innovation, but to create a culture that embraces this state of constant reinvention. So, we should think of cybersecurity as innovating our defenses and maybe even make cyber-innovation part of our innovation platform.

The second I is "intelligence." We pour untold amounts of resources into competitive intelligence, into defining, gathering, analyzing, and distributing intelligence, about our products, our customers, our competitors, and, really, any aspect of the business environment. We do it for a simple reason: to get a competitive edge. To do things cheaper, faster, better, smarter. So why do not we do the same with cyber? After all, cyber criminals are doing everything they can to get an edge into how we do business. Cybersecurity is about being one step ahead, not of industry competitors but of threat competitors. It is about protecting our highest value assets.

The third I is "investment." When we set our corporate strategies, there are countless actions we take or processes we set in motion devoting time, money, or resources in the hopes that they could be profitable. Not everything we invest in has an immediate ROI. Some of these things have a payoff in the short term; others have a longer time horizon. However, for some reason, we expect cybersecurity to pay dividends right away. Possibly because it is not generally visible to the customer. Or maybe because the initial outlay seems high.

Cybersecurity is an investment with some short and long-term payoffs. In addition, if you do not think those payoffs are real, think about this. According to Verizon [35], 58% of all of victims are categorized as small businesses. Think about funding your cybersecurity operation not just as an investment in your company's security, but in your company's longevity.

By embracing the Three I's, we can begin to overcome the inertia we face when it comes to cybersecurity, to stop thinking of it tactically and start thinking of it strategically. In the process, we can also create opportunity for our businesses. We can feel confident our IP is safe. Our customers can feel confident their data is safe. In addition, our employees and prospective employees can know their employer values their personal information as well.

Most important of all, by approaching cybersecurity as a long-term corporate strategy, our brand is safe. Because stakeholders across the spectrum, from customers to regulators to the media, will know that if there is an attack that our company has the protocols and protections in place to respond quickly and appropriately. In that sense cybersecurity is not simply a threat mitigation, but an opportunity to create a true competitive advantage, one a lot of our competitors may not yet fully embrace.

15.14 CYBERSECURITY SYSTEMS ENGINEERING

Corporate cybersecurity strategy can be accelerated through leveraging cybersecurity systems engineering, architectural frameworks, and cybersecurity IT portfolio management. To incorporate cybersecurity into corporate strategy, it is essential address cybersecurity throughout the organizations systems engineering lifecycle and artifacts [3,5].

In the face of cyber-attacks, organizations must often make quick decisions with imperfect data; however, cyber-attacks exploit vulnerabilities in people, processes, and technologies. Bayuk and

Mostashari [3] highlight where security principles should be inserted into traditional systems engineering. Integrating security requirements into the design phase of systems engineering initiatives and producing security requirements at the system level is more effective in mitigating cyber-attacks. Enterprise Architecture frameworks can be leveraged by organizations to anchor their cybersecurity strategy.

15.15 ENTERPRISE ARCHITECTURE FRAMEWORKS

Enterprise architecture frameworks utilize various views and graphics to describe the interactions of people, processes, and technologies within an enterprise [2]. Architectural frameworks portray how technical capabilities support overall business goals and provide a common taxonomy and construct that both IT and business professionals in an organization can exercise to communicate and achieve such goals.

Over the years, several architectural styles and associated frameworks have emerged. These include the Zachman Framework, the US Department of Defense Architecture Framework, and Service Oriented Architectures to include the Federal Enterprise Architecture, and the Open Group Architecture Framework. A brief description of key concepts for each model is provided below. These Enterprise Architecture frameworks can be leveraged by organizations to anchor their cybersecurity strategy.

15.16 ZACHMAN FRAMEWORK

John Zachman, an IBM researcher, first documented the need and value of having an enterprise architecture framework that tailors communications to stakeholders involved in each phase of the systems engineering process [36]. The Zachman Framework addresses the question of "Who, What, When, Where, Why, and How" a system development initiative aligns to the IT infrastructure to achieve the business goals.

Organizations need to consider who would want to attack their organization and the attackers' motives in order to anticipate what data will be targeted and how the data will be targeted in order to implement mitigations for when the organization is targeted. For example, a high school website or a small business is likely to have different attackers with different motives than government agencies, major corporations, or banks. Organizations can leverage the "Who, What, When, Where, Why and How" construct of the Zachman Framework in order to successfully plan for cyber-attack mitigation. Fig. 15.2 provides an overview of key questions organizations should consider.

Who would want to attack your organization? How much time and resources docs the cyber-attacker have to dedicate toward attacking your organization.

What data and information would a cyber-attacker want? If you know what data and information the cyber-attacker is seeking, you can proactively plan to protect this data appropriately.

When is the cyber-attacker likely to strike? If you know when the cyber-attacker is likely to strike, you can shore up your organization's cybersecurity defenses.

Who	What	When	Where	Why	How

Who would want to attack your organization?	**What** data would an attacker want?	**When** is the attacker likely to strike?	**Where** is the organization vulnerable to attack?	**Why** would an adversary attack your organization?	**How** will the adversary conduct the attack?

FIGURE 15.2

"Who, what, when, where, why, and how" to plan cyber-attack mitigations.

Where is your organization most vulnerable to a cyber-attack? Cyber-attackers exploit the weakest link in your defense and identifying where your organization is most vulnerable will be a good indicator of where the cyber-attacker will initiate the cyber-attack (e.g., people, process, and technology)?

Why would an adversary want to attack your organization? Are they seeking financial gain or executing the cyber-attack for ideological reasons? This will help you identify what data the person may be seeking.

How will the cyber-attacker launch the cyber-attack against your organization? What tactics and techniques will the cyber-attacker use? In addition to technical skill and resources, the cyber-attacker will leverage tactics to exploit the organizations weakest link. Is the organization following secure development practices? Are the organization's IT systems patched, upgraded, and the personnel appropriately trained not to fall for phishing and social engineering?

15.17 US DEPARTMENT OF DEFENSE ARCHITECTURE FRAMEWORK

The US Federal Government has reinforced the need for architectures to support business decisions through federal law. The 1996 Clinger-Cohen Act mandates that US Federal Government agencies select and manage their IT resources by leveraging enterprise architectures [37].

Furthermore, the E-Government Act of 2002 requires the development of an enterprise architecture to promote electronic government services [30]. Architecture-based decision-making provides US Federal Government agencies with a repeatable approach to communicate IT business decisions that is scalable across organizational boundaries.

The US Department of Defense Architecture Framework (DoDAF) is used by US Department of Defense agencies. DoDAF communicates the enterprise architecture through a variety of architectural viewpoints. These viewpoints include All Viewpoint, Capability Viewpoint, Data and Information Viewpoint, Operational Viewpoint, Project Viewpoint, Services Viewpoint, Standards Viewpoint, and Systems Viewpoint [30]. Researchers have demonstrated the value of leveraging enterprise architecture to enhance communications in Department of Defense systems engineering initiatives [38].

15.18 SERVICE-ORIENTED ARCHITECTURE

Service-oriented architecture (SOA) is an architectural style that has gained popularity in recent years. SOA supports the alignment between IT and business stakeholders [39]. Services are architectural descriptors that provide a common taxonomy and lexicon. Services describe "What" business goals need to be achieved, whereas "How" the service is implemented is environment specific. These architectural descriptors can be used by leaders across organizations to describe the interaction of people, processes, and technologies within an enterprise. For this reason, SOA is regarded as an agile method for driving IT and business alignment [40].

The Federal Enterprise Architecture (FEA) is the SOA implemented by most US Federal agencies outside of the Department of Defense, like the Department of Homeland Security. FEA provides a common taxonomy that enables stakeholders to communicate across organizational boundaries using a common lexicon, and supports business-based architecture analysis for government-wide improvement [37].

Likewise, The Open Group Architecture Framework (TOGAF) is an open source SOA commonly implemented in commercial organizations. TOGAF is a business driven architecture that provides a common lexicon to enable capability-based business planning across the enterprise [39].

Enterprise architecture frameworks have been found to be beneficial across government and industry as a repeatable construct to portray how technical capabilities support overall business goals to transfer architectural data into actionable information [36,38,40,41]. Enterprise architecture frameworks consist of architectural descriptors that provide a common taxonomy and lexicon that enables stakeholders to communicate across the enterprise.

15.19 CYBERSECURITY IT PORTFOLIO MANAGEMENT

A portfolio is a group of related or nonrelated projects or programs, and portfolio management is the centralized management and prioritization of projects or programs [42]. Portfolio management involves balancing the performance of IT projects/programs with both risk and return on investments (e.g., capital expenditures and operational expenditures), similar to that of financial assets. It is important for nontechnical business managers and technical IT managers to collaborate early and often on IT portfolio management as organizations often must balance the cost of security needs against competing priorities [43]. IT portfolio management techniques can be leveraged to identify an organization's current cybersecurity posture and where an organization needs to improve to mitigate cybersecurity risks.

It is critical for businesses to tightly couple the overall enterprise architecture with their business strategy and execution of IT solutions to achieve the necessary business goals [44]. IT portfolio management includes identifying which initiatives to fund and which ongoing initiatives to continue or terminate, as well as addressing necessary portfolio adjustments. Systems architecting is a decision-rich activity that involves a wide range of considerations across people, processes, and technology solutions [45]. IT portfolio management often entails conducting an analysis of alternatives to various IT investments. Smith et al. [46] stressed the importance of incorporating the opinions of key organizational decision-makers into the portfolio management process and developed an algorithm that leveraged decision-makers' opinions on portfolio elements.

Business managers need to understand how IT affects both the strategy and the bottom-line. IT professionals need to understand how IT ties into their organization's vision and business strategies to achieve the specified goals. The US federal government demonstrates the value of utilizing Enterprise Architecture frameworks to inform IT Portfolio Management and fulfill mission needs.

The Clinger-Cohen Act mandates that 27 US federal government agencies must report their IT investment goals, costs, and statuses against enterprise architecture [47]. Every year, the US Office of Management & Budget regulates the process by which 27 US Federal Agencies categorize IT investments against the FEA in support of budget appropriations processes.

As part of the Exhibit 53, IT investments are mapped to the BRM to identify opportunities for collaboration, shared services, and solution reuse. The Exhibit 53 allows for one primary service and up to four secondary service codes to be identified for each IT investment. The FEA provides a common taxonomy against which IT investments can be categorized and managed to streamline and optimize the entire IT portfolio [37].

Given the opportunity for financial gain, social effect, and business impact, sophisticated criminal networks have evolved [24]. With the development of innovative cybersecurity deterrence and mitigation tactics and techniques, new and novel cyber-attack methods will also evolve. Methods to capture necessary changes to the IT portfolio over time to meet these evolving cybersecurity needs are therefore essential [48].

The "Economic-Benefit Returns on Cybersecurity Investments: The Table Top Approach" provides an approach to calculate the economic return on investment for cybersecurity mitigation investments. In this research the authors examined a few specific cyber intrusion events and leveraged SMEs to rate the impact the intrusion events had on the customers' mission operations. The SME ratings were then categorized in the areas of prevention, detection, and quarantine. The authors contend customers should invest in the Cybersecurity mitigations associated with high impact intrusion events. The research also stresses the need to capture required cybersecurity mitigation investments in a timely manner [48].

Garvey et al. [48] recognize a need for methods for capturing the changes in dynamic Cybersecurity portfolios as technology and Cybersecurity adversaries advance and enhance over time. Garvey et al. [48] examine various cyber intrusion events and leverage subject matter expert (SME) estimates to rate the impact each intrusion has on a customer's mission. SMEs rate the impact of each intrusion event against four areas that include operations, prevention, detection, and quarantine [48].

With the development of innovative cybersecurity deterrence and mitigation tactics and techniques, new and novel cyber-attack methods will also evolve. Methods to capture necessary changes to the IT portfolio over time to meet these evolving cybersecurity needs are therefore essential [48]. Enterprise architecture frameworks are the key systems engineering methods used to describe the interactions of people, processes, and technologies within enterprises that have been found to be effective in planning cybersecurity mitigations.

One tool research proposes is a rapidly deployable model that facilitates expression of cyber-attack scenarios via enterprise architecture to develop cybersecurity enterprise architecture attack maps (CEAMs). A CEAM describes cyber-attack scenarios using architectural language, thereby providing a common taxonomy that can be leveraged across the enterprise for swift architecture-based cybersecurity decision making. Using architectural language to express cyber-attack scenarios via a CEAM provided a common taxonomy that can be used to inform IT portfolio management. Mapping the CEAM to the IT portfolio enables decision makers to perform rapid cybersecurity architecture-based decision-making supporting their business goals and mission needs [49].

15.20 SMARTER CYBERSECURITY LEVERAGING ARTIFICIAL INTELLIGENCE

AI can speed response times of underresourced security operations by analyzing massive quantities of risk data. Given the growing number and complexity of cyber-attacks, this helps security operations that are underresourced stay ahead. AI provides instant insights that can help organizations fight through the noise of thousands of daily alerts and drastically reducing response times by leveraging millions of news stories, blogs, and research papers to curate threat intelligence. Leveraging natural language processing and machine learning AI technologies, analyst can act on threats with speed and confidence [50].

AI is trained by consuming billions of data artifacts, leveraging both structured and unstructured data sources like news stories and blogs. AI leverages deep learning and machine learning techniques to improve its understanding cybersecurity threats and risks. AI identifies relationships between threats, like malicious files, suspicious IP addresses, or insiders by gathering insights and using reasoning. Security analyst can respond up to 60 times faster to cybersecurity threats as this analysis only takes a second to a few minutes depending on data set size. AI provides curated risks analysis and eliminates onerous research tasks, resulting in a reduction of time security analysts take to make strategic decisions and launch a choreographed response [50].

For example, IBM's cognitive AI, Watson for Cyber Security, enables organizations to respond to cyber threats greater speed and confidence. It provides actionable insights by learning and connecting the dots between threats with each interaction. IBM cognitive computing is an advanced type of artificial intelligences that leverages various forms of AI, including machine-learning algorithms and deep-learning networks, that get stronger and smarter over time [50].

15.21 IOT AND GROWING CYBERSECURITY RISK

In our current IoT world, daily 2.5 quintillion bytes of data are generated. The IoT will consist of up to 30 billion connected devices by 2020. Cybersecurity risks grow as the scale of IoT grows [51].

IoT is the architecture and suite of technologies needed to create, communicate, aggregate, analyze, and act upon digital information in the physical world [50]. Attackers work to infiltrate IoT deployments by identifying security weaknesses across the enterprise architecture. It is imperative that we secure by design through our cybersecurity systems engineering efforts and architect in cybersecurity mitigations across the layers of the enterprise architect.

Organizations must consider the cybersecurity lifecycle (protect, detect, respond, and recover) and implement both proactive and reactive mitigations. Organizations must build muscle memory around designing for and responding to cyber-attacks. Similar to how first responders develop strategy, train and prepare for a catastrophic event; organizations must develop cybersecurity strategies, train and prepare for the inevitable cyber-attack to help mitigate organizational catastrophe to the infrastructure, business, and brand.

Building security into an organizations IoT platform is essential to minimize risks to private data, business assets, and reputation. Ninety-three percent (93%) of consumers believe manufacturers need to do more to secure their IoT devices and 72% of companies with mature IoT programs have an appointed C-level IoT champion [53].

The Deloitte Information Value Loop is an IoT blueprint for how technologies create value and fit together. The value loop accelerates the relationship between action and data, shifting the focus from what we connect to what we enable. This enables organization to drive value more effectively and efficiently.

When thinking about how to construct the Information Value Loop, organizations should consider five key capabilities to include:

1. Create: Sensors collect data on the physical environment. For example, measuring things like device status, temperature, location, or air.
2. Communicate: Networks facilitate devices sharing data with a centralized platform or other devices.
3. Aggregate: Common standards aid data from various sources to be combined.
4. Analyze: Analysis tools spot patterns that indicate actions needed or anomalies for further investigations.
5. Act: Insights resulting from analysis either frame a choice for the user or initiate action [52].

By getting a better picture of what IoT is, organizations can learn how to best interact with IoT as well as reduce uncertainty and security risk. The value resulting from IoT data can be accelerated by extending the value loop or addressing bottlenecks [52].

In this highly connected complex IoT environment there are numerous cybersecurity threats to be considered across the cybersecurity systems engineering lifecycle from enterprise architecture to requirements, test and implementation of solutions. Four of the biggest cybersecurity threats in an IoT world are: hidden exploitable potential; forgotten and disused devices; understanding IoT attacker goals; and balancing security with user expectations [51].

Many IoT devices are designed for narrow tasks, like sensing temperature or recording movement. However, these devices have hidden exploitable potential. They run on microcontrollers and operating systems capable of doing much more in the background without impeding their primary purpose. This is a rich opportunity for attackers [51].

It is important to have an accurate inventory of devices in an IoT environment to ensure each device is adequately patched, upgraded, and integrated into the cybersecurity architecture. Like a cluttered home, sometimes old devices are simply forgotten about and not used. These devices provide an opportunity for cyber attackers to reintroduce the device to the ecosystem and leverage it as a foothold into the IoT architecture [51].

Understanding the IoT attackers and their goals as well as your organizations key assets will help you design mitigations for these attacks. Techniques like developing CEAMs designing these scenarios [49]. According NIST, "an attacker who can view the IoT device's stored or transmitted data might not gain any advantage or value from it, but an attacker who can alter the data might trigger a series of events that cause an incident" [51].

15.22 THE CHANGE MANAGEMENT CHALLENGE

How many of you think what I am saying makes enough sense, that you are willing to give making cybersecurity a part of your corporate strategy a shot? However, I suspect many of you are thinking the same thing, "Where do I start?"

I have spent a lot of time on Lean Six Sigma, because I believe you need to approach this not as a technology issue but as a change management issue. Assessing the current state, identifying pain points, and really developing that future state to mitigate those pain points. This is not about taking action for the sake of action. It is about building a process that delivers results, that is smart, specific, measurable, achievable, realistic, and timely. That is what change management is all about. In addition, the first step you need when you want to tackle any problem is to have the team in place to tackle it.

15.23 CYBERSECURITY EXPERTISE

In this case, that means you need to acquire cyber expertise. In addition, there are really two pieces to that. The first is a Chief Information Security Officer or a "CISO." A CISO plays a very different role from one organization to the next. Some come from a hardcore tech or security background. Others have business degrees. However, all of them have the ability to speak the language and mechanics of cyber, the threat environment, and solutions, but also understand business needs. This person is part of the C-Suite.

Having a CISO improves communication between Chief Executive Officers (CEOs), Chief Technology Officers (CTOs), and Chief Information Officers (CIOs), and builds understanding on how to invest strategically so you can have an effective cybersecurity posture. Businesses that worked to improve customer trust post breach by deploying a senior-level leader like a Chief Privacy Officer (CPO) or CISO, to direct customer trust initiatives, lost fewer customers minimizing the financial consequences of a breach [34].

The second, you need the help of a true cyber expert. Someone who really knows his or her stuff. For instance, at Microsoft we offer a cyber architect service. We will likely send you someone with 15 years of experience with a background in the intelligence community who has been on incident response teams. They have done the work. Maybe you will want this capability in house and hire somebody or a number of somebody has to do this for you. The point is this is not a job for your favorite management consultant. You need real expertise.

Demand for cybersecurity professionals rose to approximately 6 million globally in 2019, with a shortfall of approximately 1.5 million. Cybersecurity requires personnel with diverse skillsets ranging from Cybersecurity Manager to Reverse Engineer. The National Institute of Standards and Technology National Cybersecurity Workforce Framework [54,55] provides recommended degrees, certifications, and educational opportunities for over 100 cybersecurity job titles ranging from Cybersecurity Manager to Reverse Engineer.

Participation in formal academic programs is often needed for entry into and promotion within the cybersecurity field. For example, the US government has designated National Centers of Academic Excellence (CAE) in Information Assurance/Cyber Defense. CAE institutions offer training, 2-year degree, Bachelors, Masters, and PhD programs. The National Initiative for Cybersecurity Careers and Studies (NICSS) provides a comprehensive list of cybersecurity professional certifications [53].

15.24 ASSESS CURRENT STATE

Next, you need to assess the current state. You may be familiar with the cybersecurity framework instituted by the National Institute of Standards and Technology [12]: Identify, Protect, Detect, Respond, Recover. Identify is the assessment stage, which will determine whether you are doing anything within that framework and if you are, how well you are doing it. Much of this assessment will come in the form of software installed on your machines to pull real-time data, proactive monitoring. This helps us understand potential vulnerabilities and patterns of behavior [12].

For Protect, we will measure whether you know what your highest value assets are and are restricting access to them on a need-to-know basis. These can be everything from your intellectual property to customer and employee information to your domain credential, which is effectively controls access to all your company's data, servers, and applications. It is important to look at the most common types of attacks and the most successful, and credential theft is among the most devastating, because it is systemically woven into the architecture and difficult to remediate quickly. Many of the attacks we respond to are related to credential theft.

For Detect, the first question will be, are you monitoring your environment? Moreover, if you are, are you paying attention to any alerts when something might be amiss? You would be amazed at how many companies have protocols in place but do not actually follow through when they are alerted to a problem. That brings us to Respond, if you are monitoring your environment, do you have an incident response team and plan? Do you exercise that plan?

Lastly, Recover, do you have any protocols in place to restore capabilities that may be impaired? How you respond and recover really depends on the nature and scale of the threat. Is it enterprise-wide? We saw one prominent city in the United States endure the total destruction and deletion of all its data that meant pulling tape backups to rebuild the entire environment.

Or is it a global problem? We all heard about the entertainment company attacked across three continents simultaneously. However, another example was a multinational mining company that contracted out of the United States while its corporate leadership and data centers were spread across Europe. Remediating the attack required utilizing cyber talent located in Brazil, the United States, and Austria and to coordinate with European authorities.

We really believe this challenge requires a global approach and workforce, which is why our operations have a global footprint, from Africa to Asia, North America to Latin America, and

Europe. For this work, we conceived the intellectual property in the United States, tested it in Germany and developed our training in partnership with our peers in India.

Once the assessment stage is complete, you will work to develop a future state to put a program into action. Now, that full future state includes everything from reducing threats with identity and access management, managing mobile devices and apps, and leveraging conditional access. To increasing enterprise data protection, preventing data loss, enabling secured collaboration, and reducing malware exposure.

Now that sounds like a lot. However, the good news is, you do not have to do this alone. For instance, a cloud solution like Microsoft's Azure has unified security management and does a lot of this for you. It finds vulnerabilities and remediates them quickly. It continuously monitors the security of your machines and networks using hundreds of built-in security assessments. It uses actionable security recommendations to remediate issues before they can be exploited. It also limits exposure to threats by applying application controls and controlling access. In addition, it uses advanced, behavioral analytics and machine learning to get an edge over evolving cyber-attacks [56].

Even still, you do have some choices to make. Momentum is really important in any new strategic, complex venture. So, you really need to prioritize what is most important and prioritize some quick wins. For instance, let me suggest a big one: credential theft, which we have touched on.

You might be surprised how many people have access to domain credentials within your organization. It is kind of horrifying, really, to realize that there may be literally hundreds of people within your organization who have the ability to burn down your whole environment or give that information to someone who will. Credential theft is something that can be addressed in many organizations within six months. We often limit this ability physically, with a secure workstation not connected to the Internet.

Other parts of that future state include developing an incident response plan. Patching and upgrading, identifying any viruses or threats that need to be addressed. Moreover, beginning the process of moving data to the cloud, so more of your data will be protected by those who patch and upgrade for you.

15.25 MAKING CYBER PART OF CORPORATE STRATEGY

All this is very important stuff, but the real question is, how do you convert this from a one-time intervention with maybe a little follow-up to something you practice as a matter of habit every day? That is really where the strategic plan comes in. Yes, there are a number of additional steps we can take, but it is when we approach cyber as part of our strategic plan that it really becomes part of your business.

So, what are some of the things to look at in each strategic plan? Well, certainly we are going to regularly review what should be in the cloud. Maybe it is some assets, maybe it is everything. Maybe it is nothing. We have found that most companies like the idea of having their assets in a hybrid environment, where some of them live on premises and some of them live in the cloud.

Why does this need to be part of corporate strategy? Well, here is the funny thing about our highest value assets, they are not static. As our businesses change, so, too do our highest value assets. For instance, for a startup, at first, and maybe even for many years, our IP may be our most precious asset. However, as we begin to grow, we may develop a customer base or have more employees. Or other lines of businesses. So, part of that strategic plan process will be to review how our highest value assets change as our business changes.

During the course of developing a strategic plan, we will also review everything related to the cloud, from pricing and availability, to terms, to customer service, to who our cloud vendor is. Maybe it is Microsoft's Azure services. Or someone else. To that end, the cloud vendor we choose may be a competitive and marketing opportunity. For example, let us say we are a financial institution. Is there an advantage to putting our assets in a secure environment and letting our customers know who we use for these services?

We will also use the strategic plan process to review monitoring; can we use AI more effectively? Satya Nadella has said that "AI is going to be one of the trends that is going to be the next big shift in technology" [57]. The average total cost of a data breach for organizations that fully deploy AI security automation solutions is $2.88 million. In contrast, the cost of data breach is $1.55 million higher, $4.43 million for organizations that do not deploy automation. Specifically, implementation of AI security platforms that use machine learning, analytics, and orchestration to help human responders identify and contain breaches, saved companies an average of $8 per compromised record [34].

Businesses that use IoT devices extensively pay $5 more per compromised record on average. A mega breach of 1 million records yields an average total cost of $40 million and a mega breach of 50 million records yields an average total cost of $350 million. Other factors that decreased the cost of data breach include: use of an incident response team, extensive use of encryption, business continuity management, employee training, participation in threat sharing, artificial intelligence platforms, use of security analytics, extensive use of data loss protection, board-level involvement, and data classification schema [34].

In addition, to revisit our incident response plan as well as our whole Patch & Upgrade process. Many of these discussions have broad implications beyond cyber. For instance, when we talk about software upgrades, we also want to think about hardware, in particular, the security relationship between the two. Some companies, for example, may want to explore more frequently replacing employees' computers. After all, new machines not only give you more computing power, but also more cyber protection power. Moreover, one of the things they might weigh at their strategic review is the performance and security cost of waiting another year to refresh their machines.

At the business level, we also need to use our strategic plan to review the cybersecurity legal and regulatory environment. For instance, are we protecting assets in accordance with any new laws or rules? Then we will want to drill down and review different business needs, do different businesses within our company have different security needs? How do we address that in a thoughtful, systematic way?

We will also use our strategic plan to revisit where we started, our cyber talent management. Some of the ongoing questions will be whether we have adequate capacity, that is, enough people doing security across the board with the requisite training and expertise. From our monitoring center, to supporting our incident response plan, to supporting new data centers.

However, let us be clear: those are not the only people involved in cybersecurity. None of this work happens in an IT vacuum. Safeguarding our companies' longevity is not accomplished in a room down the hall. To be successful, it requires all of us. It requires partnership, between the various businesses, but also all the corporate functions, including legal, human resources, IT, finance, and procurement. To successfully implement a cybersecurity strategy in your company, everyone needs to know their role. Everyone needs to have skin in the game. In addition, no one should be kept in the dark.

REFERENCES

[1] National Institute of Standards and Technology, National Institute of Standards and Technology Special Publication 800-37 (Revision 1): guide for applying the risk management framework to federal information systems, The National Institute of Standards and Technology, 2010. Available from: <http://csrc.nist.gov/publications/nistpubs/800-37-rev1/sp800-37-rev1-final.pdf>, 2019 (accessed 02.04.19).

[2] National Institute of Standards and Technology, Security and privacy controls for federal information systems and organizations, NIST Special Publication 800-53, Revision 4, 2013.

[3] J. Bayuk, A. Mostashari, Measuring systems security, Syst. Eng. 16 (1) (2013) 1–14. Available from: https://doi.org/10.1002/sys.

[4] ISO/IEC, ISO/IEC 15288:2008 Systems and software engineering system life cycle process, 2008. Available from: <https://www.iso.org/obp/ui/#iso:std:iso-iec:15288:ed-2:v1:en>, 2019 (accessed 05.03.19).

[5] S. Furnell, N. Clarke, Power to the people? The evolving recognition of human aspects of security, Comput. Secur. 31 (8) (2012) 983–988. Available from: https://doi.org/10.1016/j.cose.2012.08.004. Available from: http://linkinghub.elsevier.com/retrieve/pii/S0167404812001228.

[6] C. Irvine, T. Nguyen, Educating the systems security engineer's apprentice, IEEE Secur. Priv. 8 (4) (2010) 0058–0061.

[7] R.A. Jones, B. Horowitz, A system-aware cyber security architecture, Syst. Eng. 15 (2) (2012) 225–240. Available from: https://doi.org/10.1002/sys.

[8] J. Webb, A. Ahmad, S.B. Maynard, G. Shanks, A situation awareness model for information security risk management, Comput. Secur. 44 (2014) 1–15. Available from: https://doi.org/10.1016/j.cose.2014.04.005. Available from: http://linkinghub.elsevier.com/retrieve/pii/S0167404814000571.

[9] N. Zafar, E. Arnautovic, A. Diabat, D. Svetinovic, System security requirements analysis: a smart grid case study, Syst. Eng. 17 (1) (2014) 77–88. Available from: https://doi.org/10.1002/sys.

[10] C.V. Zhou, C. Leckie, S. Karunasekera, A survey of coordinated attacks and collaborative intrusion detection, Comput. Secur. 29 (1) (2010) 124–140. Available from: https://doi.org/10.1016/j.cose.2009.06.008. Available from: http://linkinghub.elsevier.com/retrieve/pii/S016740480900073X.

[11] B. Obama, Executive order—improving critical infrastructure cybersecurity—February 12, 2013, Whitehouse.gov 2013. Available from: <http://www.whitehouse.gov/the-press-office/2013/02/12/executive-order-improving-critical-infrastructure-cybersecurity>, 2019 (accessed 18.04.19).

[12] National Institute of Standards and Technology, Framework for improving critical infrastructure cybersecurity, Version 1.1, 2018. Available from: <https://nvlpubs.nist.gov/nistpubs/CSWP/NIST.CSWP.04162018.pdf>, 2019 (accessed 05.03.19).

[13] ISO/IEC. ISO/IEC 27034-1 Information technology security techniques application security. Part 1: overview and concepts, 2011.

[14] MITRE. Common weakness enumeration (CWE), 2014. Available from: <http://cwe.mitre.org/>, 2019 (accessed 15.03.19).

[15] Shameli-Sendi, A., Cheriet, M., Hamou-Lhadj, A. (2014).Taxonomy of intrusion risk assessment and response system. Comput. Secur. 45, 1–16.

[16] R. Willison, M. Warkentin, Beyond deterrence: an expanded view of employee computer abuse, MIS Q. 37 (1) (2013) 1–20.

[17] B. Acohido, Timeline: target, Neiman Marcus Disclosures, USA TODAY 2014. Available from: <http://www.usatoday.com/story/cybertruth/2014/01/23/timeline-target-neiman-marcus-disclosures/4799153>, 2019 (accessed 22.05.19).

[18] N. Perlroth, In cyberattack on Saudi, U.S. sees Iran firing back, New York Times, 2012. Available from: <http://www.nytimes.com/2012/10/24/business/global/cyberattack-on-saudi-oil-firm-disquiets-us.html?pagewanted = all&_r = 0>, 2019 (accessed 01.04.19).

[19] J. Pepitone, New York Times hit with malicious attack, CNN Money 2014, 2013. Available from: <http://money.cnn.com/2013/08/27/technology/security/new-york-times-hacked/index.html>, 2019 (accessed 11.04.19).

[20] R. Langner, Stuxnet: dissecting a cyberwarfare weapon, IEEE Secur. Priv. 9 (3) (2011) 49–51. Available from: https://doi.org/10.1109/MSP.2011.67.

[21] K.R. Choo, Organised crime groups in cyberspace: a typology, Trends Organiz. Crime 11 (3) (2008) 270–295. Available from: https://doi.org/10.1007/s12117-008-9038-9.

[22] J. Clapper, Statement for the record worldwide threat assessment of the U.S. intelligence community senate select committee on intelligence, 2013.

[23] S. Applegate, Cybermilitias and political hackers: use of irregular forces in cyberwarfare, IEEE Secur. Priv. 9 (5) (2011) 16–22. Available from: http://doi.ieeecomputersociety.org/10.1109/MSP.2011.2.

[24] A.K. Sood, R.J. Enbody, Targeted cyberattacks: a superset of advanced persistent threats, IEEE Secur. Priv. 11 (1) (2013) 54–61.

[25] T.J. Holt, Examining the forces shaping cybercrime markets online, Soc. Sci. Comput. Rev. 31 (2) (2012) 165–177. Available from: https://doi.org/10.1177/0894439312452998.

[26] A.K. Sood, R. Bansal, R.J. Enbody, Cybercrime: dissecting the state of underground enterprise, IEEE Internet Comput. 17 (1) (2013) 60–68. Available from: https://doi.org/10.1109/MIC.2012.61.

[27] M. Simos, P. Jungles, Pass-the-hash and other credential theft and reuse: mitigating the risk of lateral movement and privilege, in BlackHat, Las Vegaas, 2013, pp. 1–36. Available from: <https://www.blackhat.com/us-13/archives.html#Simos>, 2019 (accessed 06.03.19).

[28] Microsoft, Mitigating Pass-the-Hash (PtH) attacks and other credential theft, Version 1 and 2, 2018. Available from: <https://www.microsoft.com/en-us/download/details.aspx?id = 36036>, 2019 (accessed 16.03.19).

[29] U.S. Office of the President, The comprehensive national cybersecurity initiative. <https://www.whitehouse.gov/issues/foreign-policy/cybersecurity/national-initiative>, 2009.

[30] U.S. Department of Defense, The Department of Defense Cyber Strategy. <http://www.defense.gov/Portals/1/features/2015/0415_cyber-strategy/Final_2015_DoD_CYBER_STRATEGY_for_web.pdf>, 2015.

[31] U.S. Congress, S.2588 Cybersecurity Information Sharing Act of 2014. <https://www.congress.gov/bill/113th-congress/senate-bill/2588>, 2014.

[32] European Commission, Joint Communication of the European Parliament, the Council, the European Economic and Social Committee and the Committee of the Regions: cybersecurity strategy of the European Union: an open, safe and secure cyberspace, 2013. <http://ec.europa.eu/digital-agenda/en/news/eu-Cybersecurity-plan-protect-open-internet-and-online-freedom-and-opportunity-cyber-security>, 2019 (accessed 20.05.19).

[33] Regulation (EU) 2016/679 of the European Parliament and of the Council of 27 April 2016 on the protection of natural persons with regard to the processing of personal data and on the free movement of such data, and repealing Directive 95/46/EC (General Data Protection Regulation) (OJ L 119, 4.5.2016, p. 1).

[34] IBM, Ponemon Institute's 2018 cost of data breach study: global overview, 2018. Available from: <https://www.ibm.com/account/reg/us-en/signup?formid = urx-33316>, 2019 (accessed 20.03.19).

[35] Verizon, Data breach investigations report, 2018. Available from: <http://www.verizonenterprise.com/resources/reports/rp_DBIR_2018_Report_en_xg.pdf>, 2019 (accessed 02.02.19).

[36] J.A. Zachman, A framework for information systems architecture, IBM Syst. J. 38 (2.3) (1999) 454–470. Available from: https://doi.org/10.1147/sj.382.0454. Available from: http://ieeexplore.ieee.org/lpdocs/epic03/wrapper.htm?arnumber = 5387107.

[37] U.S. Office of Management & Budget, Business Reference Model Version 3, Guidance on exhibits 53 and 300 information technology and e-government, 2013.

[38] C. Piaszczyk, Model based systems engineering with department of defense architectural framework, Syst. Eng. 14 (3) (2011) 305–326. Available from: https://doi.org/10.1002/sys.

[39] The Open Group, The SOA source book. <http://www.opengroup.org/soa/source-book/soa/soa.htm#soa_definition>, 2014.

[40] J. Choi, D. Nazareth, H. Jain, The impact of SOA implementation on IT-business alignment: a system dynamics, ACM Trans. Manag. Inf. Syst. 4 (1) (2013).

[41] B. Ge, K.W. Hipel, K. Yang, Y. Chen, A data-centric capability-focused approach for system-of-systems architecture modeling and analysis, Syst. Eng. 16 (3) (2013) 363–377. Available from: https://doi.org/10.1002/sys.

[42] Project Management Institute, A Guide to the Project Management Body of Knowledge (PMBOK® Guide), fifth ed, Project Management Institute, Inc, 2013.

[43] S.L. Garfinkel, The cybersecurity risk, Commun. ACM 55 (6) (2012) 29–32. Available from: https://doi.org/10.1145/2184319.2184330. Available from: http://dl.acm.org/citation.cfm?doid = 2184319.2184330.

[44] B. Burton, Business architecture bridging strategy and execution, Gartner Webinars, 2013. Available from: <http://my.gartner.com/portal/server.pt?open = 512&objID = 202&mode = 2&PageID = 5553&ref = webinar-rss&resId = 2301018>, 2018 (accessed 21.03.18).

[45] A.M. Madni, Generating novel options during systems architecting: psychological principles, systems thinking, and computer-based aiding, Syst. Eng. 17 (1) (2012) 1–9. Available from: https://doi.org/10.1002/sys.

[46] P. Smith, M. Ferringer, R. Kelly, I. Min, Budget-constrained portfolio trades using multiobjective optimization, Syst. Eng. 15 (4) (2012) 461–470. Available from: https://doi.org/10.1002/sys.

[47] U.S. House of Representatives. Clinger Cohen Act of 1996, Title 40, 1996. Available from: <http://en.wikipedia.org/wiki/Clinger%E2%80%93Cohen_Act>, 2019 (accessed 05.02.19).

[48] P.R. Garvey, R.A. Moynihan, L. Servi, A macro method for measuring economic-benefit returns on cybersecurity investments: the table top approach, Syst. Eng. 16 (3) (2012) 313–328. Available from: https://doi.org/10.1002/sys.

[49] M. Myauo, Expert judgment model to assess cyber-attack scenarios on enterprise architectures, PhD Dissertation, The George Washington University, 2016.

[50] IBM, AI for cybersecurity, 2019. Available from: <https://www.ibm.com/security/artificial-intelligence?cm_mmc = Search_Bing-_-Security_Security + Brand + and + Outcomes-_-WW_NA-_-ai%20and%20cybersecurity_e&cm_mmca1 = 000034XK&cm_mmca210009814&cm_mmca7 = 90817&cm_mmca8-= kwd-81226518831976:loc-190&cm_mmca9 = _k_{gclid}_k_&cm_mmca10 = {creative}&cm_mmca11 = e&msclkid = 7fb6b6223efa1b46d988eac68fadccef&utm_source = bing&utm_medium-= cpc&utm_campaign = Search%7CGeneric%20-%20CISO%20%20Artificial%20Intelligence%20LP%7C000034XK%7CWW%7CNA%7CEN%7CExact%7C10009814%7CNULL&utm_term = ai%20and%20cybersecurity&utm_content = AI_Cybersecurity_Exact>, 2019 (accessed 21.05.19).

[51] Forbes, The 7 biggest cybersecurity threats in an IoT world, 2019. Available from: <https://www.forbes.com/sites/crowe/2019/03/26/the-7-biggest-cybersecurity-threats-in-an-iot-world/#5c0bdb81648a>, 2019 (accessed 15.05.19).

[52] Deloitte, Anticipate, sense, and respond: connected government and the Internet of Things, The Internet of Things in government, 2015. Available from: <https://www2.deloitte.com/insights/us/en/focus/internet-of-things/iot-in-government.html>, 2019 (accessed 18.05.19).

[53] Microsoft, Azure IoT security, 2019. Available from: <https://azure.microsoft.com/en-us/overview/iot/security/>, 2019 (accessed 21.05.19).

[54] National Initiative for Cybersecurity Careers and Studies, National Centers of Academic Excellence (CAE), 2019. Available from: <https://niccs.us-cert.gov/formal-education/national-centers-academic-excellence-cae>, 2019 (accessed 19.05.19).

[55] National Initiative for Cybersecurity Careers and Studies, National Initiative for Cybersecurity Education Cybersecurity Workforce Framework, 2019. Available from: <https://niccs.us-cert.gov/workforce-development/cyber-security-workforce-framework>, 2019 (accessed 21.03.19).

[56] Microsoft, Strengthen your security posture with Microsoft Azure, 2018. Available from: <https://azure.microsoft.com/en-us/overview/security/>, 2019 (accessed 18.03.19).

[57] Forbes, Microsoft CEO Satya Nadella on the extraordinary potential of AI, 2018. Available from: <https://www.forbes.com/sites/bobevans1/2018/06/04/microsoft-ceo-satya-nadella-on-the-extraordinary-potential-of-ai/#43a486d1162f>, 2019 (accessed 20.05.19).

CHAPTER

ROLE OF ARTIFICIAL INTELLIGENCE AND THE INTERNET OF THINGS IN AGRICULTURE 16

Garima Singh[1], Anamika Singh[2] and Gurjit Kaur[1]

[1]*Department of Electronics and Communication Engineering, Delhi Technological University, Delhi, India*

[2]*Gautam Buddha University, Greater Noida, India*

16.1 INTRODUCTION

There will be a tremendous rise in the world population, attaining 8.5 billion by 2030 and 9.7 billion in 2050. As the population increases there is a requirement of a new method of production in agriculture. As the agriculture-based industry has come with the evolution of humanity from prehistoric times to modern days and has faithfully fulfilled the most basic need, that is, food supply. Its core mission remains the same, but now it is integrated with more complicated than ever procedure driven by multiple sociological, economic, and environmental forces. The agriculture industry has a business of $5 trillion, which represents 10% of global consumer spending, 40% of employment, and 30% of greenhouse gas emissions, which is changing rapidly over the past years to continue its pace with the world's evolution. The digital technological advancements are helping this industry to enhance its food production capacity to add value in the entire farm-to-fork supply chain and facilitating with efficient use of natural resources.

The two most popular digital technologies, that is, artificial intelligence (AI) and Internet of Things (IoT) are playing a crucial role in current and future production of agriculture and turning its as smart agriculture industry. AI and IoT connectivity is rapidly changing field applications in agriculture. With the help of AI and IoT applications in agriculture, farmers can continuously monitor their resources, water level in fields, fuel, and feed tanks that helps in improving the crop health and agriculture asset, reducing the time and labor costs. AI and IoT collectively form applications that help farmers in monitoring soil and water and managing the remote crop disease and pesticides. These technologies are making farmers aware of cattle fertility, are secure from cattle theft, and are well versed if the cattle walked or are well fed. Nowadays many companies are investing in building ingenious IoT agricultural products to optimize farm productivity [1].

For a glorious comeback of the agriculture sector in India, there is a need for creating a blueprint on integrated technology for agriculture. Through an appropriate AI and IoT application strategy, an ecosystem of enabling players, supporting government schemes to adopt smart farming technologies systematically and structurally and endorsement for implementation in progressive states, agriculture can make its comeback as a main contributor to the Indian GDP again. Smart and improved farming with digital transformation should be a focus for India today.

Artificial Intelligence to Solve Pervasive Internet of Things Issues. DOI: https://doi.org/10.1016/B978-0-12-818576-6.00016-2

16.2 ARTIFICIAL INTELLIGENCE IN AGRICULTURE

AI can be applied cross disciplinary and is bringing a revolutionary shift in the vision of the agriculture sector. Al has enabled the farmers to yield better with fewer resources with enhanced quality of production and guarantee a faster reach of crops to market [2]. The major drivers to provide IT solutions in all areas of agriculture are a technological advancement in AI and IoT. Therefore it is quite possible to use a digital technology-based solution assisted with AI to upgrade the habitation of the trodden community of farmers and gives them chance for new business entrepreneurs and uplifting the agriculture industry as the main contributor of the global economy [3]. There are many components of AI, explained in this chapter that is feasible and has potential applications in agriculture like:

- Boosting the crop yield through AI

The AI technologies are helping the farmers in their decision making that which seed and in which condition it will give a better result. Also gives the information about weather condition for the highest return [4].

- Identifying the bug hunters with AI

Companies like Rentokil are using AI technologies to kill bugs and vermin. Other companies like Accenture are also making use of android in agriculture by developing apps like PestID to find bugs. These apps used to take pictures of bugs and provide immediate solution along with the chemical information, which is to be used.

16.3 COMPONENTS OF ARTIFICIAL INTELLIGENCE REQUIRED IN AGRICULTURE

In this section, the components of AI that are very much familiar in agriculture are explained.

16.3.1 DECISION SUPPORT SYSTEM

Decision support system (DSS) is software-driven systems, which are used to collect and analyze data from various sources. DSS is mainly used to make a decision process effortless for planning, operations, management process, and also recommends the optimal solution path. DSS is used as a tool for diagnosing, assessing the risk, and also to solve complex crop production issues, in the agriculture sector [5]. Water, genetic, energy, and climate-based calculations and agronomic models are used in DSS along with human and economic inputs.

The agriculture industry is facing problems with irrigation water like a limitation on the volume of water withdrawal. DSS provides an answer to optimize agricultural water and irrigation problems. These new irrigation technologies are reducing the use of water in agriculture without compromising productivity and producing the same with a lesser amount of water [6].

For example, the software is developed in Irstea-based research institute in Montpelier named "Pilot," which is used to estimate the future productivity of crops along with irrigation and climate conditions. Through these types of DSS, models operators can prepare an irrigation schedule without requiring multiple parameters making it an important part of sustainable agriculture. Companies are trying their best to make DSS more ergonomic and user-friendly as they have already started mobile interfaces on smartphones and tablets, to make it easy to handle and can be used in real time, directly on the field [7]. These developments in DSS will allow making it user-friendly with farmers and will revolutionize the agriculture.

16.3.2 EXPERT SYSTEM USED IN AGRICULTURE

Agricultural production has progressed into a complicated business that requires the gathering and assimilation of information with knowledge from miscellaneous sources. For that reason, prevailing farmers often feel the need for agriculture specialists to remain competitive, which will provide the information for decision making. Regrettably, expert assistance in agriculture is not always accessible [8]. To alleviate this accessibility problem, expert systems have been acknowledged as a very influential technique having widespread potential in agriculture.

16.3.2.1 Characteristics of agricultural expert system

- Expert systems pretend the human reasoning around the problem area, despite pretending the area itself.
- Expert systems can solve problems through empirical or estimated methods.

The very first agriculture expert system was developed for detecting the Soybean disease problems having an exclusive feature of using two forms of decision rules and that are [9]:

1. The rules expressing experts analytic knowledge
2. The rules achieved via inductive learning commencing numerous hundred cases of disease like:

- *Diagnosis of the disorder*: This type of system is used to diagnose the disorder with plantation and provides the information to the user [10].
- *Treatment for the disorder*: After identifying the problem this system provides a cure for the diagnosed problem to the user [11].
- *Scheduling the irrigation*: This system is designed to provide the irrigation schedule mainly for plastic tunnels keeping the account for water quality and quantity followed by plant intensity and drainage efficiency [12].
- *Scheduling the fertilization*: This system was mainly designed for the fertilization of cucumber plantation. This includes fertilization requirements, that is, type of fertilization with the required quantity and the gap required between the applications [13].
- *Plant care subsystem*: This system is used to gather the information and schedules the operation required to protect the plantation from any type of expected disorder.

Some well-established examples of expert systems are given here [14].

16.3.2.1.1 CALEX expert system

The University of California has developed a blackboard-based integrated expert DSS for the management of agriculture. This expert system is used by pest control advisors, growers, consultants, and other managers.

16.3.2.1.2 MANAGE expert system

This expert system was developed to detect and prevent the disease mainly in rice crop plantation through developing a prototype, taking into consideration a few major pests and diseases and some deficiency problems limiting rice yield. This system is commonly called a rice doctor expert system.

16.3.2.1.3 LEY expert systems

Washington State University the National Weather Service and the US Bureau of Reclamation has developed this RF-telemetry based expert system. This is a computer-controlled, remote, automated, real-time weather data acquisition, and reporting system. Although this data is collected on an hourly basis, processed, and then transmitted to the NWS. During the spring season, fruit growers can protect their crops from frost with the help of this hourly monitoring expert system. This type of real-time weather monitoring and update is also important in applications like pest management, crop protection, and irrigation scheduling.

16.3.3 FORECASTING

Agriculture is full of uncertainties; there is a requirement of foresighted and informed planning forecasting system, which should be reliable and long range. Now agriculture has become a high input and cost-intensive affair that requires the judicious use of fertilizers and plant protection measures to make agriculture as profitable as it was earlier. Forecasting of different agricultural technological features is becoming very essential now because of the new pests and diseases, which are evolving as an additional threat to production. TF (technology forecasting) can also map the technological scenarios of agriculture for the future. These scenarios also facilitate the decision making power regarding the priority setting of the key agricultural component. Thus TF can make us ready for meeting emerging trends based on market pull and technology push. Assessment of food requirements in the long-term as per the future demand is of utmost importance. TF may pave the way for corrective measures for population growth versus production growth imbalance. Foresight and forecasts of the technological needs as per the emerging scenarios will enhance the sustainability and growth of agricultural systems [15].

16.4 ROLE OF MACHINE LEARNING IN AGRICULTURE

AI can be used in agriculture fields in numerous ways like machine learning (ML) is one of the biggest examples of AI in agriculture, which is used to enhance the productivity and quality of the crops without compromising with the environment. Sensors are used in modern agriculture to collect the data for understanding the environmental factors like soil, crop and weather conditions along with agricultural machines [16]. Farmers use this data to take quick results in concerning decisions. For better yielding ML with AI is applied to agriculture data by farmers. There are several agricultural applications, for example:

- Robot for agriculture
  Robots are getting programmed and developed by many companies to handle the important tasks of agriculture. The best example of ML in agriculture is using robots of crop harvesting as they work faster than human labor.
- Soil and crop monitoring

Companies are collecting data with the help of technologies and deep learning algorithms through drones. Farmers are using software to even control the fertility of the soil.

With the help of these prevailing digital technologies, farmers can save their crops and maximizes their yield. Many companies are working to design sensors, robots, and automation tools for farming. The successful examples of this automation are agricultural spray robots that see and spray which is developed by Blue River Technology. Spraying through robots is very precise which reduces the herbicide expenditures [17].

The use of AI and ML in agriculture is described in Fig. 16.1 to make it grow at an exponential rate in the coming years.

- Real-life ML example

For better understanding, the agricultural industries, companies like one in Mexico named Descartes Labs have combined technologies like ML, satellite images, sensors, and cloud computing. Many new are also getting used in agriculture by companies to identify the place and health of

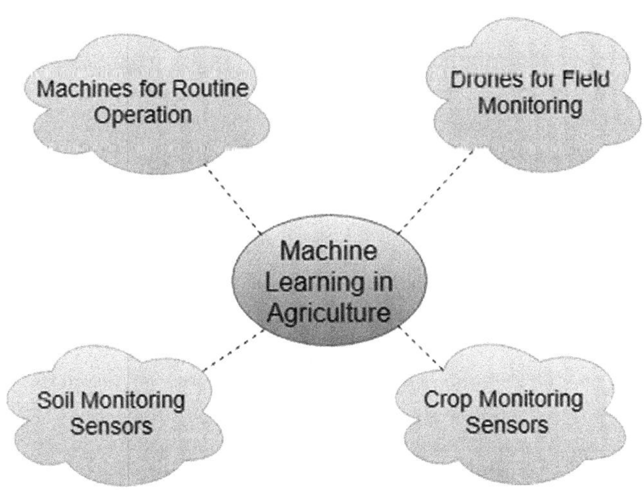

FIGURE 16.1

Machine learning in agriculture.

crops [18]. There has been a constant effort to restructure the agriculture area with the help of these types of technological initiatives.

The above given are examples of AI in agriculture that are presently getting used by farmers at a global level. The next section will explain the role of IoT in agriculture, another important prevailing technology for agriculture.

16.5 MODELS FOR FARMERS SERVICES

The following are the AI-based service models that are developed for farmers.

Chatbot: Along with agriculture these AI-based chatbots are very much famous in travel, retail, insurance, and media sector. This a virtual assistant used by farmers to get answer their queries on some specific problems in their local language. This is an ML-based application, which uses both reinforced and supervised techniques for uninterrupted and context-sensitive learning. Thus chatbot satisfies the generic problems of the farmers before any human operator intervention for any queries that are unique in nature [19].

Agri-E-Calculator: This is a smart application used by farmers to identify the most affordable and appropriate crop depending upon several factors. As the farmer selects the crop then all the required factors along with estimated results will be automatically calculated by Agri-E-Calculator. This app will provide all the data about the selected crop like the amount of water, fertilizers, seeds, and cultivation tools, labor cost will be required. It also shows the crop life cycle chart with the yield and the market scenario at the time of harvest, which gives farmers information about the profit they can buy the chosen crop. This app takes all this input data from a database of farmers and other information sources, and then processes it with ML and generates the feasibility study with estimation to help farmers in choosing their desired crop for farming.

Services for crop care: This application guides the farmers from sowing to the harvesting. This uses IoT sensors to collect this complex structured data sampled and AI is used to analyze that. PID (Proportional Integral & Differential)-based controlled mechanism is used to derive the corrective actions after the data analysis. Consequently, the corrective actions will be sent on the farmer's smartphones to prioritize the action based on severity and urgency to act upon.

Market guidance and price prediction: This service is designed to aware and safeguard the farmers from mitigation risk of price loss and market fluctuation. Based on gathered statistical data demand and predictive price information will be communicated to the farmers in the crop life cycle, which in turn helps farmers to plan the release of their yield in the market accordingly.

Insurance service and crop loan: This service is designed to facilitate the farmers with crop loans, helps in processing the loan. It also shares information like loan eligibility, limit for the proposed crop. The main advantage of this service is crop insurance to reduce the risk of crop failure due to any uncertainties or calamities.

16.6 INTERNET OF THINGS APPLICATIONS IN AGRICULTURE

The global population is growing rapidly and will reach 9.6 billion by 2050. So, possible technological advancements in agriculture are required to provide food to this growing population. Technologies like AI and IoT are best to support agricultural growth. Because to meet this increasing food demand there is a need to overcome challenges like weather and environmental conditions, climate changes, AI and IoT are the best technologies to support this. IoT in agriculture will enhance productivity and sustainability [20] (Fig. 16.2).

Perhaps one of the most interesting IoT applications in agriculture is monitoring and tracking of the cattle. Movement of every cattle can be monitored with network-connected collars and being mindful of their exact location will also prevent thefts. Moreover, fertility tracking can be accounted for optimized breeding opportunities and eating behavior and health activity can be monitored to record health issues [21].

Another interesting implementation for IoT is preventative maintenance of assets. Machinery, like pumps, generators, farm machines, and so on can now be equipped with embedded IoT sensors. The sensors alert the farmers in real time for potential failures. This minimizes unexpected costs and machine downtime due to damages caused because of neglected failures. In collaboration, all these technologies enhance the accuracy of the decisions and increase the productivity of the crops [22].

While agriculture is mostly overlooked as a consumer of IoT in India and across the globe, it has a lot more potential and stands as one of the sectors that can incur the most out of IoT. With 5G getting launched soon, the government's announcing budget allocation and investment in transforming villages into digital villages, the Indian agriculture sector stands to gain the most with the use of technologies [23].

The data about the seeds, soil, crops, livestock, farm equipment, costs and fertilizers, and water usage can be gathered by agricultural drones or sensors. With the help of IoT and advanced analytics, farmers can analyze the real-time information of temperature, weather, moisture, GPS signal, and prices as shown in Fig. 16.3. These technologies also provide insights to optimize and enhance the yield, enrich the farm planning, plan smarter decisions regarding the requirement of resource levels, when and where to distribute them to prevent waste [24].

FIGURE 16.2

Role of Internet of Things in agriculture.

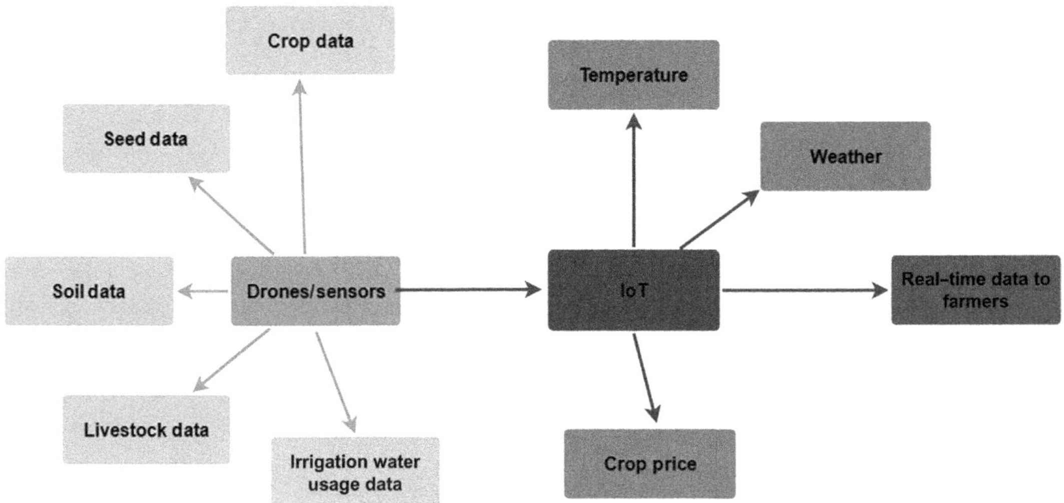

FIGURE 16.3

IoT framework in agriculture. *IoT*, Internet of Things.

16.6.1 PRECISION FARMING

The most prominent use of IoT all over the world in the agriculture sector is precision agriculture [25]. The IoT systems offer agricultural-based services like virtual optimizer PRO, soil moisture probes, variable rate irrigation (VRI) optimization, and so on. VRI is an optimization process used to maximize the productivity of irrigated crop fields along with variability in soil, which in turn enhances the yield and improves the water use efficiency. The precision farming concept plays an enormous role in their business. Gathered real-time data provides better visibility to the farmer, and the knowledge of a specific territory can be easily shared with the community [26]. Because of this, precision crop farming systems should include skills to manage the devices, security, and storage of data, and profound analysis of data coming from aerial imagery, sensors placed at fields and from remote sensing units. It will create an immediate understanding of data for the farmers and support the agro-scientists with better decision making [27].

16.6.2 DRONE-BASED TECHNOLOGY

Drones have found their maximum use in the agriculture sector, addressing the major challenges of farming and given a high-tech makeover to agriculture through real-time collection and processing of data, which helps in strategically planning for the required action.

The very important advantages of drones use in agriculture is crop health imaging in very less time along with integrated GIS mapping, which in turn enhances the capability of yield [28].

Drones give the freedom to farmers to choose the field they want to survey and can select the altitude or ground resolution from which they want data of the fields. With the help of this drone data, farmers will be able to draw useful insights on several issues like counting of plants, predicting the yield, indicates the health of the plant, height measurement of plant, nitrogen content in wheat, canopy cover mapping, and drainage mapping as shown in Fig. 16.4. The images and data collected by drones are based on thermal, multispectral, and visual conditions [29].

The following are the six ways in which drones will be used throughout the crop cycle:

1. *Field and soil study*: 3D maps are given by drones for early soil analysis. This will help in organizing the planting of seed and also collects information to precisely manage the nitrogen level and irrigation factor.
2. *Planting*: A lot of startups are working in this area and developing the drones to plant to seed, which can decrease the cost of planting to 85%. These drones hit the pods by seeds and fertilizers required for the crop.
3. *Spraying of crop*: Drones are used to scan the ground and evenly do the aerial spray on crops, which is almost five times better and faster than the traditional style.
4. *Monitoring of crop*: Drones have left them with no space for inefficient crop monitoring as they provide time-series animations. This is used in keeping track of the development of crops and tells about the inefficiencies of production, which results in better management.
5. *Irrigation*: Drones with the sensor can easily diagnose the dry portion of a field and helps in providing the necessary action.
6. *Health assessment*: By checking a crop with the help of both near-infrared light and visible, devices carried by the drone can help in keeping changes in crops and alert about their health to farmers regarding the disease if, any found. The biggest hurdle with unmanned elevated vehicles (UAVs) in becoming a reality is the use of sensors that are capable of gathering high-quality data and also the software that is required in making that high-tech dream a reality [30].

To beat the farming costs, UAVs are used to improve spatial inclusion and set up exact maps. In areas, where there are limitations on the utilization of drones—including government guidelines, low battery life, and mind-boggling expenses—tethered eye helium inflatables are utilized. These aerial sensors are used to create continuous images about the farm conditions, which are utilized to

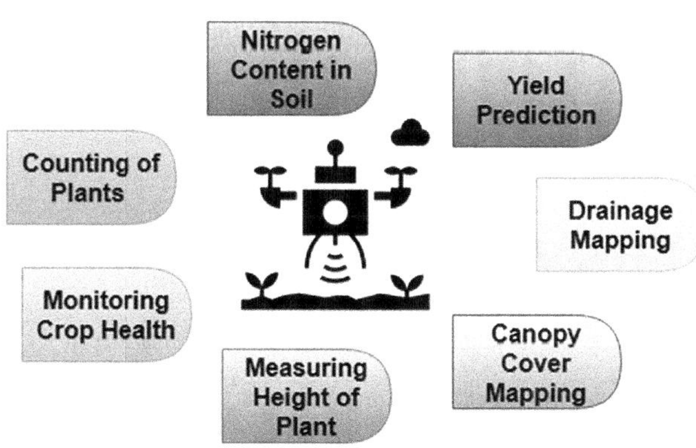

FIGURE 16.4

Drone-based technology.

further improve the information gathered by sensors on the ground. Subsequently, this methodology lessens equipment costs while encouraging the job of IoT in farming [31].

16.6.3 MONITORING OF LIVESTOCK

Through IoT applications, farmers can gather information about the location, health, and well-being of their cattle. They can use this information to classify the well-being of their livestock. For example, if there is any animal sick that can be separated from the rest of the animal to prevent the further spread of the disease in other cattle also. These IoT-based sensors provide feasibility to ranchers to locate their cattle and help in bringing the down labor costs by a substantial amount [32].

One case of an IoT framework being used by an organization is JMB North America, which is an association that gives cow monitoring solutions for cattle producers. Out of the numerous arrangements gave one of the arrangements is to help the dairy cattle proprietors watch their cows that are pregnant and going to deliver an offspring. A battery controlled sensor is placed in each livestock which provides the exact data to the farmers, making they are clear for making the decision [33].

16.6.4 SMART GREENHOUSES

Greenhouse farming is a procedure that improves the yield of harvests, vegetables, organic products, and so forth. Greenhouses control natural parameters in two different ways; either through manual intercession or a corresponding control system. Nonetheless, since manual intercession has hindrances, for example, production loss, labor cost, and energy loss; therefore these techniques are less successful. A smart greenhouse through IoT-embedded system screens insightfully as well as controls the atmosphere. In this way eliminating any requirement for human intervention [34].

The smart greenhouse uses various sensors to assess the environmental parameters according to the plant requirement that is used for controlling the environment. At that point, a cloud server is created to access the remotely available systems connected to the IoT. Inside the greenhouse, the cloud server helps in the handling of information and applies a control activity. This plan gives the ideal and financially savvy answers for the ranchers with negligible and no manual intervention [35].

16.7 CURRENTLY USED ARTIFICIAL INTELLIGENCE AND INTERNET OF THINGS TECHNOLOGIES IN AGRICULTURE

Presently, many industries are working in developing the AI- and IoT-based machinery for farming at a global level. However, in many countries, they are getting used also.

Blue river technology: This California-based startup founded in 2011 that uses computer vision, AI, and robotics to form next-generation equipment for agriculture, which will decrease the requirement of the chemical and saves costs. Computer vision is used to recognize every plant at the individual level, chooses the way to treat every specific plant and then robotics empowers the smart machines to do the necessary. These sensors are used to detect the type of weed with its correct herbicide so that the farmer can apply that on the crop incorrect amount. At that point, the correct herbicides are splashes exactly as per the encroachment area. There has been an robot developed by Blue river technology named See & Spray, which uses computer vision technology to scan and accurately spray weeds on cotton plants.

FarmBot: This Company was founded in 2011 with a vision to promote precision farming by empowering those people who are conscious of the environment. This FarmBot product falls under $4000 price and makes the owner capable of doing end-to-end farming all by himself. This bot can take care of everything required in farming from plantation of seed to detection of weed, also capable to test the soil for irrigation purposes through some open-source software system.

Harvest CROO Robotics: Company has developed a robot, especially for strawberry farming to pick and pack the strawberry crop reducing the harvest labor cost for farmers. Lack of laborers has reportedly led to millions of dollars of revenue losses in key farming regions like California and Arizona. As strawberries need to be picked within a certain period, therefore, qualifies pickers are required and so as robot. Harvests CROO Robotics organization has confidence that its invention is will help in increasing the yield, saves, and reduces energy usage and further enhances the quality.

Plant diseases diagnosis app (PEAT): PEAT is a Berlin-based agricultural startup. PEAT has developed an application called Plantix that can diagnose the possible nutrient deficiencies and defects of the soil. This is an image-based app that requires images to diagnose the diseases of plants. The smartphone is used to gather the image and then matches them with the server image to finally conclude about the health of the plant. This application uses AI and ML to solve plant diseases.

Prospera: This is an Israeli startup founded in 2014 that has revolutionized the way of farming. They have developed a cloud-based reservoir that will collect all the previously available farming data from farming sensors along with the aerial images and so on. This reservoir will be connected with the field devices to make use of this data. The Prospera device has a variety of sensors and is empowered with computer vision technology that makes it flexible to be used in fields or greenhouses. These varieties of sensors are used to take input and then find a correlation with different data labels to make predictions.

FarmBeats: Internet connectivity is a well-known problem with farmers; therefore this is a cloud computing centered system. Although Internet connectivity at farmer house is not that much fast that they can access the Big Data sets to the cloud for analysis. In addition to that, the interference sources available on the farms may disturb the connection to the cloud. FarmBeats is designed mainly for the purpose to overcome the connection disturbance between the cloud and the farm in various ways. These ways are unique IoT gateway, offline capabilities, component migration, and deep learning via edge computing. This unique feature of end-to-end IoT connectivity makes FarmBeats capable enough to provide numerous agricultural services like detection of pH, precision irrigation, yield prediction, and forecasting of microclimate, despite disturbing connection cloud computing systems. FarmBeats make farmers available with all the advantages of farming with data and can help farmers realize all of the benefits of farming with data and overpowers the implementation challenges of IoT in rural areas.

16.8 CHALLENGES WITH ARTIFICIAL INTELLIGENCE AND INTERNET OF THINGS IN AGRICULTURE

Even though AI offers tremendous opportunities and has open doors for agricultural-based application, there still exists an absence of ML solutions with cutting-edge AI arrangements in farms crosswise over most pieces of the world. Presentation of farming to external components like climate conditions, soil conditions, and the presence of bugs is in a considerable amount. So what may resemble a feasible solution while designing during the beginning of harvesting may not be an ideal one due to changes in external parameters.

A huge collection of data is required by AI systems to train their machines for a viable prediction like for huge agricultural land spatial data can be easily collected while temporal data is hard to collect. Like data specific to the crop can only be collected once in a year at the time of their yield. Then this collected will require substantial time to get mature and then develop a strong ML model. Because of this reason, AI has a lot of applications in agronomic products like fertilizers, pesticides, seeds, and so on despite in-field precision solutions [36].

Most farms are located in remote locations where internet networks probably will not be sufficiently able to encourage quick transmission speeds. Besides, communication lines might be hindered by canopies, crops, and other physical barriers. These are the factors responsible for high data transmission cost along with the slow uptake of IoT in agriculture. With the beginning of Big Data, these expenses could develop exponentially.

Currently, farmers depend upon meagerly distributed sensor networks to collect information on farm conditions, which has cost constraint as they are expensive, a sensor set cost up to 8000 USD. Therefore farmers keep depending on less propelled farming technologies, which confine their profitability.

16.9 CONCLUSION

AI and IoT technologies give farmers the facility to analyze the soil, land, and health of the crop in minimum time to get maximum yield in every season and also predict the next season weather conditions so that farmers can decide about the crop. This, in turn, reduces the labor requirement, which is the main concern in the agriculture sector. AI- and IoT-based applications help farmers by suggesting them the most suitable pesticide and the right time of sowing and harvesting before large-scale incidence of disease. Many areas of agriculture are still untouched by automatic response systems; therefore the doors of the agriculture industry are open to using upcoming technology for helping farmers by providing solutions to their queries with important guidance and suggestion. This will result in AI growth in the agricultural market. A collaborative approach is a key to success with farmers. Although, to apply AI and IoT at its full potential in the agriculture sector, there is a need to cater the issues like awareness, connectivity, and new technology fear. These issues need to be effectively managed at the farm level.

REFERENCES

[1] G. Singh, G. Kaur, Role of communication technologies for smart applications in IoT, in: Handbook of Research on Big Data and the IoT, IGI Global, 2019, pp. 300–313.

[2] H. Murase, Artificial intelligence in agriculture, Comput. Electron. Agric. 29 (1/2) (2000).

[3] C. Popa, Adoption of artificial intelligence in agriculture, Bullet. UASVM Agric. 68 (1) (2011).

[4] Y. Hashimoto, H. Murase, T. Morimoto, T. Torii, Intelligent systems for agriculture in Japan, IEEE Control. Syst. Mag. 21 (5) (2001) 71–85.

[5] A. Perini, A. Susi, Developing a decision support system for integrated production in agriculture, Environ. Model. Softw. 19 (9) (2004) 821–829.

[6] R.E. Plant, An integrated expert decision support system for agricultural management, Agric. Syst. 29 (1) (1989) 49–66.

[7] J.C.I. Ascough, H.D. Rector, D.L. Hoag, G.S. McMaster, B.C. Vandenberg, M.J. Shaffer, et al., Multicriteria spatial decision support systems for agriculture: overview, applications, and future research directions, in: Integrated Assessment and Decision Support Proceedings of the First Biennial Meeting of the IEMSS, June 2002, pp. 175–180.

[8] J. Liebowitz (Ed.), The Handbook of Applied Expert Systems, CRC Press, 1997.

[9] J.M. McKinion, H.E. Lemmon, Expert systems for agriculture, Comput. Electron. Agric. 1 (1) (1985) 31–40.

[10] G.N.R. Prasad, B.D.A. Vinaya, A study on various expert systems in agriculture, Comput. Sci. Telecommun. 4 (2006) 81–86.

[11] S.S. Abu-Naser, K.A. Kashkash, M. Fayyad, Developing an expert system for plant disease diagnosis, J. Artif. Intell. 3 (4) (2010) 269–276.

[12] A.A. Rafea, S. El-Azhari, I. Ibrahim, S. Edres, M. Mahmoud, E.N. Street, Experience with the development and deployment of expert systems in agriculture, in: IAAI, August 1995, pp. 146–155.

[13] G. Kaur, P. Tomar, P. Singh, Design of cloud-based green IoT architecture for smart cities, in: Internet of Things and Big Data Analytics Toward Next-Generation Intelligence, Springer, Cham, 2018, pp. 315–333.

[14] G.N.R. Prasad, B.D.A. Vinaya, A study on various expert systems in agriculture, Comput. Sci. Telecommun. 4 (2006) 81–86.

[15] P. Tomar, G. Kaur, P. Singh, A prototype of IoT-based real time smart street parking system for smart cities, in: Internet of Things and Big Data Analytics Toward Next-Generation Intelligence, Springer, Cham, 2018, pp. 243–263.

[16] K. Liakos, P. Busato, D. Moshou, S. Pearson, D. Bochtis, Machine learning in agriculture: a review, Sensors 18 (8) (2018) 2674.

[17] R.J. McQueen, S.R. Garner, C.G. Nevill-Manning, I.H. Witten, Applying machine learning to agricultural data, Comput. Electron. Agric. 12 (4) (1995) 275–293.

[18] S. Dimitriadis, C. Goumopoulos, Applying machine learning to extract new knowledge in precision agriculture applications, in: 2008 Panhellenic Conference on Informatics, IEEE, August 2008, pp. 100–104.

[19] P.J. Sahane, A.N. Nawathe, S.U. Kadlag, A novel approach to smart Hagen chatbot system by using data mining algorithm.

[20] Y. Shifeng, F. Chungui, H. Yuanyuan, Z. Shiping, Application of IoT in agriculture, J. Agric. Mechan. Res. 7 (2011) 190–193.

[21] I. Mohanraj, K. Ashokumar, J. Naren, Field monitoring and automation using IoT in the agriculture domain, Procedia Comput. Sci. 93 (2016) 931–939.

[22] N. Dlodlo, J. Kalezhi, The internet of things in agriculture for sustainable rural development, in: 2015 International Conference on Emerging Trends in Networks and Computer Communications (ETNCC), IEEE, May 2015, pp. 13–18.
[23] M. Stočes, J. Vaněk, J. Masner, J. Pavlík, Internet of things (IoT) in agriculture-selected aspects, Agris Online Pap. Econ. Inform. 8 (665-2016-45107) (2016) 83–88.
[24] J. Shenoy, Y. Pingle, IoT in agriculture, in: 2016 Third International Conference on Computing for Sustainable Global Development (INDIACom), IEEE, March 2016, pp. 1456–1458.
[25] S. Blackmore, Precision farming: an introduction, Outlook Agric. 23 (4) (1994) 275–280.
[26] H. Auernhammer, Precision farming—the environmental challenge, Comput. Electron. Agric. 30 (1-3) (2001) 31–43.
[27] A.M. Blackmer, S.E. White, Using precision farming technologies to improve the management of soil and fertilizer nitrogen, Australian J. Agric. Res. 49 (3) (1998) 555–564.
[28] P. Tripicchio, M. Satler, G. Dabisias, E. Ruffaldi, C.A. Avizzano, Towards smart farming and sustainable agriculture with drones, in: 2015 International Conference on Intelligent Environments, IEEE, July 2015, pp. 140–143.
[29] V. Puri, A. Nayyar, L. Raja, Agriculture drones: a modern breakthrough in precision agriculture, J. Stat. Manag. Syst. 20 (4) (2017) 507–518.
[30] B. Vergouw, H. Nagel, G. Bondt, B. Custers, Drone technology: types, payloads, applications, frequency spectrum issues, and future developments, Future Drone Use, TMC Asser Press, The Hague, 2016, pp. 21–45.
[31] P. Tomar, G. Kaur (Eds.), Examining Cloud Computing Technologies Through the Internet of Things, IGI Global, 2017.
[32] A.R. Frost, C.P. Schofield, S.A. Beaulah, T.T. Mottram, J.A. Lines, C.M. Wathes, A review of livestock monitoring and the need for integrated systems, Comput. Electron. Agric. 17 (2) (1997) 139–159.
[33] D. Guice, W. Pugh, N. Thompson, U.S. Patent Application No. 09/853,460, 2002.
[34] R.K. Kodali, V. Jain, S. Karagwal, IoT based smart greenhouse, in: 2016 IEEE Region 10 Humanitarian Technology Conference (R10-HTC), IEEE, December 2016, pp. 1–6.
[35] S.X. Chen, H.T. Ma, X. LIU, Design of a smart greenhouse system, Hebei J. Ind. Sci. Technol. 4 (2011).
[36] R. Khan, S.U. Khan, R. Zaheer, S. Khan, Future Internet: the internet of things architecture, possible applications, and key challenges, in: 2012 10th International Conference on Frontiers of Information Technology, IEEE, December 2012, pp. 257–260.

CHAPTER

INTEGRATING ARTIFICIAL INTELLIGENCE/INTERNET OF THINGS TECHNOLOGIES TO SUPPORT MEDICAL DEVICES AND SYSTEMS

17

Priya Singh[1], Garima Singh[2] and Gurjit Kaur[2]

[1]Indira Gandhi Delhi Technical University for Women, Delhi, India [2]Department of Electronics and Communication Engineering, Delhi Technological University, Delhi, India

17.1 INTRODUCTION—MEDICAL DEVICES AND HEALTHCARE SYSTEMS

Patient association with the medical system many a time required association with devices like dressing, the detector of blood pressure, syringes and detecting kits for pregnancy, pacemakers, artificial joints, instruments for surgery, and MRI and CT scanners. The technology had developed a wide variety of devices to detect, monitor, and treat patients. A major development in wireless technology, nanotechnology is leading to the innovation of medical devices that can produce, collect, test, and transfer information. The information and these devices are called the Internet of Things (IoT)—a joined system of devices, software and healthcare, and facilities. New rules, digital technology, artificial intelligence (AI), automation, and data analytics are some of the challenges and opportunities for medical devices and healthcare systems. New solutions are needed to manage data transferred and collected by digital devices This can help healthcare organizations create methods for better health outputs at affordable cost and easily accessible. Like other sectors, the health and medical device sector is positively gaining the converting nature of IoT technologies, the solution to miniaturization, software applications, and the power need of medical devices. This chapter aims to explain the drawback of conventional medical devices in Section 17.2. Section 17.3 highlights the types of medical devices. Sections 17.4 and 17.5 contribute to the need for IoT/AI in medical devices and technologies used for the same respectively. Further Sections 17.6–17.8 explains the monitoring done using AI/IoT medical devices, challenges encountered for integrating IoT/AI in medical devices, and security issues as the greatest challenge, respectively [1–34].

17.2 PROBLEMS IN CONVENTIONAL MEDICAL SYSTEMS

The conventional healthcare and medical devices have some drawbacks [1–10], which can be categorized as these five groups.

Artificial Intelligence to Solve Pervasive Internet of Things Issues. DOI: https://doi.org/10.1016/B978-0-12-818576-6.00017-4

17.2.1 RURAL POPULATION NEGLECTING

The main concern of healthcare is the neglecting the rural health needs and services. It is effectively a service implemented in urban hospitals. However, there are many rural hospitals and public health center (PHC) still the bias to work in an urban area is present, that is, doctors are not ready and comfortable to practice in a rural location. Thus instead of developing a healthcare enriched with paramedical and equipped devices rural area is neglected and serve with old quality devices and less support of medical staff to monitor.

17.2.2 SOCIAL INEQUALITY

The maintenance of health services is highly disproportional in many countries. Hilly and remote locations of the country are not served properly while in urban areas and cities the facilities are well developed. Poor people face a lot of difficulty in availing services and healthcare.

17.2.3 SHORTAGE OF MEDICAL PERSONNEL

Shortage of medical equipment and paramedic teams like doctors, a nurse, and so on is a basic problem in the health sector. Likewise, the number of dispensaries and health centers is not enough when compared to the large population.

17.2.4 MEDICAL RESEARCH

Research in the medical field needs to be aimed at treatments and drugs for tropical diseases, which are generally neglected by international pharmacy organizations due to fewer profit margins. The research needs to be directed to the right direction covering all possible diseases regardless of the profit and loss and raw should be accessible easily to make the research cheap.

17.2.5 EXPENSIVE HEALTH SERVICE

Healthcare and medical devices are quite expensive. It is hard for nomad people to arrange for the facilities and afford them easily. The cost of many important medical essentials has risen. Thus more focus should be given to the parallel systems of medicine or alternative methods should be designed to make them cheaper.

Concluding the healthcare medical device system consist of many drawbacks. These problems can be solved by the new technology and solutions and integrating them with IoT/AI for managing, monitoring and giving more accessibility to every patient needed on time [35].

17.3 CATEGORIES OF MEDICAL DEVICES

Medical devices can be broadly categorized as three types [11–17] stationary medical devices, implanted medical devices, and wearable medical devices. These categories are explained in the following text.

17.3.1 STATIONARY MEDICAL DEVICES

These types of devices consist of ultrasound or X-ray machines, computed tomography, and magnetic resonance image scanners, mammography machines, and imaging devices. They measure the physical parameters of patients and detect any faults or diseases. These expensive, technical devices, which transfer reports wirelessly to physicians and doctors, are usually used by clinics or other medical organizations. The information is connected with the patient's electronic health record devices. These devices are important to analyze and are more often integrated to other devices for healthcare to store and monitor patient information and responses to help on-time assistance and service providing.

17.3.2 IMPLANTED MEDICAL DEVICES

These types of instruments include Knee exchange and replacement devices, heart pacemakers, hip replacement, and defibrillators that analyze and provide services to cardiac conditions, stimulators of nerve, bladders, diaphragm, and many biosensors to analyze and work on distinct signals. Medically unfit people whenever need nonstop monitoring they are implanted medical devices like these. They are meant to be inside the ling being body and are inserted by surgeries.

17.3.3 EXTERNAL WEARABLE MEDICAL DEVICES

Insulin pumps used in diabetes for monitoring the level, including smartwatches and activity trackers, are an example of these devices. External wearable medical devices are helpful to patients when they are not in hospitals.

Based on research developing applications and needs of medical devices can be categorized as follows:

Chronic disease management: Companies are employing machine learning to access and keep in record patients' condition by sensors to automate the delivery of treatment using connected mobile apps.

Medical imaging: Industries are combining AI-based platforms with medical scanning devices to improve the visual clarity of image and outcomes by decreasing contact to radiation.

AI and IoT: The medical devices systems are integrating AI and IoT to monitor in a better way patient's discipline to the treatment prescribed and to improvise treatment and patient status.

17.4 INTERNET OF THINGS/ARTIFICIAL INTELLIGENCE IN MEDICAL DEVICES AND HEALTHCARE

The major issue that patients and concerned faces, mainly people living in the remote area found is unable to reach doctors and treatment during critical conditions. This leads to dreadful consequences on patients and their family minds for doctors and their services. However, with the advancement of these technologies, by using IoT/AI devices for emergency monitoring system, these issues can be sorted efficiently. IoT can help in keeping patients in continuous monitoring to keep them safe, and improving the care delivery system. Healthcare IoT/AI also increases patient satisfaction by providing easier patient and doctor interaction. The old school healthcare suffered from data management and lost the track easily thus IoT/AI is the best solution for data management and finding the solution for problems. The main reasons identified based on a literature survey to implement an IoT/AI-enabled medical device and healthcare are listed in the following text [18–24].

We already see that technology changing healthcare makes this industry less dependent on humans (and less susceptible to human error as well) and more patient oriented at the same time. The major advantages of the IoT that healthcare organizations can benefit from are as follows.

17.4.1 LOWER COSTS

Using IoT solutions and connected medical devices allows healthcare providers to monitor patients in real time. This means fewer unnecessary visits to the doctor and fewer hospital stays and admissions with the help of information storage and management.

17.4.2 BETTER PATIENT EXPERIENCE

Being connected to the healthcare system through the IoT, patients get more engaged in their treatment, and doctors improve diagnosis accuracy because they consist of all mandatory information of patient required at that time.

17.4.3 BETTER MANAGEMENT OF DRUGS AND MEDICINE ADHERENCE

IoT solutions allow hospital staff to spend less time searching for drugs, track supplies and medicine, and track hygiene practices in hospitals and effectively prevent hospital infections. Healthcare IoT monitoring solutions help people connect treatment paths and advisors to trace agreement to ordinance and cures provided.

17.4.4 REDUCED ERRORS AND WASTE

Using IoT for data collection and workflow automation is an excellent way to cut down on waste (like unnecessary tests and expensive imaging), reduce arrangement expenses, and minimize errors.

17.4.5 IMPROVED OUTCOMES OF TREATMENT

Healthcare solutions that are connected through cloud computing and use big data can provide caregivers with the ability to access real-time data. This data can be utilized to make the right decisions and to give verified treatments.

This usage of the IoT/AI in medical care is a large ecosystem as shown in Fig. 17.1 is explained as follows:

- Original equipment manufacturer, that is, manufacturer of medical devices and devices
- Medical device provider, that is, supply and marketing team

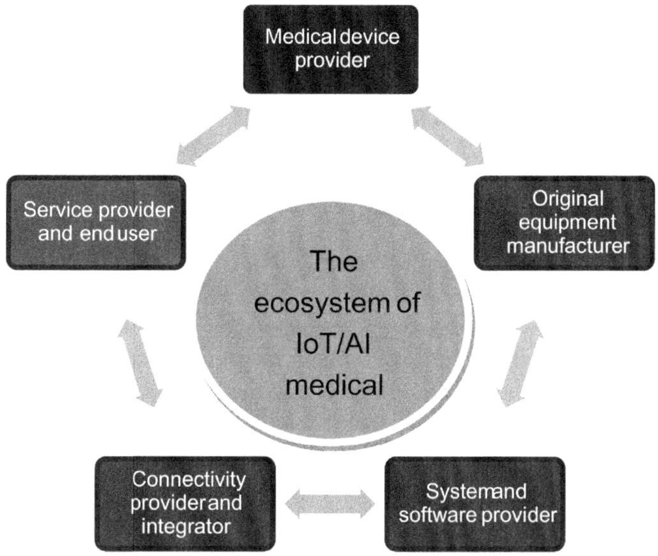

FIGURE 17.1

The ecosystem of IoT/AI medical system. *IoT*, Internet of Things; *AI*, artificial intelligence.

- System and software provider, that is, a team of programmers to encode the medical requirements
- Connectivity provider and integrator, that is, proper platform are to be designed to integrate the medical device and operating system and connect it with the communicating system to communicate, restoring the system to collect data and monitor it
- Service provider and end user, that is, hospitals and the team of paramedics to analyze the device and understand its proper usage and supply to patients on time with proper guidance

17.5 ENABLING TECHNOLOGIES OF INTERNET OF THINGS IN MEDICAL DEVICES AND SYSTEMS

Recently, communication technology and sensors and converters like transimpedance amplifier and software that integrates with these have emerged as a versatile solution in improving the medical devices and system. Main technologies enabling this development can be categorized as follows [25–30].

17.5.1 IDENTIFICATION TECHNOLOGY

An active IoT many nodes, an allotted node is authorized to generate data and access data, unaffected by its location. To make this possible it is important to effectively analyze and detect the nodes. The process of identification provides a unique identification number (UID) to every node so that unambiguous information can be exchanged by these the node. Every single resource provided a digital UID in a system like a doctor, a nurse, or any other staff is given a UID. This creates a corelation between different resources digitally. It shows the location of the resources available in the reach with the shortest transit time, this reduces the reaction time [36].

17.5.2 COMMUNICATION TECHNOLOGIES

In the IoT system, the network infrastructure is decided by communication technology. IoT networks worked on standards and transmission rates for transferring data, heterogeneous frequencies. This is further divided into two categories as long-distance technology and short-distance technology. Long-distance technology is developed to work on communication, which is the regular mode, for example, mobile phones, the Internet, and so on. Short-distance communication generally uses wireless technology, for example, Infrared Data Association, Bluetooth technology, WiFi, radio frequency identification (RFID), and so on. These type of technologies utilizes data transfer for a short distance. The characteristics of these technologies are altered for transmission rates, installation cost, distance, power consumption, the number of entities, maintenance cost, and so on based on different radio frequencies used and standards for security [37,38].

17.5.3 LOCATION TECHNOLOGY

Real-time location systems (RTLS) are used in modern location technology to locate objects. The global positioning system (GPS) is the main RTLS. This navigation system incorporating satellite dependency is used to locate objects in any complex weather condition. The GPS can help for healthcare and medical devices by locating patients, doctors, and medical facilities available nearby and so on. However, GPS or Beidou system, does not work properly for indoors because of the structural construction, this creates difficulty for the satellites to transmit the transmission signals. So it is important to design a local positioning system (LPS) that can replace the GPS and efficiently locate the indoor need with accuracy. This LPS can measure the radio signals traveling between an object and an array of receivers to locate it.

17.5.4 SENSING TECHNOLOGIES

Sensors are the most important part of IoT/AI-based system, because they are the main devices performing monitor actions in many critical conditions, collecting the data, and so on. With time there are advancements and development in sensing technologies this helps to get data continuously from patients' treatment or objects. It helps a doctor or physician to monitor the patient in critical condition like heart rate, blood oxygen saturation, and can provide emergency services. There are various types of sensors like temperature, pressure, water quality, and smoke sensors these sensors convert the nonelectrical signal to electrical signals and are then processed further by electronic devices like transimpedance amplifiers, low pass filters, decoders, and so on. All the received signals are converted into a digital signal and immediately transmitted over the network.

17.5.5 CLOUD COMPUTING

A huge amount of data generated during its operation are stored, processed, and shared. Cloud computing performs this function for IoT architecture for storage, processing critical data generated through each sensor and device. The data centers collect it and transfer it to networks. The cloud data centers can increase or decrease the capacity of computing, according to requirements. This helps in the constant monitoring of persons with disabilities for immediate services [40].

17.6 MONITORING USING INTERNET OF THINGS/ARTIFICIAL INTELLIGENCE IN MEDICAL DEVICES AND SYSTEM

An IoT/AI-based medical devices and system consists of three main layers in its architecture: information monitoring, network transmission, and application layer. The information monitoring layer mainly contains sensors. The sensors help in continuous monitoring of patients' conditioning. These collected data from sensors are transmitted with the networks, stored for future references in centers controlled by cloud computing. WiFi and any other wireless technology are applied for data transfer. The service layer of application involves the IoT provided in a hospital or clinic, providing a remote medical service there are various monitoring devices in medical devices that are of great importance for maintaining the IoT/AI-based medical devices and systems [39].

17.6.1 GLUCOSE-LEVEL MONITORING

Diabetes is one of the common age problems in old people and it is needed to continuously monitor the glucose levels of these people for providing on-time medication. With the help of the IoT system detecting glucose levels nonstop is possible in a noninvasive way. The person who is sick wears a wearable sensor that tracks their health parameters to know the glucose level; collected data is transmitted through IPV6—internet protocol network to concerned medical facility providers. The wearable tracking device contains a collector of glucose in the blood, a mobile phone and an acquisition detector in medical based on IoT to glucose level monitor.

17.6.2 ELECTROCARDIOGRAM MONITORING

In ECG, the system tracks heart rate and rhythm, multifaceted arrhythmias, myocardial ischemia, and prolonged QT intervals by monitoring and converting it into electrical signals. The ECG monitor has a wireless transmitter and a receiver. This identifies the abnormal heart activity and transfers it to a clinic or hospital concerned through a network. Various other algorithms are designed for continuous monitoring through IoT.

17.6.3 BLOOD PRESSURE MONITORING

A wearable device is designed to monitor the patient's BP. The device has a blood pressure measuring facilities, which are communicated through networks. These devices can also have an LCD for visual clarity of meter reading of BP measured.

17.6.4 TEMPERATURE MONITORING OF BODY

The variation in the degree of body temperature is used to detect homeostasis; this is an important part of medical services. It generally consists of, a sensor embedded to measure and record body temperature. The whole system contains a home gateway to monitor body this gateway uses an IR detection system to transmit the recorded data.

17.6.5 WHEELCHAIR MANAGEMENT

Various researches are done to develop smart wheelchairs with IoT/AI applications for a person with disabilities. The wheelchair consists of various technologies like wireless body area networks. It utilizes this system to detect the vibrations in the wheelchair. It tracks the situation of the patient on the electronic chair by detecting the position and surroundings and gives the command accordingly to react to movement.

17.6.6 HEART RATE

IoT is also integrated into smartphones to detect the daily heart rate of the user. Nowadays every electronic medical device can be linked to smartphones using application software. Several medical-related hardwires have been embedded and applications software have been designed for smartphones. The sensors that detect without contacting, helps to work on algorithms for image analysis in medical devices applications. A smartphone capable of detecting and recording diseases nowadays like asthma and so on.

17.7 CRITICAL ISSUES AND CHALLENGES OF INTERNET OF THINGS IN MEDICAL DEVICES AND SYSTEMS

The IoT and sensors used for medical devices is the backbone of future medical arrangements to develop an efficient system. People are adding it in their daily life to keep track of their daily health conditions [31–34]. The medical IoT systems face the following major challenges.

17.7.1 SECURITY

Nowadays people are using wearable medical devices to monitor and track their medical parameters. In this situation measures for data prevention and misuse, prevention is of main concern. Different parameters for example integrity, confidentiality, and availability of users' personal information should be protected in an IoT system. The IoT-embedded medical system must be integrated with an efficient security system.

17.7.2 MOBILITY

The medical devices should be mobile so that it does not constraint the patients or devices.

Sometimes, the selection of a proper network becomes an issue. Mobility management depends on the type of network for which it is used. These networks are categorized into three categories: service based, data based, and patient based. The data-based network divides the medical device system centered on health data. The service-based architecture is developed based on structures made by the services of organization arrangement. The sick person-based structure is developed by utilizing the structure made by people under treatment.

17.7.3 STANDARDIZATION AND SCALABILITY

All IoT devices produce large data for analyzing and collecting. The arrangement that controls these IoT devices must be scalable. The scaling of this large number of devices to make them interpretable and the management of this value-added service the management are key standardization issues at present. The vendors of the medical device do not follow maintained standards for platform-independent interfaces and protocols for devices. This creates the concern of inter portability.

17.7.4 THE APP DEVELOPMENT PROCESS

The basic steps for the development of an android app consist of four basic steps: development of the setup, coding, testing and compiling with debugging, and publishing. All other platforms also follow the basic steps for app development. However, while developing an app for the medical system the involvement of medical experts or authorized bodies is mandatory to ensure the quality of the designed app. In addition, on-time updates and installation of updates based on feedback networks without delay are required. This on-time servicing and update is a challenging task for a medical app.

17.7.5 TECHNOLOGY TRANSITION

Medical devices industry need to modernize itself with changing technology and thus develop their existing sensors and devices accordingly incorporating IoT methods with the existing network system. Thus, a seamless transformation from the old school system to the IoT-based system is a big challenge. Therefore it is required to have a backward compatibility and flexibility option in existing devices and training to the users of these devices to go on with changing methods

17.7.6 THE LOW-POWER PROTOCOL

With creating the medical devices, movable and wireless long battery life is another concern. Since these devices are critical in terms of their application thus power dissipation or battery failure cannot be afforded. A major challenge with this IoT/AI-based portable devices are to develop a low-power protocol and circuit design with least power dissipation and maximum efficiency

17.7.7 NEW DISEASES AND DISORDERS

Smartphones are considered the front wall of IoT medical devices. Every day a new app is developed to detect and analyze the diseases that are well known, but few diseases and disorders that are new are still to be considered. Research for these new diseases and disorders is essential, and the development of AI technology-oriented app to detect the disease at an early stage with the help of a case study or data stored in IoT is a new task.

17.8 IOT MEDICAL DEVICES AND SYSTEM SECURITY

In the next few years, the medical device will experience the widespread change of the IoT in the medical field and will give many new health services like E-health. These hi-tech devices consist of many personal and crucial data of the patient. Regardless of the privatization of this data, they will be connected to cloud computing and global level network thus they can be accessed by anywhere and at any time. Thus these emerging trends of utmost concern are to be protected by the attackers and hackers.

For the successful implementation of IoT medical devices, it is mandatory to provide high security, identify the critical aspects, and analyze them for safety concerns. The security system's important aspects as shown in Fig. 17.2 are explained below. The main aspects of security are listed as requirements of security challenges faced by security, a model of threat, and a taxonomy for the attack.

17.8.1 SECURITY REQUIREMENTS

Medical devices with IoT/AI also deal with the same security issues as in normal communication systems. Thus to maintain secure services, following security requirements need to be emphasized:

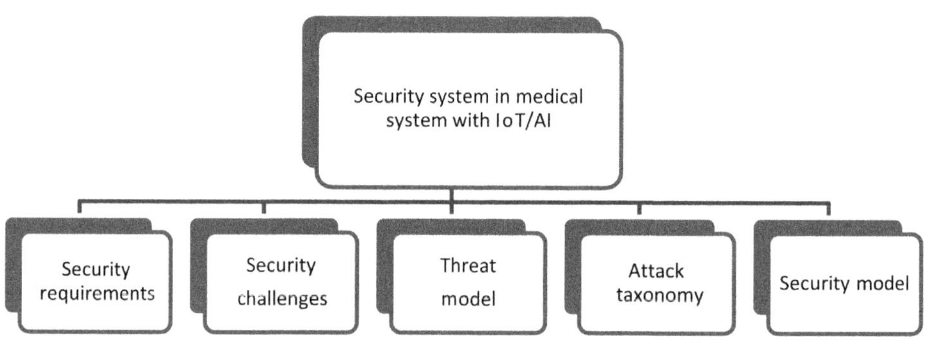

FIGURE 17.2

Security system aspects of AI/IoT medical devices and networks. *AI*, Artificial intelligence; *IoT*, Internet of Things.

confidentiality, integrity, authentication, availability, data freshness, nonrepudiation, authorization, resiliency, fault tolerance, and self-healing. These requirements are explained in short.

17.8.1.1 Confidentiality

Unauthorized users cannot access the data. Moreover, eavesdroppers cannot see the confidential messages.

17.8.1.2 Integrity

It stands for maintaining the quality and content of data while in transit and it is not altered. Stored data and contents are not compromised in terms of any parameters.

17.8.1.3 Authentication

The communicating side's accessibility is maintained by providing an authorized right to see or read the data.

17.8.1.4 Availability

It ensures the survival of IoT medical services, that is, cloud services, local service, or global service to legal parties whenever asked, that is, after full authentication only.

17.8.1.5 Data freshness

I mean data and its key should be stored as fresh data so that it does not become volatile with power and can be accessed. Since every IoT medical network has few time-variant systems, so it is important to be sure that information is fresh. This implies that every information is new and makes it sure that no false data overlap old information (Fig. 17.3).

17.8.1.6 Nonrepudiation

It means a node of the network cannot deny transferring the message which is once transmitted.

17.8.1.7 Authorization

It makes sure that nodes that are authorized can only access nodes for resources and network services.

17.8.1.8 Resiliency

If any medical service is being compromised or attacked, in that case, a security service should be developed to protect these nodes/networks/devices from attacks.

17.8.1.9 Fault tolerance

During processing the data there could be a lot of mistakes, for example, glitch in the software, a compromise of the device, and a failure of the device in this situation also a security system should be present to protect them in their respective ways.

17.8.1.10 Self-healing

When any medical device in the system fails or becomes energy deficient, the remaining or connected devices should have some initial level of security; this is known as self-healing.

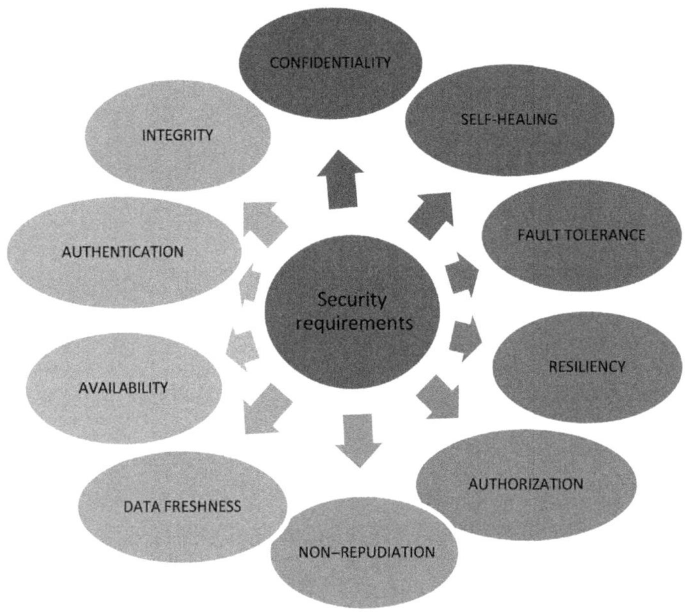

FIGURE 17.3

Security requirements of a medical device with Internet of Things/artificial intelligence.

17.8.2 CHALLENGES FACED FOR SECURITY

Since the IoT security needs are not protected like conventional security methods, strong solutions should be found to counter or solve new challenges faced by the IoT. Some of the challenges are listed as follows.

17.8.2.1 Computational requirements

IoT using medical devices is integrated with processors of less-speed. CPU speed is slow in these types of applications due to large data size availability. These devices do not perform complex algorithms and expensive operations; instead, they just act as actuators or sensors. Thus developing a solution for security that decreases the resource consumption and helps in increasing security is a complex task (Fig. 17.4).

17.8.2.2 Memory limitations

Generally, IoT medical devices consist of low memory integrated into the device. These devices operate through an integrated operating system (OS), an application binary, and system software. Thus its memory allotted is not compatible to run complex security protocols.

17.8.2.3 Limitations of energy

A general IoT-based medical system has few medical devices of low-power battery like temperature detector of body and blood pressure sensors. These devices store power by turning on the

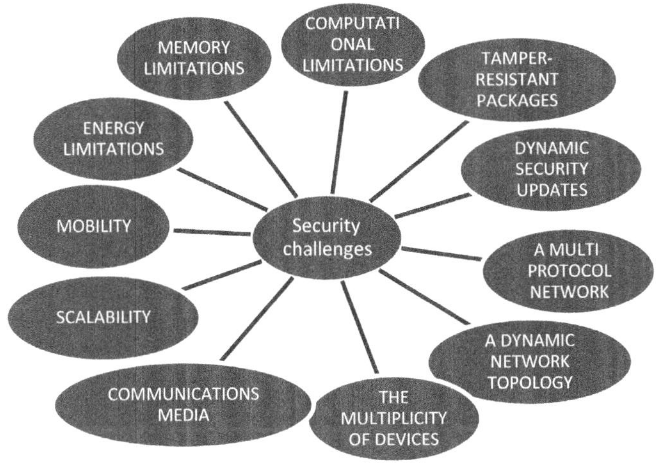

FIGURE 17.4

Security challenges for IoT/AI-based medical devices. *IoT*, Internet of Things; *AI*, artificial intelligence.

mode of energy saving when they are not working. They work at a decreased speed when no important work is assigned to work on asses. Thus, the power-limited characteristics of the IoT-based medical devices create it difficult to find the security solution, which is also energy efficient.

17.8.2.4 Mobility

Typically, medical devices mobile. These medical devices are linked by the IoT service suppliers to the Internet. Such devices may be connected to a home network or office network depending on the location of the user. All types of networks on distinct security methods and settings. So designing a mobile and portable security protocols that alter according to the type of network used is highly challenging.

17.9 SCALABILITY

With time the IoT devices have increased, this results in a large number of devices connected to the global network. Thus, developing scaled defense arrangements without lowering defense needs to develop into a difficult task.

17.9.1 COMMUNICATIONS MEDIA

Medical devices are linked to two types of networks global and local networks most of the time by the means of wireless networks. Wireless siphon properties of this system create a conventional wired security system less secure. Thus, it is challenging to design an alternate defense for a protection system that can secure wireless channel as well as and wired channel properties [41].

17.9.2 THE DEVICES MULTIPLICITY

Medical devices in an IoT-based medical hobnob are of many types, some PCs are full-edged and RFID tags with low end. The application of these devices changes depending on their capability in algorithm, memory, power, and embedded software. Thus developing a security system that works for all types of devices is challenging.

17.9.3 A DYNAMIC NETWORK TOPOLOGY

A medical device can link to an IoT-based medical hobnob anytime and anywhere. A well as it can exit the network with notification or abruptly. This property of integrating medical devices to IoT makes them dynamic. Thus designing a model for changing hobnob topography is challenging.

17.9.4 A MULTIPROTOCOL NETWORK

A medical device sometimes transmits or receives data with other devices through the local network with a proprietary network protocol or some time with the IP network. Thus the specialist design of security that is compatible with multiprotocol communication is difficult to design.

17.9.5 DYNAMIC SECURITY UPDATES

To counter potential threats, security protocols should be up-to-date. Thus regular updating of defense spots is needed for IoT medical devices. However, developing such a small interval mechanism will be power consuming for the changing installation of these defense spots, which makes it a difficult task.

17.9.6 TAMPER-RESISTANT PACKAGES

Physical defense for protection is the most necessary task of the IoT-based medical devices. A hacker tries to disturb the node's physical security and thus can extract encoded information, alter programs, or exchange those with virus nodes. Tamper-resistant packaging defends mechanism is a must against these types of attacks, but this is very difficult to implement with energy limitations.

17.9.7 A THREAT MODEL

IoT medical devices and systems are damaged attacked in the security model due to a lot of surfaces that are delicate and need to be protected. The first type consists of expanded networks, cloud services, and networks. The second type is high communication in the IoT-based devices, services through cloud, networks, and applications. The last one includes hardware and software in the device dead end. These threats may attack from outside or inside the network. If a hack starts from a medical node in a close hobnob, then this is a serious attack. It is challenging to detect the virus or effecting device within the close system. In addition, the adverse virus may hack a medical node and hobnob system directly or indirectly and can utilize connected types of power devices or IoT sources like laptops, smartphones, and tablets to dig deeper into the network of interest (Fig. 17.5).

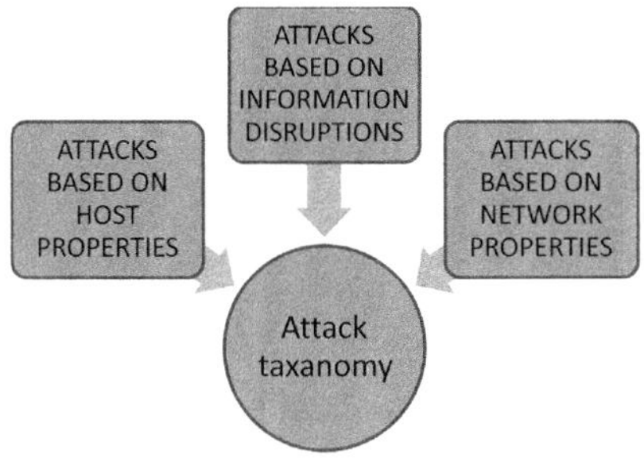

FIGURE 17.5

Attack taxonomy on AI/IOT integrated medical devices. *AI*, artificial intelligence.

17.9.8 AN ATTACK TAXONOMY

IoT, Internet of Things. The IoT is evolving with time and many new IoT medical devices are being developed. An attacker can create a different type of threat of security to hack or disrupt present and tomorrow IoT-based medical nodes and systems. Many attacks are predictable and tangible, while many are unpredictable. This section classifies current potential threats based on three key characteristics—information, host, and network properties

17.9.8.1 Information disruptions-based attacks

The message stored or ready to transfer or in the way can be altered or manipulated by a hacker to send a disrupted message and disturb the integrity of the information. These attacks are classified as follows.

17.9.8.1.1 Interruption

An attacker inserts denial-of-service signals to the network to cancel the communication link in the protocol. This type of attack weakens the availability and functionality of network service and nodes ownership.

17.9.8.1.2 Interception

A hacker spy on personal medical data transferred in a memo to corrupt data confidentiality and privacy.

17.9.8.1.3 Modification

An attacker controls illegal passage to medical messages and changes them to generate bewilderment in true events and part in the IoT-based medical network.

17.9.8.1.4 Fabrication

An attacker forces information by inserting false messages to weaken the data authenticity and mislead participants.

17.9.8.1.5 Replay

An attacker replays information already existing to disturb its freshness. This creates misleads and confused innocent entities and thus working of the device is altered.

17.9.8.2 Attacks based on host properties

Three types of attacks can be fired based on host properties.

17.9.8.2.1 Compromise by user

An attacker trade-offs the patient's medical devices and hobnobs by tampering or stealing. This type of attack provides delicate data such a cryptographic keys passwords, and very personal user information.

17.9.8.2.2 Compromise on hardware

An attacker changes physical devices and can obtain on-device codes or program, data and keys thus can reprogram the device with infected or wrong codes.

17.9.8.2.3 Compromise on software

A hacker jumps for advantage operating systems, applications, system software weakness, and glitches and insists IoT medical devices to corrupt and malfunction.

17.9.8.3 Attacks based on network properties

This type of attack emerges in two forms: protocol and layer specific.

17.9.8.3.1 Compromise of standard protocol

An attacker reflects the device from protocols standardized like application protocol and networking standards protocol and inserts virus to hamper the availability of offers, the privacy of message, integrity, and authenticity.

17.9.8.3.2 Network protocol stack attack

Layer of the protocol of active group for the IoT hobnob has many drawbacks that can lead an attacker to misuse or launch infectious actions. The performance of IoT medical system layers needs to be improved in terms of longevity, security, and link under the different surrounding scenario, the high defense should be present at each stack protocol layer.

17.9.9 SECURITY MODEL

IoT medical system is still developing and will continue for some time. Thus it is not possible to detect each coming threat and vulnerability due to the developing phase. A stratified networking model of protected IoT. Credible and calculable problems—these types of defense ideas should have the capability to alleviate unseen or not calculable problems that are yet to emerge. To gain this ambition in terms of protections, relevance should be developed with progressive properties and AI. That is with experienced outcomes the problems should be detected initially and the feedback system for the least error is needed to be designed. Along with the development of medical equipment, apparatus, networks, and applications in time if a mugger identifies a different type threat then security services need to be adept to detect the new threat with smart calculations or AI.

17.10 CONCLUSION

AI systems are developed to analyze the thinking powers to solve and facilitate complex algorithms or redundant /repetitive tasks and thus giving advanced new insights and making users concentrate on a different area of services. AI/IoT can gain data and encrypt it into messages and process it logically, store it in cloud networks for further access. The chances of mistakes are less, thus complex processing is done with accuracy and on time thus it makes the device faster and accurate. Thus it provides professionals more trust to rely on the devices and increase their bandwidth of services for patients by focusing on treatments. These IoT/AI systems have made a new world of possibilities for the medical device industry, by helping doctors in providing right and effective service more quickly. While technology is a necessary defense mechanism it is also a gate to the new threat, it is as challenging to make sure every user/professional gets to know the ways they need to take to protect medical data in their daily activities. That is technology understanding is needed for the user also. Hospitals should provide more focus on training employees to recognize the attacks on or through devices and problems it can create.

REFERENCES

[1] F.H. Alqahtani, The application of the Internet of Things in healthcare, Int. J. Comput. Appl. 180 (2018).

[2] M.S. Shinde Sayali, P. Vaibhavi, A survey paper on Internet of Things based Healthcare System, in: International Advanced Research Journal in Science, Engineering and Technology National Conference on Innovative Applications and Research in Computer Science and Engineering (NCIARCSE-2017) AGTI's Dr. Daulatrao Aher College Engineering, Karad, Vol. 4, Special Issue 4, January 2017.

[3] S.M. Riazul Islam, D. Kwak, H. Kabir, M. Hossain, K.-S. Kwak, The Internet of Things for health care: a comprehensive survey, IEEE Access, 1 June 2015.

[4] S.C. Padwal, S.V. Kurde, Long-term environment monitoring for IoT applications using wireless sensor network, International Journal of Engineering Technology, Management and Applied Sciences 4 (2016) .

[5] S. Warren, R.L. Craft, Designing smart health care technology into the home of the future, 25 March 1999.

[6] R. Singh, A proposal for mobile E-care health service system using IoT for Indian scenario, J. Net. Commun. Emerg. Technol. 6 (1) (2016).

[7] A. Khanna, P. Misra, The Internet of Things for medical devices prospects, challenges and the way forward, Wireless Commun. Mobile Comput. (2019).

[8] L. Yu, Y. Lu, X.J. Zhu, Smart hospital based on Internet of Things, J. Netw., 7(10), October 2012 .

[9] A. Mathew, F. Amreen S.A., P. H.N., A. Verma, Smart disease surveillance based on Internet of Things (IoT), Int. J. Adv. Res. Comp. Commun. Eng. 4(5) (2015).

[10] K.U. Sreekanth, K.P. Nitha, A study on health care in Internet of Things. IJRITCC, 4, (2), Feb. 2016, pp. 44 -7, doi:10.17762/ijritcc.v4i2.1759.

[11] P.A. Laplante, N. Laplante, The Internet of Things in healthcare: potential applications and challenges, IT Prof. 18 (2016) 2–4.
[12] M.B. Blake, An Internet of Things for healthcare, IEEE Internet Comput. 19 (2015) 4–6.
[13] P. Po Yang, O. Amft, Y. Gao, L. Xu, Special issue on the Internet of Things (IoT): informatics methods for IoT-enabled health care, J. Biomed. Inform. 63 (2016) 404–405.
[14] Y. Yin, Y. Zeng, X. Chen, Y. Fan, The Internet of Things in healthcare: an overview, J. Ind. Inf. Integr. 1 (2016) 3–13.
[15] P.L.L. Benny, I. Henry, Y. Guang-Zhong, Transforming health care: body sensor networks, wearables, and the Internet of Things, IEEE Pulse 7 (2016) 4–8.
[16] S. Madakam, R. Ramaswamy, S. Tripathi, Internet of Things (IoT): a literature review, J. Comput. Commun. 3 (2015) 164–173.
[17] K. Ashton, That Internet of Things Thing, RFID J. 22 (2009) 97–114.
[18] N. Gershenfeld, R. Krikorian, D. Cohen, The Internet of Things, Sci. Am. 291 (2004) 76–81.
[19] L.M.R. Tarouco, L.M. Bertholdo, L.Z. Granville, L.M.R. Arbiza, F. Carbone, M. Marotta, et al., Internet of Things in healthcare interoperability and security issues, in: IEEE International Conference Communications, IEEE, Sydney, 2012.
[20] F. Mattern, C. Floerkemeier, From the Internet of computers to the Internet of Things, in: K. I.P. Sachs, P. Guerrero (Eds.), From Active Data Management to Event-Based Systems, Springer, Berlin, Heidelberg, 2010, vol. 6462, pp. 242–259.
[21] I. Lee, K. Lee, The Internet of things (IoT): applications, investments, and challenges for enterprises, Bus. Horiz. 58 (2015) 431–440.
[22] P.A. Williams, A.J. Woodward, Cybersecurity vulnerabilities in medical devices: a complex environment and multifaceted problem, Med. Devices (Auckl.) 8 (2015) 305–316.
[23] D. Miorandi, S. Sicari, F. De Pellegrini, I. Chlamtac, Internet of Things: vision, applications and research challenges, Ad Hoc Netw. 10 (2012) 1497–1516.
[24] M. Busch, C. Hochleitner, M. Lorenz, T. Schulz, M. Tscheligi, E. Wittstock, All in targeting trustworthiness for special needs user groups in the Internet of Things, in: M. Huth, N. Asokan, S. Čapkun, I. Flechais, L. Coles-Kemp (Eds.), Lecture Notes in Computer Science, Springer, Berlin, 2013, pp. 223–231.
[25] R. Peng, M.L. Sichitiu, Angle of arrival localization for wireless sensor networks, in: Proceedings of the Third Annual IEEE Communications Society on Sensor and Ad Hoc Communications and Networks, IEEE, Reston, VA, 2006.
[26] K.K. Tremper, S.J. Barker, Pulse oximetry, Anesthesiology 70 (1989) 98–108.
[27] J. Gubbi, R. Buyya, S. Marusic, M. Palaniswami, Internet of Things (IoT): a vision, architectural elements, and future directions, Future Gener. Comput. Syst. 29 (2013) 1645–1660.
[28] D. Mulfari, A. Celesti, M. Fazio, M. Villari, A. Puliafito, Achieving assistive technology systems based on IoT devices in cloud computing, EAI Endorsed Trans. Cloud Syst. 1 (2015) 1–6.
[29] A.-M. Rahmani, N.K. Thanigaivelan, T.N. Gia, J. Granados, B. Negas, P. Liljeberg, et al., Smart e-health gateway: bringing intelligence to Internet-of-Things-based ubiquitous healthcare systems, in: Proceedings of the Annual IEEE Consumer Communications and Networking Conference, IEEE, NV, January 2015.
[30] S. Agrawal, D. Vieira, A survey on Internet of Things, Abakós, Belo Horiz. 1 (2013) 78–95.
[31] F.J. Riggins, S.F. Wamba, Research directions on the adoption, usage, and impact of the Internet of Things through the use of big data analytics, in Proceedings of the 48th Hawaii International Conference on System Sciences, HI, 2015.
[32] J. Yu, M. Kim, H.C. Bang, S.H. Bae, S.J. Kim, IoT as applications: cloud-based building management systems for the Internet of Things, Multimed. Tools Appl. 75 (2015) 14583–14596.

[33] T. Suzuki, H. Tanaka, S. Minami, H. Yamada, T. Miyata, Wearable wireless vital monitoring technology for smart health care, in: Proceedings of the Seventh International Symposium on Medical Information and Communication Technology, IEEE, Tokyo, 2013.
[34] C. Liu, C. Yang, X. Zhang, J. Chen, External integrity verification for outsourced big data in cloud and IoT. A big picture, Future Gener. Comput. Syst. 49 (2015) 58–67.
[35] S. Goel, P. Tomar, G. Kaur, A fuzzy based approach for denoising of ECG signal using wavelet transform, Int. J. Bio Sci. Bio Tech. 8 (2) (2016) 143–156.
[36] C. Dores, L.P. Reis, N.V. Lopes, Internet of Things and cloud computing, in: Iberian Conference on Information Systems & Technologies, IEEE, Barcelona, 2014.
[37] D. Tomar, J.P. Bhati, P. Tomar, G. Kaur, Migration of healthcare relational database to NoSQL cloud database for healthcare analytics and management, Healthcare Data Analytics and Management, Academic Press, 2019, pp. 59–87.
[38] G. Kaur, N. Gupta, E-health: a new perspective on global health, J. Evol. Technol. 15 (1) (2006) 23–35.
[39] Y. Ma, J. Rao, W. Hu, X. Meng, X. Han, Y. Zhang, et al., An efficient index for massive IoT data in a cloud environment, in: 21st International Conference on Information and Knowledge Management, ACM, Maui, 2012.
[40] D. Kedia, G. Kaur. Examining different applications of cloud-based IoT, Examining Cloud Computing Technologies Through the Internet of Things, IGI Global, 2018, pp. 125–146.
[41] G. Singh, G. Kaur. Role of communication technologies for smart applications in IoT, Handbook of Research on Big Data and the IoT, IGI Global, 2019, pp. 300–313.

CHAPTER

MACHINE LEARNING FOR OPTICAL COMMUNICATION TO SOLVE PERVASIVE ISSUES OF INTERNET OF THINGS

18

Dushyant Singh Chauhan and Gurjit Kaur

Department of Electronics and Communication Engineering, Delhi Technological University, Delhi, India

18.1 INTRODUCTION

Nowadays as a huge number of real-time devices and sensors are being deployed, there is a need of low latency and high data rate network. So, Internet of Things (IoT) is a solution that provides a pervasive network paradigm offering transparent services. Apart from high data rate and low latency, the major consideration in an IoT-based optical communication system are how to efficiently utilize the spectrum, and how different network technologies coexist with the current technologies. The aforementioned requirements can be achieved by incorporating neural networks, artificial intelligence, and machine learning (ML) to tackle with the massive amount of information generated by various sensors and IoT devices to make efficient and accurate decisions.

With the help of artificial intelligence-based methods the information is extracted from the raw data, and data patterns are formed which helps end devices. ML, is another important and emerging area, which has obtained a great recognition and success in various fields including digital image processing, audio and speech recognition, computer graphics, natural language processing, computer vision, intelligent control, and decision making. It has also been introduced in optical communication and optical networks. Various ML algorithm finds application in solving issue related optical networks like traffic prediction, efficient routing, networking problems, security, and resource allocation [1].

18.2 MACHINE LEARNING TECHNIQUES

ML is a branch of artificial intelligence that allows any applications to give output more accurately in terms of prediction by fetching correct data, machines can itself know how to solve a specific problem. By the additional support provided from statistical analysis tools and complex mathematical tools, ML provides help to machines, which are able to perform a given task that have been conventionally solved by humans. In the field of networking, the idea is to automatize the complex problems related to network designing and its operation [2]. Fig. 18.1 shows the classification of machine leaning along with the various algorithms used for predicting the outcome more accurately.

Artificial Intelligence to Solve Pervasive Internet of Things Issues. DOI: https://doi.org/10.1016/B978-0-12-818576-6.00018-6

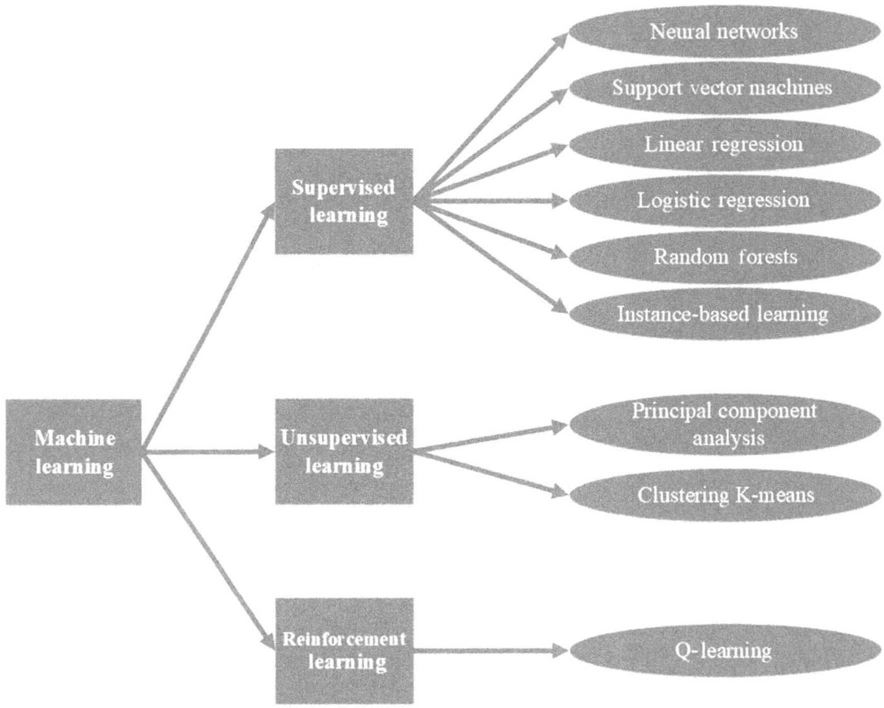

FIGURE 18.1

Various machine learning algorithms.

18.2.1 SUPERVISED LEARNING

Supervised learning is a method of predicting the value of output variables when a vector of input variables is given. Some of the applications of these types of learning methods are object recognition and spam detection. Depending upon the output variable, as in regression problem the output is a continuous variable while in classification problem the output is a discrete variable. There are two datasets available for training. One is of input variables of N samples and another dataset corresponding to output variables. A function is constructed based on various learning methods that predicts the output variable corresponding to input variables. Furthermore, supervised learning is divided into two main categories on the basis of fixed parameters and parameters dependent on training set, as parametric models and nonparametric models, respectively.

18.2.2 UNSUPERVISED LEARNING

In unsupervised learning, when the output is not available and the test data present in the database has not been classified or categorized, then the aim of unsupervised learning is to create a model rather than prediction based on the test data. Unsupervised learning methods are more powerful in comparisons with supervised learning methods. Without any training, this learning technique can group the unclassified data on the basis of similarities, differences, and patterns. Unsupervised learning algorithm can be further categories as:

- **Principle component analysis:** This type of analysis is used before supervised learning algorithm for preprocessing of data.
- **Clustering:** It is used to make subgroups from the test data present in the database.

18.2.3 SEMI SUPERVISED LEARNING

In this type of learning, the feature vector present in the dataset contains combination of both unlabeled and labeled examples. Typically, this combination has large amount of unlabeled data in comparison with labeled data. Firstly, using clustering technique the unlabeled data is grouped into small clusters and then labeled data is further used to label remaining data present in the database.

18.2.4 REINFORCEMENT LEARNING

ML has another subfield known as reinforcement learning, in this the decisions has been made sequentially, that is, the output depends on the present input and the future input depends on the output of the previous input. Reinforcement learning is different from both supervised and unsupervised learning on the basis of how it interprets inputs.

18.3 MACHINE LEARNING TECHNIQUES USED IN OPTICAL COMMUNICATION

ML finds many applications not only in the field of optical communication but also in the field of optical networking. ML finds application at different layers (physical and network) as well as at cross-layers also. Examples of ML at physical layers include monitoring of bit error rate (BER), which can affects directly network layer also in terms of routing. Another motivation for using ML is that, due to the optical fiber nonlinearities, closed form expression for the optical channel cannot be obtained. This has implications for the predictions in performance in terms of Q-factor, BER, and demodulation of signals in optical communication system [3]. Fig. 18.2 shows the various application of ML in optical communication.

ML can be used to solve different types of problems. Fig. 18.3 shows the basic blocks involves in designing a ML-based solution for an optical network. Data collection refers to generation and gathering of right information and creates vector based on the characteristics. Feature engineering is used to reduce dimensionality of the input vectors associated with the characteristics whereby reduce the computational complexity and increases the accuracy. Then ML algorithm analyses the input and output data to train a model, which helps in predicting the future output.

18.4 MACHINE LEARNING IN PHYSICAL LAYER

In any optical communication network there are various challenges involves at physical layer that need to be taken care off like estimation of quality of transmission (QoT) and optical performance

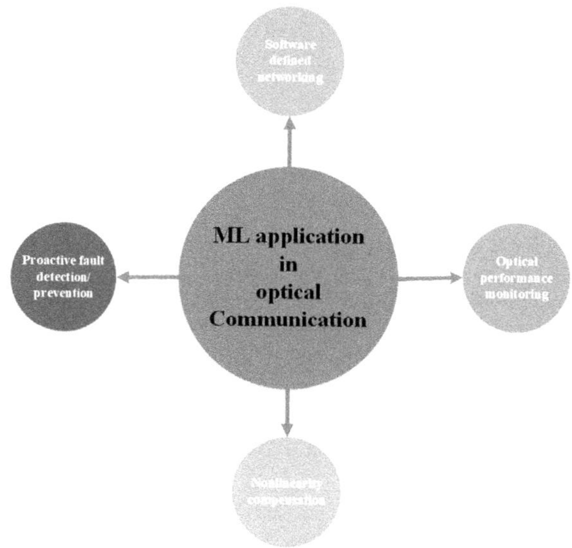

FIGURE 18.2

Some machine learning applications in optical communication.

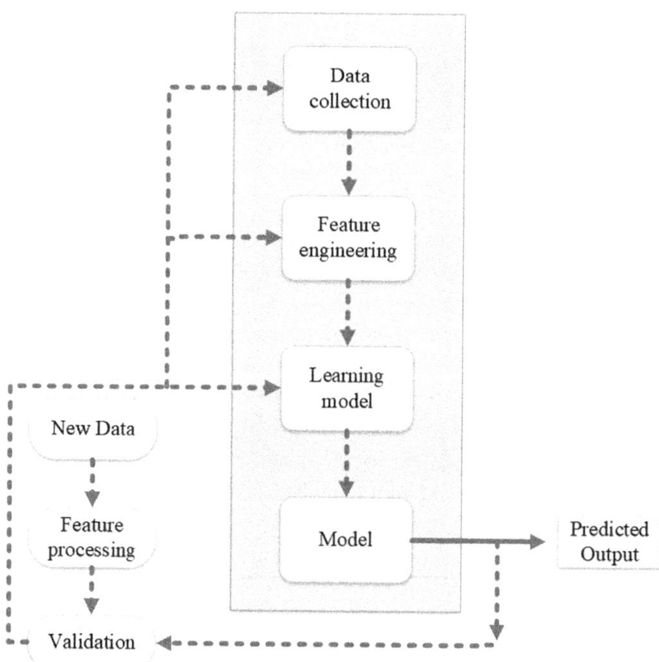

FIGURE 18.3

Building blocks of machine learning-based model for problem solving.

monitoring, when performing lightpath provisioning. Various cases at physical layers were discussed in more detailed form with the help of ML.

18.4.1 QUALITY OF TRANSMISSION ESTIMATION

Estimation of QoT is the performance measurement that has been taken directly from field and data is collected from receiver side with the help of various installed optical monitors or on ligthpath characteristics in terms of quality metrics like Q-factor, BER, and optical signal to noise ratio (OSNR). The estimation of QoT is generally applied in two cases:

- First, case is unestablished lightpaths, in predicting the transmission quality based on the measurements of past data collected from already established ones.
- Second, with the aim of identifying malfunctions and faults, the QoT of already deployed lightpaths is monitored.

For the unestablished lightpaths, QoT prediction relies on intelligent tools that are sufficient enough to predict, whether a ligthpath will meet the desired quality of service like BER, OSNR, and threshold value of Q-factor. This type of problem is generally based on binary classification-based problem in which the classifier outputs either yes or no depending on the various ligthpath characteristics like modulation format (MF), number, and length of links used for transmission. Another approach known as case-based reasoning, which typically depends on the preservation of a knowledge database where various information related to the measured Q-factor of deployed link, along with the selected wavelength, route, total number, and total length is stored. Once a request is generated for new traffic, the most comparable one (in terms of Euclidean distance based) is fetched from the stored database and a prediction is performed on the basis of comparing the Q-factor associated with the predefined system threshold stored in database. The performance of the case-based reasoning technique can be greatly affected by the maintenance and correct dimensioning of the database, various algorithms are available to keep database up to date by removing unwanted entries from the database. The trade-off between effectiveness of the classification-based performance, computation time, and database size is broadly studied in [3], the method is shown to outperform various ML algorithms like Random Forests, J4.8 tree, and Naive Bayes. In [4] an experiment is performed for real test bed and good results have achieved. Another approach based on database is suggested in [5] to minimize the uncertainties on design margin and parameters associated with optical network, the field data stored in central repository base on software defined network controller. Then, Q-factor measurement tool is used to produce an estimation of field-measured signal to noise ratio (SNR) values based on prediction of the unknown network parameters. Using gradient descent algorithm, the predicted values keep on updating in the database. This prediction of SNR continues, once the difference between measured and estimated SNR is minimized and fall below a certain threshold value, the parameters corresponding to the estimate SNR gets saved in the database and new design margins is produced which helps in fulfilling future demands. Using numerical simulation, the trade-off between variation of the estimated field SNR error and database size are evaluated. Similarly, with reference to multicast transmission, a network is trained using neural network (NN) in Ref. [6] using various features shown in figure along with the layer which is used to predict whether the Q-factor value exceed a given system threshold or not. In addition, NN is trained with small groups of data, to regularly update the prediction model.

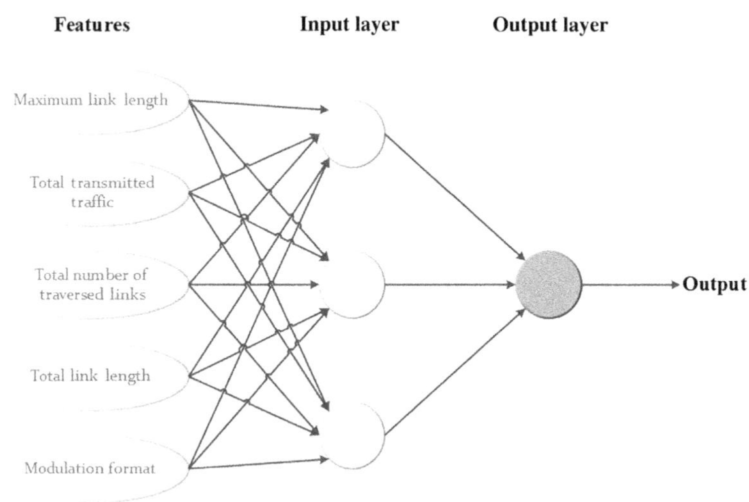

FIGURE 18.4

Various features as input to the classifier.

To avoid overfitting problem, a dropout technique is used in training phase. In case of unestablished lightpaths, a binary classifier based on random forest is adopted in Ref. [4] to predict the value of BER. The classifier takes a set of features as input shown in Fig. 18.4, which is to be adopted for transmission.

18.4.2 OPTICAL AMPLIFIERS CONTROL

The operating point in any erbium-doped fiber amplifier (EDFAs) affects their gain flatness and noise figure, which have a significant effect on the QoT. So, by means of ML algorithm the autonomous tuning of the operating point can be accomplished based on the signal input power. Most of the existing literature rely on evaluating performance metrics experimentally as an initial amplifier characterization (e.g., gain flatness, noise figure, and gain control accuracy). These characterization results are then represented within the operation region as a set of separate values. In the implementation of EDFAs, microcontrollers are unable to obtain the noise figure and gain flatness values. Moreover, exact and accurate measurement takes lots of time. To address the issue of gain flatness and noise figure, ML-based algorithms can be implemented and mapping function of measured points is extended over nonmeasured points. So, NN algorithm is adopted for interpolation the mapping function. Conversely, a cognitive approach, which is best suited for the dynamic networks. Once the new request is generated, the previous database of gain flatness and amplifier gains is updated along with the measured value of SNR and lightpath characteristics. So, the incoming lightpath request is retrieved on the basis of highest degree of similarity, and a new option for choosing again is generated. Then, with the help of new array of gains, the OSNR value is obtained with the help of simulation and the new value of OSNR is stored in the database. After this, the combination of best OSNR is selected and used in the EDFA when the new lightpath is requested.

18.4.3 MODULATION FORMAT RECOGNITION

In any optical network MF is a very important parameter, but this parameter is manually set by the designer depending upon the network. So, by the application of ML-based algorithm including neural networks, and k-means clustering the autonomous (i.e., without any information requiring from the transmitter) MF is recognized in the receivers and this issue can be addressed. At the receiver side, depending on OSNR value, the performance of six unsupervised clustering-based algorithms are compared to differentiate between five different MFs like 8-QAM, 16-QAM, QPSK, BPSK, and 8-PSK in terms of running time and true positive rate. Silhouette coefficient, is used in some algorithms to predict the number of clusters, which evaluates the degree of closeness among intra and inter distances. The advantage of Stokes space (i.e., not affected by phase and frequency) is taken to represent DP-QPSK, DP-BPSK, and DP-8-QAM. NN is used for next-generation heterogeneous fiber optic network, by minimizing the number of neurons placed in the hidden layers of NN model with respect to various features. In Ref. [5], a combination of genetic algorithm and NN is used to improve the efficiency during the training phase.

18.4.4 NONLINEARITY MITIGATION

In any optical communication system data-rate distance product is one of the important performance characteristics. Due to the losses occur in the fiber, amplification of optical signals is essential. More number of amplifiers need to be installed to increase the overall transmitted distance in any optical network. Optical amplifiers contain noise and these noise gets added. To maintain good SNR value, an increase in optical signal power of source is required. However, due to the increase in optical power at source creates nonlinearities (Kerr effect) which leads to noise [i.e., nonlinear interference (NLI) noise]. The impact of NLI is in detection of symbol, so to perform optimum symbol detection ML-based approaches are used. In general, symbol detection in an optimized manner is the primary task of the receiver.

The optimized symbol detection in the situation of circularly symmetric Gaussian distribution noise is performed by reducing the distance (Euclidean) between all possible symbols of the constellation alphabets and the received symbols. So, the decision boundaries are linear for this type of detection. For memoryless nonlinearity, like in-phase and quadrature nonlinearity, driving electronics nonlinearity, and nonlinear phase noise, the noise related with the symbol may get distorted and shape of constellation diagram gets distorted. So Euclidean distance-based symbol detection is no longer valid and the knowledge of likelihood function of full parameterization information is necessary. So, ML-based algorithms, like Kernel density estimator, support vector machines (SVM), Gaussian mixture models, and k-nearest neighbors can be employed to determine the likelihood function. Fig. 18.3 shows various ML-based algorithm for nonlinearity compensation. In Ref. [6] by employing ML-based Gaussian mixture model, a gain of almost 3 dB is achieved in the input power, for 14G baud and DP 16-QAM transmission over a distance of 800 km link, which is a dispersion compensated link. Moreover, k-nearest-neighbors classifier is adopted on the basis of weighted distance to compensate system losses in different dispersion conditions, with 16-QAM

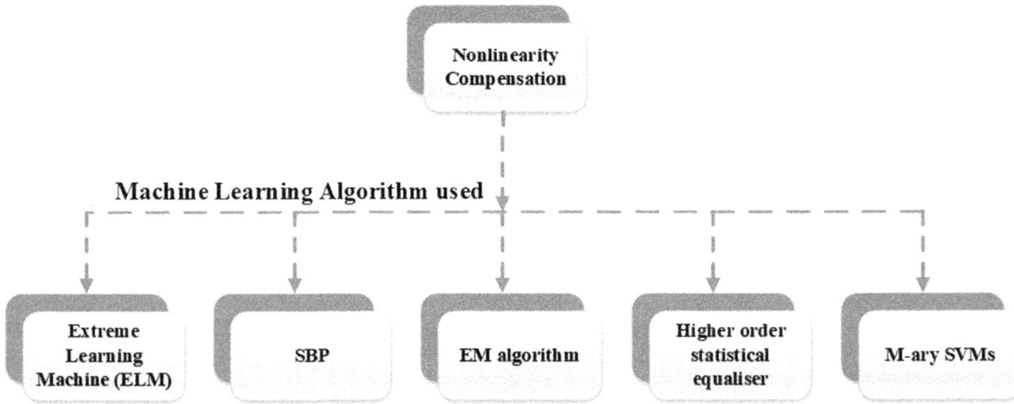

FIGURE 18.5

Machine learning algorithms used for nonlinearity compensation.

transmission, whereas for nonlinear equalization NNs are proposed with number of symbols equal to number of neurons in 16-QAMOFDM transmission. In training phase an advanced equalizer named as extreme learning machine (ELM) equalizer is used to reduce the computational complexity. ELM is based on neural networks in which by the help of generalized matrix inversion, the weights of IP/OP mapping error can be minimized. Fig. 18.5 shows various ML algorithms used for nonlinearity compensation.

18.4.5 OPTICAL PERFORMANCE MONITORING

NNs are best ML algorithms used to perform monitoring in optical networks, as they are capable of learning complex mapping. This mapping is between optical fiber channel parameters (i.e., OSNR, polarization mode dispersion, baud rate, and polarization-dependent loss) and the extracted features from the symbols [7]. The parameters that are used as input to the NN model can be derived by means of various approaches based on feature extraction. Fig. 18.6 shows some techniques for feature extraction.

Manually feeding input features to the neural network is relatively simple because only single hidden layer is present. The advantage of single hidden layer problems is that it requires a less amount of training data as compared to multi hidden-layer problems. Fig. 18.7 shows various ML algorithm for optical performance monitoring.

18.5 MACHINE LEARNING IN NETWORK LAYER

There is a demand of ML at network layer also. Once there is a failure in network, restoration of existing network or a requirement of new lightpath is evolved. So, these kinds of failure require complex and fast decision. Since operator must take care of the traffic generated due to newly

FIGURE 18.6

Feature extraction techniques.

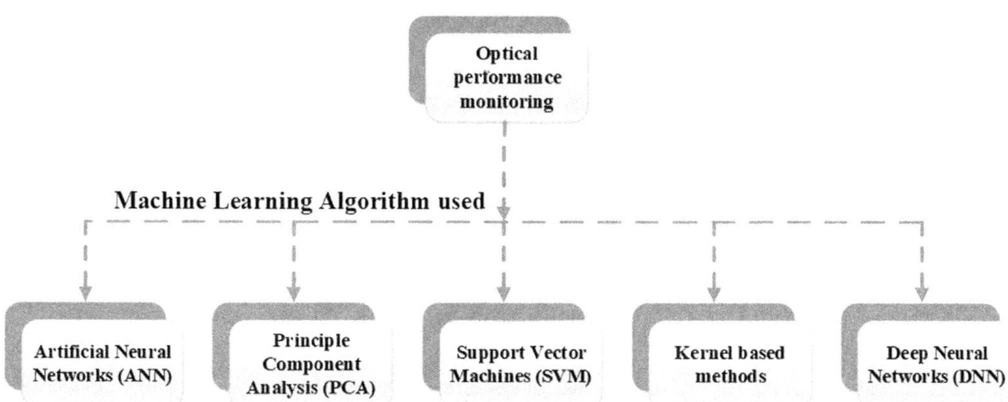

FIGURE 18.7

Machine learning algorithm used for optical performance monitoring.

inserted users. For an effective network operation, an assessment of number of users and their service required to users is necessary, as it helps in reducing the cost to lay down a network by deploying resources with sufficient margins and avoid additional bandwidth for network resources. In addition, at network layer, we identify four main use cases [8].

18.5.1 TRAFFIC PREDICTION

Traffic prediction at network layers plays a very important role in today's increasingly diverse and complex networks. Its role is important specifically in optimization and planning for network resources. Since the inherent property of ML is to train the existing model with the help of current dataset and predict the output, so ML can be efficiently used to predict the behavior of traffic. For example, in Ref. [9] a method of supervised learning algorithm known as autoregressive integrated moving average (ARIMA) method which is applied on data based on time series. By using virtual topology reconfiguration, ML algorithms are used for traffic prediction. This is a prediction method that gives good results on applying time series data, so it is best suited for the applications like virtual topology reconfiguration and traffic predictions. By using ARIMA model, decision maker and network planner (DMNP) module is used for predicting traffic. The DMNP then work together with various other modules to perform virtual topology reconfiguration. Since, input data must be in time series data form, so that virtual topology should handle the change in the network traffic with respect to time. More appropriately the inputs of the model are real time traffic data observed just before the current period. In addition to this in a low operational expenditure-based application, a low complex ARIMA model is preferred. To achieve good accuracy the model will become more complex, so the ML technique is chosen in such a way that the trade-off between complexity and accuracy does not affect. A prediction model based on NNs is for the generation of source-destination traffic matrix is given in Ref. [10]. This prediction matrix is used by a network model known as decision maker model to check whether present virtual network topology need to be change or not. NNs are best suited not only for the variation in input traffic but also for the accuracy in predicting the output traffic on the basis of past data due to its better adaptability. In most of the research ML algorithms focuses on traffic prediction based on estimation of traffic pattern using call data record. By utilizing the set of real data from network service provider and clustering-based algorithms, useful information can be fetched that helps in utilizing the network resources better. The call data record matrix consists of various information of the user. Fig. 18.8 shows call data record matrix.

Another matrix known as point of interest matrix, this matrix gives the information about various regions, which are visited mostly corresponding to each and every base station. So, both the matrix which contains input information are applied to ML-based clustering algorithm that is known as nonnegative matrix factorization (NNMF) and a variation called as collective NNMF. The algorithm divides output in two different nonnegative matrices, first gives the information about various types of traffic patterns and second shows the similarity between the base stations with reference to traffic patterns. Genetic algorithms can also be used for virtual topology design, with the help of adaptable fitness function. For future solutions principle of reinforcement learning may be used to update the fitness function using the previous solution for virtual topology design.

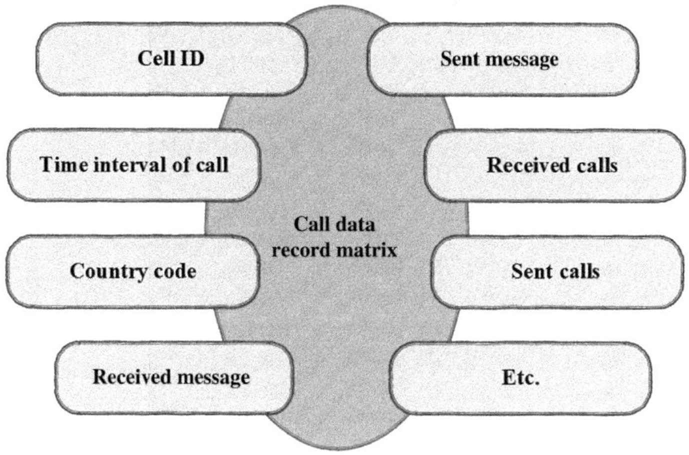

FIGURE 18.8

Call data record matrix.

18.5.2 FAULT MANAGEMENT

When there is an abnormality in the network, there is a requirement of detection, isolation, and correction, which is termed as fault management. It requires administrator or network operators to have a collective information of the whole optical network, including various devices, and applications using the networks. However, expecting this kind of knowledge is practically not possible. So, ML algorithms can be applied to detect/predict/locate exact location of fault in the network depending on the training data. Generally, ML-algorithm-based fault detection approaches uses various supervised learning techniques based on input data. Networking is misused to locate the exact fault location [11]. If localization of fault is not achieved, so additional information of lightpath is required, which increase the number of variables and the dimensions of the routing matrix gets increased. On the basis of load on the network, the number of receiving nodes essential to ensure exact localization is calculated. Similarly, inputs are applied to the Bayesian network in terms of received power and BER, which evaluates whether a fault occurs in the lightpaths and try to recognize the cause, based on specific characteristics of the measurement patterns. The accuracy of Bayesian based classifier shows in [12] that only 0.8% of the tested case were misclassified. Some basic ML techniques for fault detection is shown in Fig. 18.9.

18.5.3 TRAFFIC CLASSIFICATION

Network operators need to perform various management activities and network operation, so traffic classification is an important factor. These activities include security and intrusion detection, capacity planning, quality of service, and performance monitoring. So, another emerging area of ML-based application at network layer is traffic classification. In [13], in an optical burst-switched network, a model is defined that detects various forms of packet losses. Then these losses are classifying the packet loss data using a Hidden Markov Model and Expectation Maximization algorithms

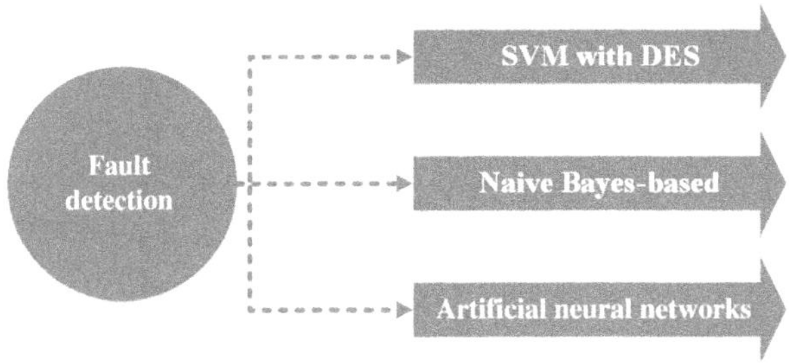

FIGURE 18.9

Machine learning techniques for fault detection.

as contention and congestion losses. In a second example a NN model is trained to categorize traffic in data center network. The input feature vector matrix includes information regarding IP address, port of destination and source, and transport layer protocol. In addition to this, to increase the speed and performance top 30 TCP segments corresponding to packet size and set of intraflow timings are used out of four packets of traffic. In Ref. [14] the foremost output of NN used is the classification of elephant and mice flows in the data center. This type of NN model used is known as multilayer perceptron. This multilayer perceptron is relatively easy to implement which includes four hidden layers.

18.5.4 PATH COMPUTATION

The common problem in optical communication is selection of efficient path on the basis of various parameters of physical and network layer. At physical layer various parameters like OSNR, QoT, and MF were predicted using ML-based algorithms. The primary objective is the decision of the selection of best optical path suited for communication among various alternatives. So, this computation process is a cross-layer method with ML-based algorithms at different layers. From the network layer perspective, in Ref. [15] a method based on wavelength and path selection is implemented to minimize burst lost probability for selection/computation of path in optical burst switching network. This problem is solved by using Q-learning and it is framed as a multiarm bandit problem. In optical burst switching network, by pulling one of the variables, the burst loss probability is minimized, which acts as a reward for the path selection insource-destination pair. This multiband arm problem is a traditional problem that lies in the category of reinforcement learning. These types of problems can be solved by using dynamic programming, learning automata, and Gittins indices. However, Q-learning algorithm shows definite convergence as compared to various other methods of path computation using ML. In Ref. [16] Fuzzy C Mean Clustering algorithm is used for quality of service aware path computation. When this algorithm is used in software defined optical networks, it achieves improvement in network performance in comparison with non-cognitive control plane. At physical layer this algorithm predicts the best possible parameters from the set of applied input parameters (like lightpath lengths, traffic request, BER, OSNR, and MF). The membership score is the degree of closeness between each lightpaths and physical layer parameters. With the help of this information of membership score, the real time decision is taken to setup the lightpaths [17].

18.5.5 RESOURCE MANAGEMENT

Network performance can be improved with the help of two important keys know as network adaption and resource management. Efficient routing and traffic scheduling are some of the important issues related to efficient recourse management and these issues can be framed as a decision-making problem. Science there are many challenges due to several factors, like noisy inputs and various system environmental conditions. Arbitrary parameter assignments are easy to understand by people based on experiences, but this type of method is lacking in optimization. So, ML is capable of finding solution due to its ability to describe the inherent relationships between the inputs and outputs of network systems without human intervention. By using the existing router architecture, network traffic patterns of each router in the network act as an input and give the next node as output. In network scheduling and routing, these achievements unleash the ability of the ML-based methods. Reinforcement-based learning attains great results in many problems related to optical communication [18].

18.5.6 CONGESTION CONTROL

An important network operation is congestion control, which helps to increase the speed of number of packets, which are entered in the network. It ensures the stabilization of network, acceptable packet loss ratio, and efficient resource utilization. The most common and well-known congestion control protocol is known as transmission control protocol. Several improvements in congestion control mechanisms have been evolved like named data networking, and delay-tolerant networks. Irrespective of these congestion control mechanisms, various limitations like packet loss classification, queue management, congestion interference, and congestion window update. So, ML is applied to improve the congestion control in different optical networks. Bharathi et al. [19] deals with the classification of contention and congestion losses in an optical burst switching (OBS) network. By simulating the data with the help of OBS modules, a new feature is derived from the existing losses, which is known as number of bursts between failure. Two ML-based classifiers are generated based on packet losses for clustering and hidden Markov model. These classifiers integrate two TCP variants that keep a low control overhead for providing better performance (e.g., higher throughput and fewer timeouts).

18.6 CONCLUSION

Since ML have been successfully used in various optical communication and networking area. This chapter provides knowledge of application of ML at various layers of an optical network including physical and network layer. With the unprecedented growth in the degree of traffic, network becomes more complex and more resources are required. The traditional approaches are not enough to deal with these kinds of issues, so ML not only helps in traffic prediction, classification, and routing but also provides exposure of fault management and security attacks. By incorporating advanced ML algorithms and huge amount of input data available from network monitoring. The future optical network is capable to automatize an existing optical network by efficiently utilizing the network resources.

REFERENCES

[1] Y. Zhao, Y. Li, X. Zhang, G. Geng, W. Zhang, Y. Sun, A survey of networking applications applying the software defined networking concept based on machine learning, IEEE Access. 7 (2019) 95385–95405.

[2] V. Havlíček, A.D. Córcoles, K. Temme, A.W. Harrow, A. Kandala, J.M. Chow, et al., Supervised learning with quantum-enhanced feature spaces, Nature. 567 (7747) (2019) 209–212.

[3] J. Wass, J. Thrane, M. Piels, R. Jones, D. Zibar, Gaussian process regression for WDM system performance prediction, in: 2017 Optical Fiber Communications Conference and Exhibition (OFC), IEEE, 2017, pp. 1–3.

[4] F. Musumeci, C. Rottondi, A. Nag, I. Macaluso, D. Zibar, M. Ruffini, et al., An overview on application of machine learning techniques in optical networks, IEEE Commun. Surv. Tutor. 21 (2) (2018) 1383–1408.

[5] J. Zhou, P. Fan, Modulation format/bit rate recognition based on principal component analysis (PCA) and artificial neural networks (ANNs), OSA Continuum. 2 (3) (2019) 923–937.

[6] D. Zibar, O. Winther, N. Franceschi, R. Borkowski, A. Caballero, V. Arlunno, et al., Nonlinear impairment compensation using expectation maximization for dispersion managed and unmanaged PDM 16-QAM transmission, Opt. Express. 20 (26) (2012) B181–B196.

[7] Z. Wan, Z. Yu, L. Shu, Y. Zhao, H. Zhang, K. Xu, Intelligent optical performance monitor using multi-task learning based artificial neural network, Opt. Express. 27 (8) (2019) 11281–11291.

[8] P. Tomar, G. Kaur (Eds.), Examining Cloud Computing Technologies Through the Internet of Things, IGI Global, 2017.

[9] T. Alghamdi, K. Elgazzar, M. Bayoumi, T. Sharaf, S. Shah, Forecasting traffic congestion using ARIMA modeling, in: 2019 15th International Wireless Communications & Mobile Computing Conference (IWCMC), IEEE, 2019, pp. 1227–1232.

[10] A. Azzouni, G. Pujolle, NeuTM: a neural network-based framework for traffic matrix prediction in SDN, in: NOMS 2018-2018 IEEE/IFIP Network Operations and Management Symposium, IEEE, 2018, pp. 1–5.

[11] D. Rafique, T. Szyrkowiec, H. Grießer, A. Autenrieth, J.P. Elbers, Cognitive assurance architecture for optical network fault management, J. Lightwave Technol. 36 (7) (2018) 1443–1450.

[12] M. Ruiz, F. Fresi, A.P. Vela, G. Meloni, N. Sambo, F. Cugini, et al., Service-triggered failure identification/localization through monitoring of multiple parameters, in: ECOC 2016 42nd European Conference on Optical Communication, VDE, 2016, pp. 1–3.

[13] M. Shafiq, X. Yu, A.K. Bashir, H.N. Chaudhry, D. Wang, A machine learning approach for feature selection traffic classification using security analysis, J. Supercomput. 74 (10) (2018) 4867–4892.

[14] H. Rastegarfar, M. Glick, N. Viljoen, M. Yang, J. Wissinger, L. LaComb, et al., TCP flow classification and bandwidth aggregation in optically interconnected data center networks, J. Optical Commun. Netw. 8 (10) (2016) 777–786.

[15] Y.C. Huang, J. Zhang, S. Yu, Self-learning routing for optical networks, in: International IFIP Conference on Optical Network Design and Modeling, 2019, Springer, Cham, pp. 467–478.

[16] G. Baggio, R. Bassoli, F. Granelli, Cognitive software-defined networking using fuzzy cognitive maps, IEEE Trans. Cognit. Commun. Netw. 5 (3) (2019) 517–539.

[17] D. Tomar, P. Tomar, G. Kaur, Deep learning in Big Data and Internet of Things, in: S. Minz, S. Karmakar, L. Kharb (Eds.), Information, Communication and Computing Technology. ICICCT 2018. Communications in Computer and Information Science, vol 835, Springer, Singapore, 2019.

[18] P. Tomar, G. Kaur, Green and Smart Technologies for Smart Cities, CRC Press, 2019.

[19] L. Bharathi, N.S. Priya, B.S. Sathish, A. Ranganayakulu, S.J. Rao, Burst rate based optimized IO queue management for improved performance in optical burst switching networks, in: 2019 Fifth International Conference on Advanced Computing & Communication Systems (ICACCS), IEEE, 2019, pp. 168–172.

CHAPTER

IMPACT OF ARTIFICIAL INTELLIGENCE TO SOLVE PERVASIVE ISSUES OF SENSOR NETWORKS OF INTERNET OF THINGS

19

Akanksha Srivastava[1], Mani Shekhar Gupta[2] and Gurjit Kaur[1]

[1]*Department of Electronics and Communication Engineering, Delhi Technological University, Delhi, India*

[2]*Department of Electronics and Communication Engineering, National Institute of Technology, Hamirpur, India*

19.1 INTRODUCTION

In present scenario smart intelligent system is used to connect different networks like smart automation system network, smart grid system network, smart transportation system network, smart home system network, smart infrastructure systems network, smart water management system networks, and others. This smart intelligent system has minimized the size of the world. The main concept behind this idea is using the Internet of Things (IoT), which is strongly associated with sensors networks [1]. This complete physical structure is fully linked with information and communication technology (ICT) where various hardware, microprocessors, microcontrollers, embedded systems, and software are used for monitoring, processing, and management purpose, and with the help of distributed sensor networks, the interconnected device's network sends useful information and control instructions [2]. A wireless sensor network (WSN) is made by a group of various sensor nodes that performs detection of various physical parameters like temperature, pressure, heat, traffic, weather conditions, lights, and so on [3]. WSNs are observed as the most innovative data collecting systems, to create an ICT system, which facilitates to enhance the efficiency and reliability of overall systems. WSNs are more flexible than wired sensor networks because they can easily deploy anywhere. With the advancement in the sensors technology now WSNs have become an integral part of the IoT [4,5]. WSNs are integrated with the IoT system, in which each sensor nodes are connected through the Internet dynamically and practice it to perform and achieve their work. WSNs are considered as a long-term performance network for information gathering for IoT operation. Table 19.1 represents the data of the total number of connecting devices to IoT from 2015 to 2025.

Artificial Intelligence to Solve Pervasive Internet of Things Issues. DOI: https://doi.org/10.1016/B978-0-12-818576-6.00019-8

Table 19.1 Number of Connected Devices to IoT from 2015 to 2025 [6].

Year	Total Number of Connected Devices
2015	15.41 billion
2016	17.68 billion
2017	20.35 billion
2018	23.14 billion
2019	26.66 billion
2020	30.73 billion
2021	35.82 billion
2022	42.62 billion
2023	51.11 billion
2024	62.12 billion
2025	75.44 billion

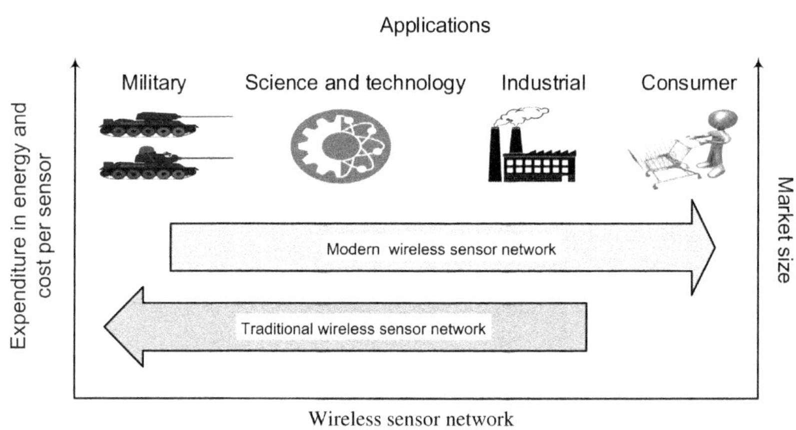

FIGURE 19.1

Applications of wireless sensor network from past to present.

19.2 SENSOR NETWORK TECHNOLOGY

Advancement in technology in the field of electronics and communications have focused on to develop small size, low-priced, low-power, multifunctional sensor nodes that transfer information in short distances. These small sensor units perform following functions sensing, data signaling and processing, and communicating with other components. Fig. 19.1 represents the market size and expenditure on WSNs from past to present.

19.2.1 HISTORY OF SENSOR NETWORKS

For different services usage of sensor network is not new. At the time of Cold War, sound surveillance system was used for detecting the Soviet submarines [7]. This system consists acoustic sensors for detecting process. Now National Oceanographic and Atmospheric Administration is adopting this technology for studying and sensing the various phenomena takes place inside the oceans. "Aerostats" works as a sensor is adopted by air defense radar networks [8]. In the direction of advancement in sensor field, Defense Advanced Research Projects Agency of United States have formed Advanced Research Project Agency in 1969, which designed a framework of networking technologies for connecting several research centers and universities [9]. Nowadays the most important application of sensor network is industrial automation.

19.2.2 CHARACTERISTICS OF SENSOR NETWORK

- A sensor network is defined as the "group of nodes" that senses and controls the different parameters of environment and placing a proper interface between computers, persons and the environment.
- Sensor nodes are basically two types of wired sensor network that include Ethernet, cables and WSN having routers, gateways, actuator nodes, sensor nodes, and clients.
- In the monitoring area huge number of sensor nodes are deploying randomly, by networks in self-organizing manner.
- Sensor nodes collect the data and send to next sensor node through hopping [10]. In the process of multihop routing data is processed and handled by multiple nodes and then through the gateway data is finally reached to the destination via satellite and Internet.
- As the technology upgraded, cost of WSNs components has reduced dramatically on the other hand its applications are spreading in all fields like military areas, industrial area, medical field, traffic monitoring, and commercial fields.

Fig. 19.2 represents the architecture of WSN.

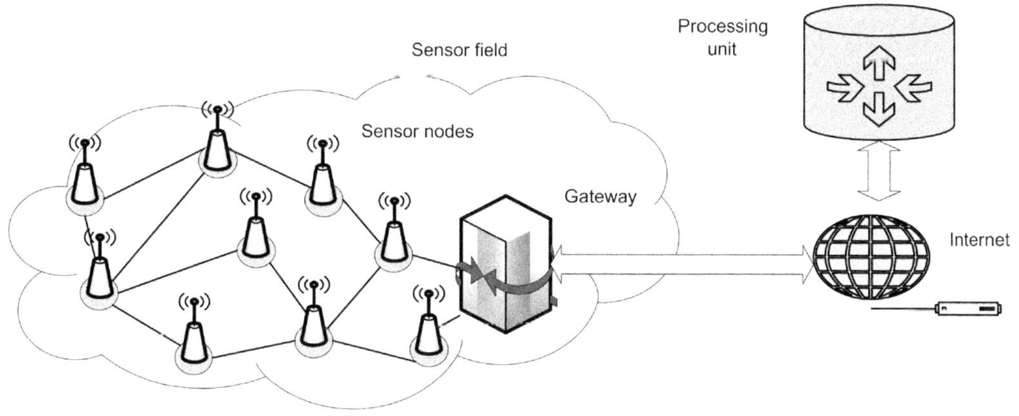

FIGURE 19.2

Architecture of wireless sensor network.

19.2.3 TOPOLOGIES

In WSN two main components are sensor nodes that gather information and the other one is the gateway that connects this network to the Internet for information transmission. Fig. 19.3 represents the placement of sensor nodes and processing [11]. This process is initialized by broadcasting the status of each node to the other nearby nodes. After that sensor nodes detect its nearby other nodes and establish an organized network according to some topologies like bus, tree, star, mesh, and so on represented in Fig. 19.4.

Now the appropriate path is selected after considering the various parameters and situations and transmitted the sensed data on that path. WSNs are powered by batteries, so the transmission distance of WSN nodes is short. The range of outdoor communication for sensor network is almost 700–1000 m in direct line of sight while in case of indoor communication this range is reduced up too few meters only. To improve the network coverage, multihop transmission method is used by the sensor network. In a network a node can work as a transmitter and as a receiver both. For data communication the first node that starts the communication is consider as a source node, which transmit the data to its nearby node and nearby node further transmit this data to its nearby node on way toward the gateway. This process is continuously repeating until the does not received at the destination through the gateway. Therefore the sensor networks are self-adapting, self-organizing, and unstable and low nodes energy transmission.

19.2.4 DATA AGGREGATION

Data aggregation is defined as the merging of multiple copies of information into a single copy, which is more effective and fulfill the quality of experience (QoE). The concept of data aggregation not only reduces the energy consumption but also provides accurate and precise information [2].

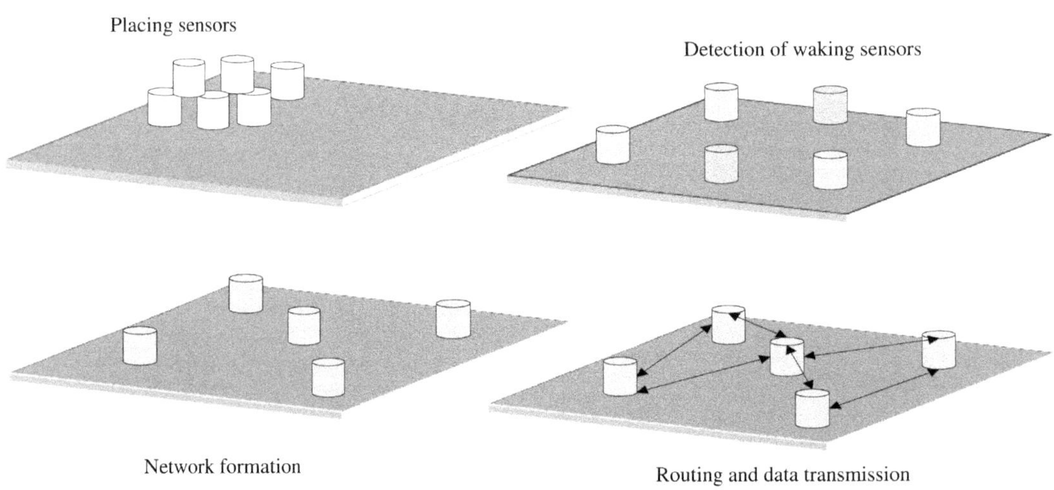

FIGURE 19.3

Data transmission process of wireless sensor networks.

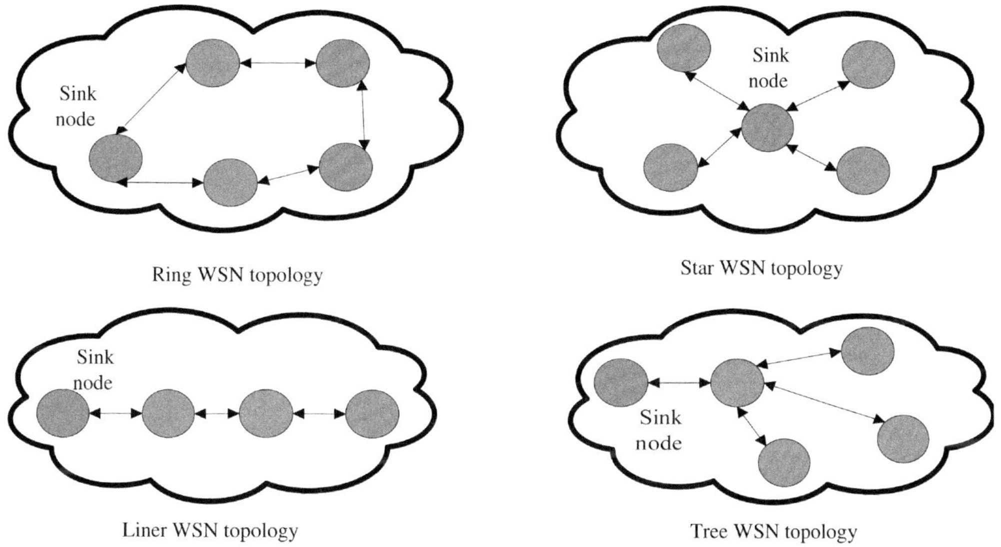

FIGURE 19.4

Various topologies used for WSNs organization [12]. *WSN*, Wireless sensor network.

The energy required for data transmission is much higher than that energy required for data processing in sensor networks. Significantly data aggregation can minimize data redundancy, but the drawback is it also reduces the robustness of the sensor network system.

19.2.5 ACCESS TECHNOLOGIES OF SENSOR NETWORK

There is a significant progress in the access technologies of sensor network, which completes essential requirements of the WSN applications. On the basis of speed and range access technologies are divided into four types of personal area network, local area network, metropolitan area network, and wide area network [13]. Some leading access technologies based on sensor network are WLAN standard IEEE 802.11 for IoT applications, Bluetooth 4.0 focused on medical WSN, IEEE 802.15.4e for industrial WSN. Fig. 19.5 represents the upgradation of wireless generation and technologies, which are associated with sensor network.

19.3 PERVASIVE ISSUES RELATED TO SENSOR NETWORKS

Today WSN is a developing technology that includes various issues, challenges, aspects, and layers to support the communication system. Some of its major pervasive issues are discussed in this section.

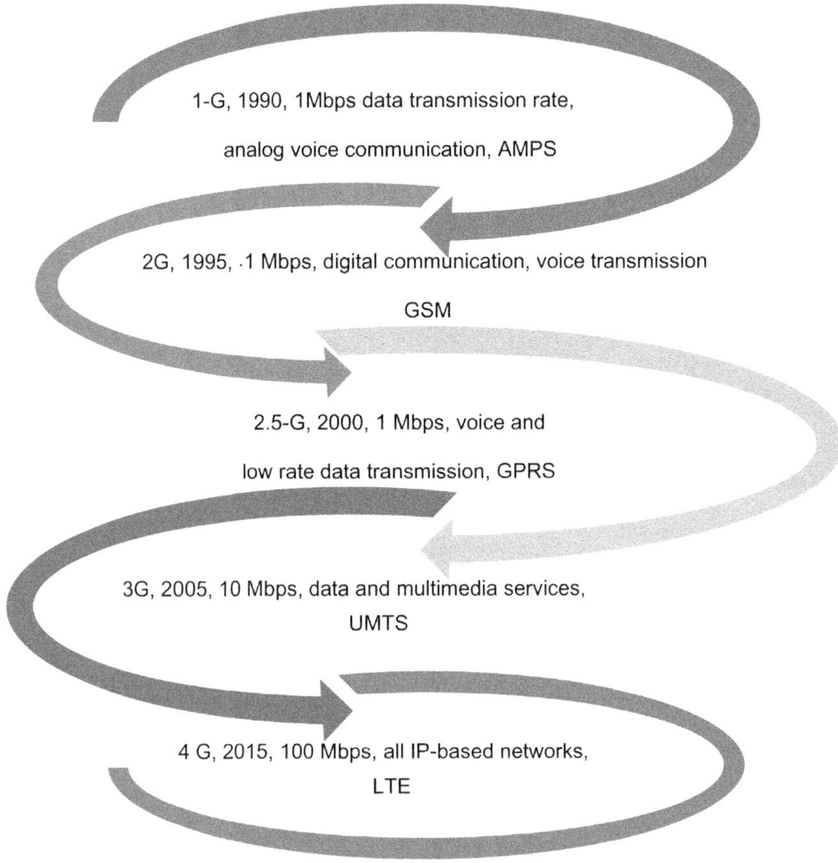

FIGURE 19.5

Upgradation of wireless generation.

19.3.1 REQUIREMENT FOR AN ARCHITECTURE FRAMEWORK

There are several issues of WSN but the most important one is "range of sensor network." Sensors are sensing the various parameters and provide this data for IoT operation. The range of "parameters" that are sensed, detected, and manipulated by the IoT-based sensor network having a significant role. WSN sensing parameters, including pressure, flow, temperature, speed, and others physical quantities, having multidimensional heterogeneous features. Table 19.2 represents selection of different system qualities on the basis of architecture.

19.3.2 COEXISTENCE WITH OTHER TECHNOLOGIES

As the number of applications of WSN is increasing in military, in commercial and in industry, more demanding performance requirements are expected at the same time. The performance of WSN can improve with the coexistence of other technologies. The present access technology

Table 19.2 Selection of System Qualities on the Basis of Architecture.

Architectural View						
System Quality	Parameters	Deployment	Information	Coexistence	Operational	Functional
	Performance	High	Medium	High	Low	Medium
	Scalability	Low	High	Medium	Low	High
	Availability	High	Low	Medium	Medium	Low
	Security and privacy	High	Medium	Medium	Medium	Medium

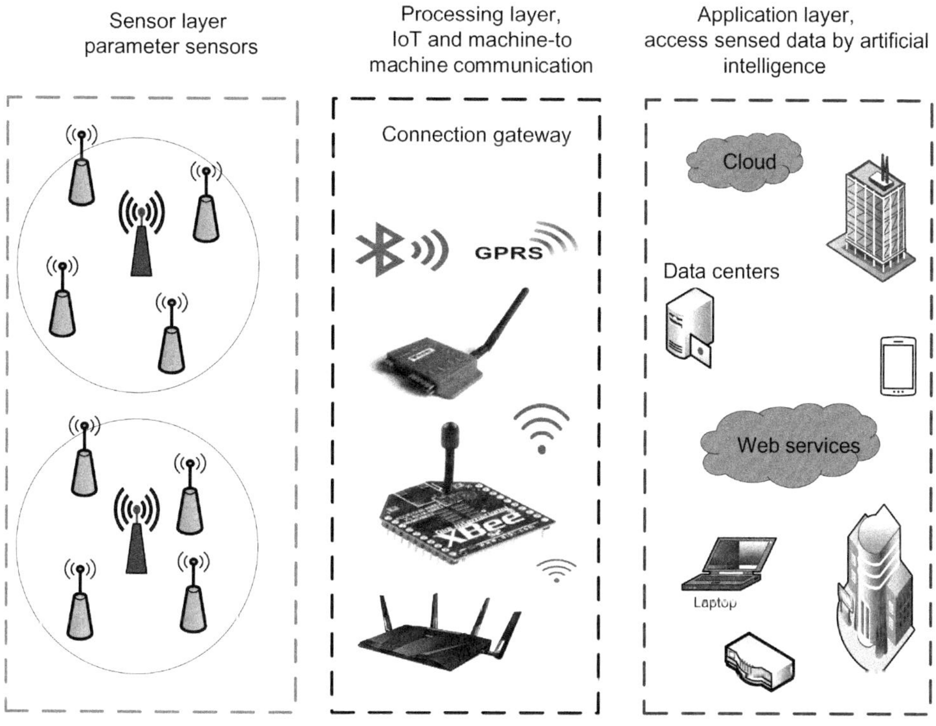

FIGURE 19.6

Network optimization by artificial intelligence technique.

generally adopts such approaches like unutilized resource allocation among multiple users, frequency division and time division, IoT, artificial intelligence (AI), and so on, to guarantee reliable transmission [14] (Fig. 19.6).

19.3.3 REAL-TIME OPERATION AND MANAGEMENT

Generally, WSNs are used to observe, gather and process data of the objects in the network-oriented areas and forward this data toward the observers for online and offline investigation in less real-time requirements, like pollution measurement, meter reading, weather condition, environmental monitoring, and so on. [15]. Hence WSN research concentrations on how to expand sensor network reliability and decrease power consumption. Therefore with the nonstop expansion of the infrastructures of the cities, the coverage area of network is also increasing, so real-time data transmission is requirement of present time.

19.3.4 MORE SECURITY AND PRIVACY

Presently WSNs link the information and communication system strongly with different layers of infrastructure. Due to the virus threats infrastructure damage like transportation, energy, power, and national security may take place. WSN is generally more exposed to several security threats as the unguided is more susceptible to security attacks than the guided communication channel [16].

19.4 ROLE OF ARTIFICIAL INTELLIGENCE TO SOLVE PERVASIVE ISSUES OF SENSOR NETWORK

AI plays a vital role to solve pervasive issues of sensor network because the raw data collected by the sensors for the IoT devices needs to be processed by the AI. As the sensor network complexity is continuously increasing the requirement of the AI is also increasing in IoT devices. In data communication process extracting the required information from the raw data is a difficult task [17]. The major advantage of deep learning and machine learning is having the capability of extracting required information from a set of data in very small set of weights [18]. Nowadays number of connected devices with IoT are increasing and these devices are connected with internet so to manage this network, quality of service, and QoE an intelligent algorithm is necessary. The best option to minimize the power consumption is neural network. A metal oxide uses AI for sensing the different gasses.

19.5 FEATURES OF ARTIFICIAL INTELLIGENCE IN THE INTERNET OF THINGS REVOLUTION

AI is technique that accomplish a large number of smart tasks without human presence like face and voice recognition, decision-making, language translation, and so on. On the other hand, IoT comprises a group of interconnected devices that performs various operations like accurate communication link establishment between transmitter and receiver, data transfer through the network. Recently IoT devices entered into daily lives of human being and make their live more comfortable and easier. These devices are connected through internet and continually monitor the daily live activities of human being like user requirements, user behaviors, user's personal information, and user preferences, and so on [19]. Therefore there are many organizations and enterprises, which stores user data and information but still they are unable to use these data intelligently. This is the big hindrance in the advancement and growth of the IoT. In this situation AI, helps to glean desired information from the bulk of the data. It permits to scrutinize the information and make sense out of it. In this way AI plays a vital role in the revolution of IoT.

19.6 CONCLUSION

The latest emerging technologies like the IoT, WSN, and AI are not single technologies, but these technologies provide output after processing various algorithms and logics. For data communication, these technologies represent a complex architecture because of the integration of various technologies from the initial layer (physical layer) to the uppermost layer (application layer). As discussed in this chapter there is a vast set of issues, applications, and challenges of WSNs. To solve these issues with the help of AI is an important topic for research for different industries and research organizations. AI technology facilitates to improve the performance of IoT-based sensor networks by using machine learning, deep learning, Big Data, and so on.

ACKNOWLEDGMENT

The authors would like to thank the Women Scientists Scheme-A under the Department of Science and Technology Government of India for its financial support of this work under File No: SR/WOS-A/ET-154/2017.

REFERENCES

[1] Lazarescu, T. Mihai, Wireless sensor networks for the internet of things: barriers and synergies, 2017 Components and Services for IoT Platforms, Springer, 2017, pp. 155–186.

[2] P. Goyal, G. Kaur, A review on high speed and low power CMOS optical interconnects, Int. J. Comp. Network. Wireless Commun. 6 (1) (2016) 1–6.

[3] L. Mainetti, L. Patrono, A. Vilei, Evolution of wireless sensor networks towards the internet of things: a survey, in: 2011 SoftCOM 2011, 19th International Conference on Software, Telecommunications and Computer Networks, IEEE, 2011, pp. 1–6.

[4] M. Kocakulak, I. Butun, An overview of wireless sensor networks towards internet of things, in: 2017 IEEE Seventh Annual Computing and Communication Workshop and Conference (CCWC), IEEE, 2017, pp. 1–6.

[5] B. Ahuja, G. Kaur, Throughput analysis of cooperative spectrum sensing for cognitive radio network, Int. J. Comp. Network. Wireless Mobile Commun. 5 (6) (2015) 1–8.

[6] Statista, Internet of Things (IoT) connected devices installed base worldwide from 2015 to 2025 (in billions). <https://www.statista.com/statistics/471264/iot-number-of-connected-devices-worldwide/>, 2019.

[7] The Evolution of Wireless Sensor Networks, Silicon Laboratories, Inc, China, 2013. <https://www.silabs.com/documents/public/white-papers/evolution-of-wireless-sensor-networks.pdf>.

[8] G. Kaur, D.M. Saxena, N. Gupta, Reliability of OCDMA MAN system using wavelength-time matrix encoding and decoding, ICFAI Univ. J. Electr. Electron. Eng. 2 (1) (2009) 14–23.

[9] V. Jindal, History and architecture of wireless sensor networks for ubiquitous computing, Int. J. Adv. Res. Comput. Eng. Technol. 7 (2) (2018) 214–217.

[10] A.A. Alkhatib, B. Alhameed, G. Singh, Wireless sensor network architecture, in: 2012 International Conference on Computer Networks and Communication Systems (CNCS 2012), IACSIT Press, Singapore, 2012, vol. 35.

[11] M.S. Gupta, A. Srivastava, K. Kumar, Seamless vertical handover for efficient mobility management in cooperative heterogeneous networks, in: Proceedings of the Second International Conference on Data Engineering and Communication Technology, Springer, 2019, pp. 145–153.

[12] X. Yu, P. Wu, W. Han, Z. Zenglin, A survey on wireless sensor network infrastructure for agriculture, Comp. Standard. Interfaces, 35 (1) (2013) 59–64.

[13] H. Tharon, WANs, connectivity, and computer literacy: an introduction and glossary, Comp. Composit. 9 (3) (1992) 41–58.

[14] J. Tan, S.G.M. Simon, A survey of technologies in internet of things, in: IEEE International Conference on Distributed Computing in Sensor Systems, 2014, pp. 269–274.

[15] B. Ahuja, G. Kaur, Study of soft decision fusion based cooperative spectrum sensing in cognitive radio, in: Proceedings of National Conference on Recent Advances in Microwave Antennas for Cognitive Radios and Wireless, Galgotias College of Engineering and Technology, Greater Noida, 2013, pp. 248–252.

[16] S. Babar, A. Stango, N. Prasad, J. Sen, R. Prasad, Proposed embedded security framework for internet of things (IoT), in: IEEE Second International Conference on Wireless Communication, Vehicular Technology, Information Theory and Aerospace & Electronic Systems Technology (Wireless VITAE), 2011, pp. 1–5.

[17] J. Gubbi, R. Buyya, S. Marusic, M. Palaniswami, Internet of Things (IoT): a vision, architectural elements, and future directions, Future Gen. Comput. Syst. 29 (7) (2013) 1645–1660.

[18] A. Mishra, G. Kaur, VASNET model with warning message protocol, in: Proceedings of the First International Conference on Innovations and Advancements in Information and Communication Technology, March 2012, pp. 346–351.

[19] M. Thurfjell, M. Ericsson, P. de Bruin, Network densification impact on system capacity, in: 2015 IEEE 81st Vehicular Technology Conference (VTC Spring), IEEE, 2018, pp. 1–5. 27 November 2016.

CHAPTER

PRINCIPLES AND FOUNDATIONS OF ARTIFICIAL INTELLIGENCE AND INTERNET OF THINGS TECHNOLOGY

20

Harshit Bhardwaj, Pradeep Tomar, Aditi Sakalle and Uttam Sharma

Department of Computer Science and Engineering, University School of Information and Communication Technology, Gautam Buddha University, Greater Noida, India

20.1 INTRODUCTION

The integrated feature of science and computer, which allows systems programs or computers to take decisions independently and in an intelligent manner and with a solution to problem, is artificial intelligence (AI) [1].The primary goal of AI systems is to discover, which increases people's performance and productivity overtime.

Machine learning (ML) and deep learning (DL) include artificial intelligent technology [2] tools, which provide an analysis report to enhance the clarity of planning, reasoning, thought, problem solving, and learning. Advantages of AI technology are:

- Artificial education intelligence [3] makes a valuable input to people. This is a complex problem, divided into subunits and the solution for each subunit will be found. The device can be a machine or an individual who tries to solve the problem. The new idea explains that educational cognitive science has built an instructor utilizing computer programming and that the teacher is observing students' problem-solving skills.
- In the area of AI, the expert framework [4] is commonly utilized. Script corrector and script monitor is the most common application. They work by checking orthographs and grammatical errors as a proofreader and provide any recommendation to get the best article. Throughout 80% of its manufacturing process, the robotics expert program is commonly used. It reduces labor costs, reduces error, and achieves maximum output in a low time because no lunch or break is required by the robot. The person takes hours to finish a painful job, but the robot can do that in just a few minutes.
- Applied AI robotics [5] is most appealing and human resource efficient. The robots are designed to conduct a routine function that increases productivity and is used effectively. The machines are special in their absence, space exploration, and does whatever function is dangerous for people. The more sophisticated work of robotics is performed by integrating crash sensors; camera and ultrasound cameras to allow them to see, hear, and touch. The device is designed for space travel and is sustainable and physical.

Artificial Intelligence to Solve Pervasive Internet of Things Issues. DOI: https://doi.org/10.1016/B978-0-12-818576-6.00020-4

- Human reasoning delays ideas and struggles with implicit thinking. In addition to emotional care, a computer is programmed to give an independent analysis and successful assessment.
- In day-to-day lifestyles AI has been applied and develops effectively, utilizing technology as project design, skill trainings, data interpretation problem management throughout areas like connectivity, time management, health services, safety measures, traffic control, transactions, shopping, and preparation.

On the other hand, the Internet of Things (IoT)[6] relates to the rising network of any physical thing or objects with which, Internet connectivity can be made using IP addresses of those objects. The IoT [6] offers a wide range of tools, which uses embedded hardware to connect and engage with the exterior environment, all over the Internet [7]. Examples are connected security systems, thermostats, cars, electronics, lightings, warning clocks, vending machines, and so on [8]. Advantages of IoT technology are:

- **Efficient resource utilization:** We can increase efficiency for resource utilization also monitoring of natural resources by knowing the functionality of each device.
- **Minimize human effort:** Human efforts can be minimized using devices of IoT, which interact and communicate with each other for performing lot of task in our day-to-day life.
- **Save time:** Saves our time by minimizing efforts. Time is the primary factor that can save through IoT platform.
- **Improve security:** System that is interconnected at all levels is more effectual and secure.

20.2 BACKGROUND

The AI this is an important concept, and when it is complemented with computers with ample facets of the smart problem-solving skill, then the human and the computer can do more together than one or the other. Alternatively, the number of devices connected to the web nowadays is experiencing a very explosive rise. We had previously only Internet-linked personal computers (PCs) and mobile devices, but now the Internet of objects, that is, millions of devices contribute to the IoT idea of connecting items to the Internet. The advancement of IoT contributes to the concept of a machine to a machine communication, which means that there is now a possibility for two machines to communicate with each other and for the use of remote access to any data that was previously inaccessible through a private server. Different scientists in various fields have applied AI and IoT and facilitated their research.

Farrugia et al. [9] aims at stressing the importance in medical information technology and biomedicine of the work of AI. Within this huge area of medical science, we have analyzed recent AI work with attention to medical diagnosis. A variety of smart computation methods have been considered, from rules-based expert systems (ES) and thriving logic to neural networks and genetic algorithms for medical diagnosis. Hydrocephalus is known to cause headaches as a health condition. The author has studied a prototype of what is currently being developed as a mobile device application for the collection of hydrocephalus patients and eventually recommending the creation of a professional mobile application program to help clinicians diagnose, evaluate and manage the disorder.

Yeasmin [10] analyzes the benefits of medical AI. It looks at how medical intelligence serves the medical sector and how the use of this common concept is influenced by the diagnosis of illnesses, patient treatments, and error reduction, and how the patient is practically present. Finally, it shows how medical research can be affected by AI.

Pirbhulal et al. [11] proposed the power-ON-(EEOOA) algorithm for transmitting medical data between medical system sources and destination nodes. In addition, the Internet of Medical Things (IoMT)-for eHealthcare is built in this report. One easy way to coordinate, track, and prescribe healthcare applications is the IoMT. Given the limited and resource-design of medical equipment in IoMT, dealing with immense energy drain problems would be of considerable importance. The proposed algorithm in terms of energy savings is also compared to the baseline.

AI and IOT have also a big impact on the field of agriculture. The ES focused on smart agriculture systems were developed by Shahzadi et al. [12]. The IoT concept in this system consisted of sending the data to the server in order to make appropriate decisions by the actuators in the field.

The automatic irrigation system was developed by Gutiérrez et al. [13] and used the GPRS module as a communication device. The device is built into a portal that regulates the amount of water. Water saving was 90% higher than conventional irrigation systems. It has been demonstrated.

For the sensing and control of irrigation from a remote location, Kim et al. [14] used a distributed wireless network.

The apps in which Gondchawar and Kawitkar pose themselves [15] are highlighted are the smart remote GPS-based robot for tasks like spinning, spraying, detecting rain, scaring, watching over birds or livestock, and so on.

The use of robot in agriculture was explored by Katariya et al. [16]. The robot is designed to follow the white line path where a work is required, with a black or brown surface. The device is used to spray fertilizer, drop plants, provide water, and ploughing.

Dholu et al. [17] proposed cloud-based IoT application in the field of agriculture. Precision agriculture is essentially a term that insists on properly providing resources at and for the exact period. These resources may include water, electricity, pesticides, and so on. These resources. The advantages of IoT have been used to introduce precision agriculture. The basic concept is to sense all the necessary farming parameters and take the necessary decision to power the actuator. These agricultural parameters include soil humidity, temperature and relative humidity, and intensity of light. According to the sensor–sensor calculation, the irrigation valve is worked based on readings on soil moisture, the fogger valve is triggered based on relative moisture readings, and so on.

Patil et al. [18] do the analysis and review of the current state of the agricultural system. The climate change and the precipitation over the past decade have been unpredictable. Because of this in recent times, many Indian farmers have adopted climate-smart methods called smart farming. Smart agriculture is an IoT-automated and driven information technology. In all wireless environments IoT is evolving rapidly and extensively. Remote monitoring system is introduced in conjunction with Internet and wireless communications. The main aim is to gather information about the farm production environment in real time, providing convenient access to farm facilities like short message service warnings and weather pattern advisory services, crops, and more.

In the educational sector, AI and IoT also have an influence. Florea et al. [19] talk about four perspectives on the relationship between IA and education: how IA can improve education; how to personalize learning experience; how to assist teachers in their efforts and how education in IA can be developed in order to develop the workforce needed to cope with this current technological revolution; and how to strengthen education and education.

Saxena et al. [20] explores the idea of an individual's three preferred learning styles: visual, auditive, and kinesthetic. The standard method therefore consists of classifying students according to their types and allocating them assignments according to their preferred types. The program proposed implements an IoT-based system that categorizes students automatically and records their results, task, history of the submission, and so on. It will help minimize the workforce of faculty members and boost student success by tailoring activities according to the innate style, from 5%–42%.

Akiyama et al. [21] suggested an educational program that was tailored for the creation and propagation of ideas of an IoT system, which could be accomplished also by liberal art students. For the students in child education and childcare, we have applied the program and have analyzed the results in this study.

20.3 EXISTING TECHNOLOGIES IN ARTIFICIAL INTELLIGENCE AND INTERNET OF THINGS

Some of the few top technologies [22], [23], [24], [25] of AI and IoT, rocking the world are listed in the following text, with their short description:

20.3.1 NATURAL LANGUAGE GENERATION

Generation of natural language (NL) [26] method is popularly referred to as "language processing" by psycholinguists to turn all structured data to NLs. Basically, the generation of NL should be seen like a procedure that evolves thoughts into words. For example, if a child sees a butterfly in a garden, he may think about it differently. However, when the infant in its NL explains the method of learning (mother language, the cycle can be named the phase of the NL).

20.3.2 NATURAL LANGUAGE UNDERSTANDING

The comprehension of the NL is the opposite of the generation of NL [27]. This approach is more related to NL comprehension. In this case the child will view the information given to him in different ways when it comes to the butterfly instead of being shown. The boy takes a photo of a butterfly moving in the garden based on this understanding. If the interpretation is correct, the method (NL comprehension) may be believed to have worked.

20.3.3 SPEECH RECOGNITION

Speech acknowledgment [28] technology is used for conversion of human speech to a computerized-accessible format with AI. This process is useful and acts as intermediate in the interaction between people and computers. The machine will understand human expression in several NLs utilizing speech recognition technologies. This also allows the computer to interact with people faster and smoother. Just say, for instance, that the child was asked in the first example, "What are you like?" When humans interact normally with humans. On the basis of the data (knowledge) already present in his brain, when a babe listens to the human speech sample. The child draws appropriate inferences and eventually gives an idea of the nature of the image.

20.3.4 MACHINE LEARNING

In the AI area, ML is an important technology [29]. The aim of ML is to teach a computer to understand and thought independently. In addition, ML utilizes several complicated computers testing algorithm. During the method, a series of classified or unclassified training data for a specific or general domain are presented to the computer. The computer then analyses and saves the data for potential use. The computer uses stored implications to draw the requisite assumptions and provide a response if it considers any further sample data of the subject it has studied. Let us assume, for starters, that a set of toys was shown in the first case. The infant communicates to the training dataset (toys) and discovers how the toys function (using his senses like contact, feel, and so on). Such features can differ from toy scale, color, form, and so on. The child stores inferences on basis of his observations and uses it to differentiate among toys with which he may have future meetings. Therefore, the kid has understood can be inferred.

20.3.5 VIRTUAL AGENTS

Virtual agents [30] are a result of a system aimed at creating a physical and successful inequality for men. Digital assistants are commonly used for customer care sector to use the blend of AI, ML, and so on to perceive the consumers complaints and concerns. The sophistication and techniques used to build the agent are subject to a clear understanding by virtual agents. Similar tools are frequently used in several applications, including chat bots, associate networks, and so on. Such devices will humanely interact with people. In the aforementioned cases, the child will use mixture of person's already acquired information, language processing and other required "resources" to interpret the individual if the baby is called a virtual agent and interacts with unfamiliar participants. After the experience is that, the child draws encounter-based assumptions and can effectively address based the participant's questions.

20.3.6 EXPERT SYSTEMS

The ES are computer programs that are programmed to address difficult problems at the level of exceptional human intellect and experience in a particular area. In AI, ES [31] uses a previously stored information in a knowledge base and makes decisions. Systems works on reasoning skill and the predefined "if-then" rules.

20.3.7 DECISION MANAGEMENT

AI-based systems with abilities of interpreting and giving predictive models by converting data are proven revolutionary for modern decision management systems [32]. Enterprise-level applications at a very large number have utilized expert system. In this example child is using his reasoning abilities based on knowledge data, hence considered as a decision management system, effectively managing his decisions.

20.3.8 DEEP LEARNING

DL [33] also a subset of ML using artificial neural networks (ANN). The complexity of network and the number of levels of data transform refer the term "deep." News Aggregation (sentiment-based), Computer Vision, Automated Translations, and so on are some of the applications of DL.

20.3.9 TEXT ANALYTICS

Text analytics [34] in simple is the text structure analysis. Text analysis is a part of AI systems, which is used for text analytics for interpreting the meaning of text. Fraud detection systems and security issues has application for such systems.

20.3.10 BIOMETRICS

Through biometric techniques [35], human comportment, and physical features of the structure and the shape of the body are defined, assessed, and analyzed. This enables natural interactions between people and computers, including touch and is broad in the field of market research. Following firms are all biometrically involved in this fie ld. 3VR, Agnitio, Affective Sensory, FaceFirst, Synqera, and Tahzoo.

20.3.11 NATURAL LANGUAGE PROCESSING

NL processing [36] utilizes text processing, by computational methodologies and ML, to recognize the structure of the sentences as well as their context and purpose. Text mining and NLPs for safety systems and fraud detection are being used. These are also used to retrieve unstructured data with a vast array of automatic helpers and software. Service providers including Basis Software, Coveo, Indio, Knime, Linguamatics, Expert System, Sinequa, Synapsify, and Stratifyd are among others.

20.3.12 DIGITAL TWIN/ARTIFICIAL INTELLIGENCE MODELING

Digital twin [37] is a technological framework that ties the break between digital and physical structures. For example, General Electric (GE) builds AI employees to track their aircraft engines, locomotives. Their projects supply field for companies, which use twin-digital and AI-modeling, Akselos is used to safeguard essential structure and supply; Dynamics has built a SaaS system for managing the supply of raw materials in dynamic, highly dispersed manufacturing environments.

20.3.13 CYBER DEFENSE

Cyber defense [38], a networking defense mechanism designed to prevent, identification, and prompt reaction to technology and information attacks or incidents. The Breach Index spotted more than two billion records broken during 2017, and 76% of the sample documents have been accidentally lost, and 69% are infringements of identity theft. In tandem with ML methods, repeating neural networks, which can process input sequences, can be used to generate guided learning technologies. Both companies in AI-enabled cyber defense market, including Darktrace, which incorporates behavioral analytics and advanced mathematics to automatically identify suspicious activities within enterprises.

Deep instinct is another cyber defense venture, a deep technology initiative recognized as the Nvidia's Silicon Valley Most Disruptive Startup, which defends endpitals, computers, and mobile devices of businesses.

20.3.14 COMPLIANCE

Conformity compliance [39] is an acknowledgment or assurance that an individual or an entity complies with and retains a substantial business with the provisions of agreed procedures, legislation, rules or terms, and conditions of contract. There are several instances of the use of AI in practice worldwide. For example, solutions from NLP can scan regulatory text and a cluster of keywords to match its patterns way that identifies changes relevant for an organization. However, with the deeper learning used as a framework for increasingly complex business rules, the number of financial practices reported as possible examples of money laundering can be minimized.

Compliance.ai is a company that blends regulatory documents with a business functions, and with its patented analysis software raise consumer acceptability levels while minimizing theft and manual repair are all organizations that work in this field.

20.3.15 KNOWLEDGE WORKER AID

Although certain people are rightly concerned about AI replacing employees at work, note that AI technologies will greatly help workers, particularly those engaging in knowledge work. Yes, information automation [40] is identified as the two most disruptive emerging trends. As a diagnostic tool, staff gradually consider AI as their medical and legal practice, which relies heavily on wise personnel. In this sector, there are more and more companies working on technology. Kim Technologies has the aim of empowering knowledge workers with little or no experience in the IT programming of tools for creating a novel workflow and document-creating process with the help of AI. Kyndi is another one whose platform is designed to support the information processing of knowledge workers.

20.3.16 CONTENT CREATION

The development of content [41] is now comprising materials like images, commercials, white papers, forums, infographics, and other visual or written tools, which add to the online world. Brands like USA Today, Hearst, and CBS utilize AI for their material. Wordsmith is a digital analysis platform that utilizes NLP to deliver news stories based on data from the earnings.

20.3.17 PEER-TO-PEER NETWORKS

Peer-to-peer networks [42] are generated in the cleanest sense by linking two or more several PCs and sharing resources. However, peer-to-peer networks can even be utilized by cryptocurrencies, and they are able to find solution for some of the most demanding problems in the world by gathering and analyzing large quantities of information to entrepreneur. NanoVision is another team that utilizes peer-to-peer networks and AI approach to cure medical threats. It is a start-up that offers molecular data to consumer user with cryptocurrencies.

20.3.18 EMOTION RECOGNITION

Emotion recognition [43] lets software "read," using advanced images or audio data processing the emotions on a human face. Now we are at the stage of catching "microexpression" or implicit linguistic signs, and a vocal intonation that reflect the emotions of the individual. Lawyer can use this tool through interviews to seek to find certain details about someone. Nevertheless, it also has several communications uses. In this sector there are more and more start-ups. Audio source that explains the behavior of an individual, including how optimistic, frustrated, moody, or excited, are beyond verbal study. Viso uses emotional video analytics to inspire and enhance the consumer experience, new product ideas and new upgrade and the emotion AI of Affectiva is used to use face coding, emotional recognition from face to voice data in the sports, auto, robots, school, healthcare, and other areas.

20.3.19 IMAGE RECOGNITION

Image recognition [44] is a digital image or video process to identify and detect an object or feature, and AI is increasingly being highly effective in using this technology. AI can search for images on social media platforms and equate them to several datasets to determine which ones are important in image search. The technology for image recognition can also be used in the identification of license plates, diagnosis of diseases, analysis of clients and opinions, and users "faciles." For detecting near duplicate image and for categorizing pictures Clarifai an image recognition system finds application. In order to detect close duplicates and find similar uncategorized pictures, Clarifai offers picture detection system for clients. SenseTime is one of the leading suppliers of payment and image analysis services for the authentication of bank cards and other applications in this field.

20.3.20 MARKETING AUTOMATION

The marketing departments have so far learned from AI and for good reason AI is put in the field. Fifty-five percentage of marketers are confident that AI has an improvement in the social media influence of their industry what a declaration. Companies can improve engagement and increase efficiency in marketing automation [45] in order to increase revenues quickly.

Some of the emerging IoT technologies are as follows.

20.3.21 BIG DATA CONVERGENCE

IoT not only underlines altering the way people live and do business, but it also keeps an eye on data production. Big Data systems [46] tend to support large-scale storage requirements and carry out the essentials for the maximum benefits of IoT to be derived. This is the latest IoT trends we face and will soon see in major mode. Big Data and IoT are strongly related and nowadays there are several new devices chosen for a fair share of the results. It can monitor computational needs and storage although working with larger data this retains certain information, so IoT will from now on focus on its large-data intersection.

20.3.22 DATA PROCESSING USING EDGE COMPUTING

The fundamental weakness in IoT is that appliances are placed behind the network's firewall. System protecting might be simple, but it requires much more to protect IoT systems. The protection between network communication and software applications that link to computers must be embedded in us. IoT's cost efficiency and performance of data processing produce the best success. In all autonomous including self-drive vehicle and smart traffic light, speedier data processing is popular. The solution to this problem is Internet of Things Trends is named edge computing [47]. The Cloud typically beats edge computing in terms of speed and efficiency. We both realize, and edge computing, that higher computation means lower latency. Edge computing data processing with the cloud will be necessary to boost IoT.

20.3.23 GREATER CONSUMER ADOPTION

In the next 10 years, when you shift away from the consumer IoT, you will see a major IoT change [48] just like Lily Robotics "market flops." The growth in the funding of the consumer IoT will decline and the industrial IoT infrastructure and platform will continue in the future. It will take time to develop this IoT trends. Build IoT architecture, Veniam, BetterView, and the Swift navigation companies, like this one, will be seen to solve difficulty in insurance, transport, agriculture, or telecommunications, and a vibrant decrease in capital expenditure will be seen.

20.3.24 "SMART" HOME DEMAND WILL RISE

We have seen IoT apps grow with the idea of clever home technology in the past and this will continue soon so that the home becomes more interactive. People would not control the machines, but the systems tell the people what to do. Developments or developments in the IoT industry are the hot topic of today's smart home Internet Things [49].

20.3.25 THE HEALTHCARE INDUSTRY EMBRACES INTERNET OF THINGS

When retailers are benefited from customer interaction, wearable devices (like bands, chips) are used by the healthcare sectors [50], and these industries are in constant but stable development.

20.3.26 AUTO MACHINE LEARNING FOR DATA SECURITY

Developers today are working on new methods, by using blockchain like technology, where citizens can easily share data. Most industrial companies already learn to build the trust and embrace the concept of ML and will adapt their activities through process performance to prevent downtime. The teaching of the auto ML toolset [51] will also be highly automated and becoming common.

20.3.27 INTERNET OF THINGS—MASSIVE GROWTH COMING

IoT devices usually get more details and data regarding apps and consumers compared to the rest of the industry, and by 2020 IoT devices would hit about nearly 31 billion worldwide. Today, we consider IoT devices to be the main monitoring and tracking functionality. We can simply presume it will enlarge to the fullest in accordance with the IoT technology trends. IoT normally collects or collects so many data, and the essential information is collected by AI. We will soon see the IoT systems operating for behavior and allow the technicians to make informative recommendations.

20.3.28 BLOCKCHAIN FOR INTERNET OF THINGS SECURITY

Soon there will be decentralization, self-governance, self-healing, and smartness in a wide range of market, financial and policy systems, markets, and sectors. Several start-ups are developing their territories on top of the IOTA Tangle (IOTA is a centralized ledge built to monitor and perform IoT system transactions) for the creation of modules and products for businesses free from SaaS or cloud charges. The centralized and monolithic computer structures should be ready to break apart into the function and micro resources dispersed to the decentralized computers, machines, and appliances. Someday, IoT can enter infrastructure, health care, sales, banking, and other fields that are impossible today. Such trends in IoT technology will make significant differences for us [52].

20.3.29 BETTER DATA ANALYTICS

AI is an advanced training program that detects patterns rapidly. Patterns in IoT technologies should boost the management of data and make it easier to protect them. It also provides information from the important data to help prioritize our lives. Obviously less IoT [53] incorporation is very clever, attentive, and self-learning.

20.3.30 SMART CITIES BECOMING MAINSTREAM

More systems and instruments have been adopted by States to take advantage of open data collection methods. We will soon see IoT trends that will allow for pioneering data exchanges by future-oriented cities [54] that will allow the public and private enterprise access to and mix of data.

20.3.31 BLOCKCHAIN AS EFFICIENT BACKEND

While mass adoption, scalability, safety, and costs are very important; you will see many technical problems with which IoT is kept back. Most software is designed to gather a large amount of information, and it seems more profitable to offer it here. The main issue here is the unified backend network used to power the IoT apps. So, confidence is now a bigger problem. Blockchain is important for proper information security and safety as one of the main IoT development trends. If we are going to have a range of intelligent devices in our houses, we need to address the consumers trust issues because they are at greater risk and the danger may overshadow their identification. Nevertheless, one of the latest IoT developments of the next year is to find the "blockchain as an important backend." The blockchain is here are the side and the encryption issue of storing an overwhelming amount of data in one place is decentralized.

20.3.32 PERSONALIZATION OF THE RETAIL EXPERIENCE

IoT is very helpful in retails supply chain management [55]. Sensors and many other smart technologies are helping in tailoring shopping experience making it more comfortable, and accurate. We can make our own settings and personalize our trade with changing IoT Trends. Advertisements and notifications for discount and sale of useful and usually bought product from frequently visited shop and site is and easy task for IoT, but getting an indoor map for a shop is extra ordinary task. This IoT expertise Trend will ensure the better integration of personalized retail experience, which ultimately can bring up a new era of shopping.

20.3.33 PREDICTIVE MAINTENANCE BOOST UP BY INTERNET OF THINGS

In future owner will be informed by home about plumbing leaks, issues in electrical connection, any appliance failure. Using IoT, the sensors applied in public are as and transports like factories, airport, planes, and cars, will soon enter our home. The system using IoT in home will act same in your absence and presence at the time of any issue. There will be more secure and smart cars, homes, and public places in future.

20.3.34 CLOUD COMPUTING: THE FUTURE OF INTERNET OF THINGS

One of the most important safety patterns will be data protection. An Internet-connected computer can be dangerous in many ways, because the personal information can be readily available for violators or spyware. Investors, thieves, and so on, can hold the database of infrastructures for smart home, autonomous cars, and wearable devices. The importance of these IoT trends needs to be given.

20.3.35 SOFTWARE-AS-A-SERVICE BECOMES THE NORM

While analyzing IoT Trends, Software as a Service [56] for potential industries is one of the main topics. With the aid of low entry prices, the SaaS rapidly rises to the top because it is also the popular IT games company and we will soon be able to see the comprehensive company's spouse. Apps as a service can render people's lives better than ever among all these developments of IoT technology.

20.3.36 IOT SECURITY AWARENESS AND TRAINING

The industry is big, so protection and preparation are required. The curriculum includes basic knowledge of advantages and danger, as well as other protection advice [57]. Proven security should be built, safety incorporated, transparency promoted throughout the Internet with the assistance of appropriate training.

20.3.37 CREATION OF UNIFIED FRAMEWORK FOR INTEGRATION

The lack of the unified IoT framework [58] is a challenge for the IoT industry as it cooperates. There is no single cooperative forum for the firms. All the IoT developments I mentioned previously include a unifying mechanism since that is the only way to safeguard the market. Those are the programs have high performance that also helps to maintain the typically necessary data intensive process.

20.3.38 ENERGY AND RESOURCE MANAGEMENT

The control of energy [59] relies on a better understanding of energy consumption. The goods that can be used in electric panels typically come on the market and can track household energy consumption. These IoT technologies can easily be built into resource management to make people's lives smoother and more secure.

For submitting mobile notifications, push notification may be enabled if the energy levels exceed. Certain functionality may also be added, like indoor temperature control, sprinkle control, and so on.

20.3.39 A SHIFT IN VOICE CONTROL: FROM MOBILE PLATFORMS TO MANAGING INTERNET OF THINGS ECOSYSTEMS

Currently the talks are going on and adapting to the changing security environment is also progressing ahead, and when their cyber superlative plan fails, people are jumbled about what the leading developers will do. The mobile-based developments in IoT should extend and become administration of the IoT ecosystem [60].

20.3.40 STAY AWARE AND ALLIED: FURTHER EXPANSION OF INTERNET OF THINGS

The IoT techniques true implementation is being sought by TEC executives, development companies, and businesses [60].The increased business value need and tests show the way toward an improved number of sensors embedded into the new digital goods. Many industries and companies today will see the IoT as a wall magic to draw the customer, grow products, and improve user experience according to all these IoT patterns.

20.4 FUTURE OF ARTIFICIAL INTELLIGENCE AND INTERNET OF THINGS

The fact that citizens can use smart devices, ingest smart capsules that measure the effect of medication on their medical health, stay the smart houses, and so on, sounds like a science fiction, but it is more about this work. Everything will be intelligent and Internet related. At its core, AI is ML, and the intelligent use of that data and information. The future work is impacted by innovations that can be made from such information. Today's IoT applications are helpful for identifying patterns as they recognize places where "traction" has proved its value and where consumer and industrial capital are now going. When AI is paired with IoT, "smart machines" simulate intelligent behavior to make informed decisions not impacted by human behavior. Both research departments must function together to build something of great importance. We are going to see an intelligent internet movement. Boosted productivity results from the tandem innovation of both AI and IoT. As the technology matures and progresses, accuracy will continue to increase as it becomes more intelligent and robust when processing different variables. The proverbial ocean of possible uses of IoT + AI, are in the fields of vehicles and vacuums, and they are: security and access devices—organizations like ACT (access control technologies) now advocate the usage of key fob technology to unlock doors and utilize devices with respect to strictly IoT applications. AI, also in organizations with well under a thousand workers, may be used as the foundation for assessing daily access habits of different staff or positions and employee groups-giving visibility into possible workplace designs and possibly detecting malicious behavior (using the same technologies utilized in outsourcing identification through current information security). Although we have not been able to identify big fob/access technologies that incorporate AI or predictive analytics, we should expect that improving fob technology and acceptance may lead to security insight (particularly with large corporations that analyze data from multiple locations).

Another application is Emotional Analysis and Facial Recognition—The Facial Recognition, even in the past 5 years, has seen several huge leaps and strides, and it seems likely that its implementations have not been used from surveillance through marketing. Companies like Kairos also have brandished clients like Nike and IMB on their websites for advertisement applications. The glossing of details from customer responses to goods and ads is perhaps never simpler with a camera on virtually every device and smartphone produced today. The self-tagging of Facebook is an illustration of which most people are acquainted—and many market models and implementations are yet to be created.

20.5 CONCLUSION

AI has been increasingly conscious of the essence of knowledge and has given a remarkable variety of implementations in diverse fields throughout its brief lifespan. The comprehension of human thought and of the essence of intellect in general has been improved. In the other side, despite technical advancements and customers' ability to combine technologies like mobile phones with household appliances, IoT is practically infinite in its future. WiFi has allowed people and machines to be linked to the ground, air and sea. IoT connects in the network and ties the device. IoT often creates massive data; however, AI is the technology that allows vast amounts of data to be accessed and beneficial. The link of IoT and AI systems is mutually beneficial. The coexistence of both innovations' profits from lots of environments and market niches. Finally, IoT and AI are strong and will make the company more intelligent. So, once these two innovations are merged, businesses will do much more digital change. Lots of places will reap the benefits of the two technology's coexistence. Naturally, the synthesis of AI and IoT is not a jump through the park; it needs not only heaviness, but also specific abilities and knowledge. However, combined these two revolutionary innovations have a huge effect on businesses to increase income and work more efficiently.

REFERENCES

[1] J. Togelius, J. Juul, G. Long, W. Uricchio, M. Consalvo, What Is (Artificial) Intelligence, MIT Press, 2018.

[2] P. Michelman, What to Expect from Artificial Intelligence Technology, MIT Press, 2017.

[3] M. Kassab, J. Defranco, J. Voas, Smarter education, IEEE IT Professional, 2018.

[4] M. Eremia, C. Liu, A. Edris, Expert Systems, Wiley-IEEE Press, 2016.

[5] M. Broman, P. Broman, Human-robotics & AI interaction: the robotics/AI legal entity, in: IEEE International Symposium on Technology and Society (ISTAS), 2017.

[6] Q. Hassan, Introduction to the Internet of Things, Wiley-IEEE Press, 2018.

[7] H. Geng, Internet of Things and Cloud Computing, Wiley Telecom, 2017.

[8] Q. Hassan, A Tutorial Introduction to IOT Design and Prototyping with Examples, Wiley-IEEE Press, 2018.

[9] A. Farrugia, D. Al-Jumeily, M. Al-Jumeily, A. Hussain, D. Lamb, Medical diagnosis: are artificial intelligence systems able to diagnose the underlying causes of specific headaches?, in: Sixth International Conference on Developments in eSystems Engineering, IEEE, 2013.

[10] S. Yeasmin, Benefits of artificial intelligence in medicine, in: Second International Conference on Computer Applications & Information Security (ICCAIS), IEEE, 2019.

[11] S. Pirbhulal, W. Wu, S. Mukhopadhyay, G. Li, A medical-IoT based framework for eHealth care, in: International Symposium in Sensing and Instrumentation in IoT Era (ISSI), IEEE, 2018.

[12] R. Shahzadi, M. Tausif, J. Ferzund, M.A. Suryani, Internet of things based expert system for smart agriculture, Int. J. Adv. Comput. Sci. Appl. 7 (9) (2016) 341–350.

[13] J. Gutiérrez, J.F.V. Medina, A.N. Garibay, M.A.P. Gándara, Automated irrigation system using a wireless sensor network and GPRS module, IEEE Trans. Instrum. Meas. 63 (1) (2014) 1–11.

[14] Y.J. Kim, R.G. Evans, W.M. Iversen, Remote sensing and control of an irrigation system using a distributed wireless sensor network, IEEE Trans. Instrum. Meas. 57 (7) (2008) 1379–1387.

[15] N. Gondchawar, R.S. Kawitkar, IoT based smart agriculture, Int. J. Adv. Res. Comput. Commun. Eng. 5 (6) (2016) 838–842.

[16] S.S. Katariya, S.S. Gundal, M.T. Kanawade, K. Mazhar, Automation in agriculture, Int. J. Recent. Sci. Res. 6 (6) (2015) 4453–4456.

[17] M. Dholu, K. Ghodinde, Internet of Things (IoT) for precision of agriculture, in: Second International Conference on Trends in Electronics and Informatics (ICOEI), IEEE, 2018.

[18] K. Patil, N. Kale, A model for smart agriculture using IOT, in: International Conference on Global Trends in Signal Processing, Information Computing and Communication (ICGTSPICC), IEEE, 2016.

[19] A. Florea, S. Radu, Artificial intelligence and education, in: 22nd International Conference on Control Systems and Computer Science (CSCS), IEEE, 2019.

[20] A. Saxena, K. Shinghal, R. Misra, A. Agarwal, Automated enhanced learning system using IoT, Fourth International Conference on Internet of Things: Smart Innovation and Usages (IoT-SIU), IEEE, 2019.

[21] K. Akiyama, M. Ishihara, N. Ohe, M. Inoue, An education curriculum of IoT prototype construction system, in: Sixth Global Conference on Consumer Electronics (GCCE), IEEE, 2017.

[22] M. Kassab, J. Defranco, J. Voas, Smarter education, IEEE IT Professional, 2018.

[23] P. Michelman, What to Expect from Artificial Intelligence Technology, MIT Press, 2017.

[24] M. Eremia, C. Liu, A. Edris, FACTS Technologies, Wiley-IEEE Press, 2016.

[25] T. Forester, Artificial Intelligence and the Fifth Generation, MIT Press, 1985.

[26] C. Barros, E. Lloret, Analysing the influence of semantic knowledge in natural language generation, in: IEEE Conference, 2017; L. Stephen, D. Geert, K. Andreas, P. Frank, A non-biological AI approach towards natural language understanding, in: IEEE Future Technologies Conference, 2016.

[27] L. Stephen, D. Geert, K. Andreas, P. Frank, A non-biological AI approach towards natural language understanding, in: IEEE Future Technologies Conference, 2016.
[28] A. Prodeus, K. Kukharicheva, Training of automatic speech recognition system on noised speech, in: IEEE International Conference on Methods and Systems of Navigation and Motion Control (MSNMC), 2016.
[29] A. Nayak, K. Dutta, Impacts of machine learning and artificial intelligence on mankind, in: IEEE International Conference on Intelligent Computing and Control (I2C2), 2018.
[30] Z. Zhang, D. Weng, H. Jiang, Y. Liu, Y. Wang, Inverse augmented reality: a virtual agent's perspective, in: IEEE International Symposium on Mixed and Augmented Reality Adjunct (ISMAR-Adjunct), 2018.
[31] K. Ishii, S. Hayami, Expert systems in Japan, IEEE Expert. (1988).
[32] L. Sui, Decision support systems based on knowledge management, in: IEEE Proceedings of ICSSSM '05. 2005 International Conference on Services Systems and Services Management, 2005.
[33] A. Nandi, H. Ahmed, Deep Learning, Wiley-IEEE Press, 2019.
[34] G. Wilcock, Introduction to linguistic annotation and text analytics, Morgan Claypool (2009).
[35] H. Shrobe, D. Shrier, A. Pentland, Behavioral Biometrics, MIT Press, 2018.
[36] A. Farzindar, D. Inkpen, Natural language processing for social media, Morgan Claypool (2015).
[37] N. Mohammadi, J. Taylor, Smart city digital twins, in: IEEE Symposium Series on Computational Intelligence (SSCI), 2017.
[38] E. Seker, H. Ozbenli, The concept of cyber defence exercises (CDX): planning, execution, evaluation, in: IEEE International Conference on Cyber Security and Protection of Digital Services (Cyber Security), 2018.
[39] X. Yulin, J. Caijun, Y. Jie, Compliance control for grasping with a bionic robot hand, in: IEEE Chinese Control and Decision Conference (CCDC), 2016.
[40] A. Brobert, Learners as knowledge workers-some implications, in: IEEE FIE'99 Frontiers in Education, 29thAnnual Frontiers in Education Conference, Designing the Future of Science and Engineering Education, Conference Proceedings, IEEE Cat. No. 99CH37011, 1999.
[41] C. Pedersen, J. Togelius, G. Yannakakis, Modelling player experience for content creation, IEEE Trans. Comput. Intell. AI Games (2010).
[42] L. Ramaswamy, B. Gedik, L. Liu, A distributed approach to node clustering in decentralized peer to peer networks, in: IEEE Transactions on Parallel and Distributed Systems, 2005.
[43] A. Revathy, P. Shanmugapriya, V. Mohan, Performance comparison of speaker and emotion recognition, in: IEEE 2015 Third International Conference on Signal Processing, Communication and Networking (ICSCN), 2015.
[44] F. Shih, Image Representation and Description, Wiley-IEEE Press, 2010.
[45] C. Yee, M. Aziz, S. Hasan, Applying instant business intelligence in marketing campaign automation, in: IEEE Second International Conference on Computer Research and Development, 2010.
[46] A. Uta, A. Varbanescu, A. Musaafir, C. Lemaire, A. Losup, Exploring HPC and big data convergence: a graph processing study on intel knights landing, in: IEEE International Conference on Cluster Computing (CLUSTER), 2018.
[47] M. Satyanarayanan, W. Shi, Overview of Edge Computing, IEEE, 2018.
[48] J. Yun, Y. Ahn, N. Sung, J. Kim, A device software platform for consumer electronics based on the internet of things, IEEE Trans. Consum. Electron. (2015).
[49] H. Geng, Smart Home Services Using the Internet of Things, Wiley Telecom, 2017.
[50] H. Geng, Home Healthcare and Remote Patient Monitoring, Wiley Telecom, 2017.
[51] G. Mcgraw, R. Bonett, H. Figueroa, V. Shepardson, Security engineering for machine learning, IEEE Comput. (2019).
[52] Q. Hassan, Blockchain Based Security Solutions for IoT Systems, Wiley-IEEE Press, 2018.

[53] H. Geng, Security of IOT Data, Wiley Telecom, 2017.
[54] H. Geng, Smart City Architecture and Planning, Wiley Telecom, 2017.
[55] Singla, V., Malav, I., Kaur, J., Kalra, S., Develop Leave Application Using Blockchain Smart Contract, IEEE, 2019.
[56] K. Kaushik, S. Dahiya, Security and privacy in IOT based E-business and retail, IEEE International Conference on Systematic Modeling & Advancement Research Trends (SMART) (2018).
[57] H. Shrobe, D. Shrier, A. Pentland, Data Security and Privacy in the Age of IOT, MIT Press, 2018.
[58] Q. Hassan, A Low-Cost IOT Framework for Landslide Prediction and Risk Communication, Wiley-IEEE Press, 2018.
[59] Y. Gupta, S. Bajoria, R. Singh, O. Swathika, IOT based energy management system with load sharing and source management features, in: IEEE Uttar Pradesh Section International Conference on Electrical, Computer and Electronics (UPCON), 2017.
[60] N. Bouhai, I. Saleh, Introduction to the Technologies of the Ecosystem of the Internet of Things, Wiley Telecom, 2018.

Index

Note: Page numbers followed by "*f*" and "*t*" refer to figures and tables, respectively.

C

D

E

K

L

M

T